# SIGNAL AND IMAGE PROCESSING WITH NEURAL NETWORKS A C++ SOURCEBOOK

## Timothy Masters

WILEY
**JOHN WILEY & SONS, INC.**
New York   Chichester   Brisbane   Toronto   Singapore

Publisher: K. Schowalter
Editor: Diane D. Cerra
Managing Editor: Micheline Frederick

Designations used by companies to distinguish their products are often claimed as trademarks. In all instances where John Wiley & Sons, Inc. is aware of a claim, the product names appear in initial capital or all capital letters. Readers, however, should contact the appropriate companies for more complete information regarding trademarks and registration.

This text is printed on acid-free paper.

This publication is designed to provide accurate and authoritative information in regard to the subject matter covered. It is sold with the understanding that the publisher is not engaged in rendering legal, accounting, or other professional service. If legal advice or other expert assistance is required, the services of a competant professional person should be sought.

Portions of the source code were originally published in *Practical Neural Network Recipes in C++* by Timothy Masters.
Copyright © 1993 by Academic Press, Inc.

*Library of Congress Cataloging-in-Publication Data:*

Masters, Timothy.
    Signal and image processing with neural networks : a C++ sourcebook /
    Timothy Masters.
        p.        cm.
    Includes index.
    ISBN 0-471-04963-8 (paper/disk)
    1. Neural networks (Computer science)  2. C++ (Computer program language)
    3. Signal processing—Digital techniques  4. Image processing—Digital techniques.
    I. Title.
    QA76.87.M373  1994                                    94-6975
    621.382'2'028563—dc20                             CIP

Printed in the United States of America
10 9 8 7 6 5 4 3 2 1

*This book is lovingly dedicated to my parents, Charles and Lois Masters. Their nurture and encouragement throughout my childhood, and their unselfish support of my education, are what made this all possible.*

# Preface

Signal and image-processing problems account for a high percentage of research activity today. Good solutions are available for some of these problems. But in many cases, there is no known optimal solution. The best that can be done is to use methods that have good mathematical credentials for related problems and hope for the best. The results are often less than satisfactory. On the other hand, it is well known that neural networks can be startlingly effective at solving problems for which no clear-cut solution method exists. They are also exceptionally robust against noise, and they are immune to violations of assumptions that would cripple many traditional methods. This has led to their increased use in a wide variety of disciplines. Therein lies the mission and the purpose of this book. It is not a book on traditional signal and image-processing techniques. Nor is it a neural network text. Rather, it is a book about a marriage—a marriage between two disjoint fields whose union provides a whole new storehouse of approaches for solving signal and image processing problems.

A principal focus of this text is the multiple-layer feedforward network (MLFN). This model has become extremely popular for both classification and prediction. However, there is one difficulty with using the traditional MLFN in many signal and image-processing applications. These problems are often best approached in the frequency domain or in phase space. When this is done, the fundamental data on which the network operates is usually complex. The implication is that when the data is presented to a traditional neural network, half of the network's inputs are real numbers and half are imaginary. This is perfectly legal, and good results can usually be obtained with this approach. However, the intuition balks when there is a mismatch between a problem and its solution tool. In such cases, it would seem more virtuous to employ neurons that themselves operate in the complex domain. And indeed, it turns out that far more than just intuition is pleased when the network's domain is matched to the problem's domain. This book will present a wide variety of problems for which complex-domain networks significantly outperform their real-domain counterparts. They almost always train faster and are more resistant to becoming trapped in local minima. But most importantly, they have a strong tendency to generalize better. Suppose that one trains a real-domain network and a complex-domain network (having equal numbers of weights) on identical training data. When these two networks are tested using an independent validation data set, the complex-domain network will almost invariably have significantly lower error than the real-domain network.

A large section of this book is dedicated to the design and training of complex-domain MLFNs. All necessary equations are presented and justified. Highly commented code fragments for all aspects of computation are supplied. Complete source code for a working program is provided on disk.

Some of the most popular signal and image-processing algorithms are reviewed in this book. Those algorithms that are particularly suitable for marriage to complex-domain neural networks are emphasized. Source code for efficient implementations of the Gabor transform and for several versions of the Wavelet transform is given. The reader is also provided with advice and guidance on choosing the parameters for these algorithms when faced with specific problems. Complete source code for the radix-2 FFT is given, along with advice and algorithms related to some common frequency-domain normalization techniques. Methods for handling paths in phase space and normalizations appropriate for perimeter-based shape input are covered in detail. Convolution algorithms that operate in the time domain are presented first, as they lie close to the mathematical development. Fast algorithms that operate in the frequency domain are then shown, as they are more practical for large problems. The underlying thrust is always toward using these algorithms in conjunction with neural networks, and the interface between traditional methods and neural networks is discussed in detail.

Image processing as a generalization of signal processing receives considerable attention. The Fourier transform, Gabor transform, and Morlet wavelet transform are all generalized to two dimensions. Fast filtering with the two-dimensional DFT is covered. Some special image-processing techniques are also studied. Numerous methods for characterizing two- and three-dimensional shapes based on moments are discussed in detail. Variables that can be computed without segmenting the image are also given.

The final sections of many chapters in this book are devoted to a variety of contrived and real-life applications that benefit from the techniques presented earlier in the book. Many of the applications are at home in the complex domain, and processing is in accordance with this fact. A significant amount of space, however, is devoted to closely related problems that are better served by real-domain processing. This helps the reader to learn how to distinguish between the two cases. The fact that complex-domain networks are almost always superior will be clearly illustrated, so that the reader is encouraged to begin exploring them. At the same time, the reader will be shown that they are not always the best solution. Guidance for making that choice will be given through meticulously designed applications.

All of the program code in this book is in C++, as this language combines high-level, object-oriented structure with nearly the speed and compactness of assembler. An attempt has been made to use designs unique to C++ only when they are vital. Thus, much of the code can be easily adapted to C compilers. Vector and Matrix classes, as well as operator overloading, have been deliberately avoided. Such specialized constructs contribute little or nothing to program efficiency and complicate the lives of programmers who wish to use these algorithms in other languages. Strictly C++ techniques have been used only when they significantly improve program readability or efficiency. Because the operation of all programs is thoroughly discussed in the accompanying text, translation to other languages should not be difficult.

All of the C++ code in this book and on the accompanying disk has been tested for strict ANSI compatibility. No special proprietary libraries are ever referenced. The author uses Symantec C++ version 6.1. The code has also been compiled with Borland C++ Professional 4.0. It is anticipated that any ANSI C++ compiler should be able to compile the code correctly.

# Contents

# 1

# The Role of Neural Networks in Signal and Image Processing

A vast array of algorithms exists for processing time-series and image data. The development of many of these algorithms was inspired by specific applications (such as radar), and many years of experience have tweaked their operation to near perfection. In fact, some of the most prominent of such algorithms can be shown to be mathematically optimal for their tasks under certain assumptions. Therefore, this book will not dare to issue a sweeping indictment of traditional methods, with the goal of replacing them all by neural network algorithms. That would be senseless. Nevertheless, there are many reasons for considering neural network solutions to problems for which other solutions already exist. This chapter will examine some of those reasons. And, of course, there are many problems for which *no* effective traditional solution exists. Neural networks are often surprising in their ability to solve such problems.

## Why Use Neural Networks?

Let us start by answering a question that is somewhat more narrow than that posed by the title of this section. Why should we study neural network solutions to problems for which mathematically optimal solutions apparently already exist? Many of the practical examples presented in this book fall into that category. Superficially, it might seem to be a waste of time. But there is a very simple reason: the harsh reality of real-life problems. It is unlikely that the *exact* problem facing a particular researcher will be described in any textbook or scholarly paper. Similar problems may abound, and it may be that the researcher will be able to adapt the optimal traditional solutions in whatever ways are needed. However, such adaptation frequently destroys optimality. Sometimes the degradation is slight, and sometimes it is not. Sometimes what seems like a trivial departure from assumptions leads to the closest traditional solution being rendered invalid. The fact that there exists an excellent traditional solution to a similar problem is irrelevant if the reader's variation on the problem precludes use of that traditional solution. In such cases, the inclusion of neural networks in the solution can save the day.

Many algorithms that are mathematically optimal make assumptions that may be unreasonable in practice. The most common assumption is that contaminating noise follows a particular distribution. In many cases of practical interest, violation of those assumptions is poorly tolerated. Neural networks, on the other hand, are quite tolerant of unusual noise distributions. It may well be that when they are used on real-life problems, they outperform their "mathematically optimal" counterparts. Life would be much smoother if all physical phenomena were linear, with Gaussian error. The author is not acquainted with many such phenomena.

Another reason for considering neural network alternatives to theoretically superior traditional methods is that the neural network may be faster. Many applications of signal and image processing must operate in real time. If the "ideal" algorithm comes up with its answer too late, it's worthless. Neural networks are not only fast, but they are also intrinsically parallel. If a single-CPU implementation is not fast enough, chances are that a multiple-CPU approach will be acceptable. They are easy to implement in parallel. Even if another solution exists and can be made to execute on parallel processors, the implementation of a parallel version of that solution may be difficult.

Finally, there is always the possibility that there is no nice traditional solution to anything resembling the problem at hand. One can always delegate a clever person to analyze the problem and attempt to design a custom solution. On the other hand, neural networks are famous for finding solutions to baffling problems. It is nearly always cheaper and easier to drop the problem into the lap of a neural network and let it find a solution in whatever way it finds best.

## Choosing a Neural Network Model

There are hundreds of different neural network models described in the literature. On first glance, it might seem that the selection process would be difficult, given the number of possibilities. Luckily, that is not the case. Relatively few of the available models have a time-tested performance record. While it is always possible that some obscure model may provide the best solution to a particular problem, it is also likely that one of the few common models will provide an excellent solution. Therefore, this section will focus on the models that have formed the foundation of the author's storehouse. If the favorite model of any reader is omitted, that should not be interpreted as casting aspersions on that model. The vast majority of the preprocessing and integration techniques presented later in this text should work well with any reasonable neural network model, so readers should not feel constrained to the particular models mentioned here.

## Supervised versus Unsupervised Training

The first question to be answered is how training is accomplished. There are two primary methods. *Supervised* training uses known examples to train the network. This is the most common situation. Perhaps the researcher is examining photomicrographs of a manufactured material to determine whether or not samples are defective. Numerous examples of satisfactory and defective material are supplied, with the status of each sample known at training time. The network would be trained in supervised mode. This means that as each sample is presented to the network, its condition (defective or satisfactory) would also be told to the network.

The author has found two neural network models to be particularly effective for problems that can make use of supervised training. These are the *multiple-layer feedforward network* (MLFN) and the *probabilistic neural network* (PNN). There are many other contenders, but the author has found that one or both of these two models are as effective as anything else out there.

The MLFN is the "standard" neural network model. It was arguably the first practical neural network, and it has maintained its lead ever since. The original description of it can be found in Chapter 8 of [Rumelhart *et al.*, 1986]. Many details on its implementation and use are given in [Masters, 1993]. This text presents a generalization of the MLFN in which all neurons operate in the complex domain. That version will be shown to be superior to the strictly real version for a wide variety of problems. The MLFN program, supplied in both executable and source form, implements this network.

The other leading contender for supervised training applications is the PNN. It was first described in [Specht, 1990], although his implementation is mathematically identical to a statistical algorithm that appeared in [Meisel, 1972]. Donald Specht's contribution was in discovering how that extremely powerful statistical algorithm could be broken down into a collection of simple processes that are able to be executed in parallel. His neural network version of the Parzen-Bayes classifier finally brought to this algorithm the fame that it deserved. A detailed examination of this model appears in [Masters, 1993].

The choice between using the MLFN versus the PNN is usually very easy because they have extremely different properties. The chief disadvantage of the MLFN is that excruciatingly long training periods are often required. Some of the examples provided in this text required on the order of one hundred hours of training time. That problem is offset by the fact that execution of the trained network is among the fastest of all known models, as well as being intrinsically parallel. Real-time applications are often best served by the MLFN model.

The PNN is just the opposite. In its most basic form, training is, for all practical purposes, instantaneous. Unfortunately, execution time can be very slow. Even though the PNN algorithm is easily implemented on parallel processors, it is nearly always out of the question for applications in which speed is of paramount importance. Also, the PNN requires a relatively large amount of memory. In some cases this may be a difficulty.

The PNN has one clear advantage over the MLFN. It can be shown to be (asymptotically) a mathematically optimal classifier. Also, its method of operation is well understood and backed up by rigorous mathematics. For applications in which this sort of knowledge is vital, the PNN may be the best choice. Analysis of the MLFN is difficult, and sometimes nearly impossible. There are not many implementations of MLFN solutions in which the designers of the network fully understand how the network gets its answers. There are some tools for gleaning a few hints, but the task is not easy.

Another advantage of the PNN is that a byproduct of its computations is Bayesian posterior probabilities. For applications in which confidence levels are important, this is a tremendous plus. They are easily obtained, and they rest on a solid mathematical foundation.

Unfortunately, the PNN has a severe intrinsic limitation. It is fundamentally a classifier. It is designed exclusively for classification problems. General function mapping can be coaxed out of it as is demonstrated in [Schioler and Hartmann, 1992]. Even autoassociative versions exist as discussed in [Masters, 1993]. However, these are of questionable merit. Applications that do not involve classification probably will not be served well by PNN models.

The final decision criterion involves the training set. The PNN is unusually picky about having a thorough training set. It often does not generalize as well as the MLFN. If the training data is expensive to collect, and hence sparse, it would probably be better to use an MLFN. On the other hand, the PNN is better able to handle outliers. If it is to be expected that the training set will include wild points that cannot be reliably removed, the PNN is less likely to be adversely impacted by this abnormal data.

So far, this section has been devoted to deciding between the two principal supervised training models. There is another possibility. It may be that true class membership for the training cases is not known. Then we must resort to *unsupervised* training. The neural network is given the burden of deciding how to separate the training cases into unique classes. The best neural network model for handling this problem is most likely a member of the *Kohonen* family. The definitive source for information on that family of neural networks is [Kohonen, 1989]. A much more readable description of the simplest version of the Kohonen network can be found in [Caudill, 1990]. A detailed implementation of the model is given in [Masters, 1993]. A theoretical discussion of the Kohonen learning algorithm is given in [Lo, Yu, and Bavarian, 1993]. A very nice description of two popular versions of the model is in [Maren, Harston, and Pap, 1990].

In the interest of completeness, it should be mentioned that there is a third type of training that lies somewhere between supervised and unsupervised methods. In *reinforcement learning*, the neural network is allowed to react to each training case. It is then told whether its reaction was good or bad. The author has not yet seen any practical applications of this procedure, so it will not be discussed further. However, the algorithms described in this text should certainly be applicable to reinforcement learning situations.

## Real-Domain versus Complex-Domain Networks

Suppose that after having studied the decision criteria discussed earlier, the reader has selected an MLFN model to solve his or her problem. The next choice is whether the network is to be constructed with real-domain or complex-domain neurons. Nearly all traditional neural networks operate strictly in the real domain. However, it has recently been discovered that when an application's data is inherently complex, as could be the case for processing frequency-domain or phase-plane data, performance of MLFNs can be significantly improved by generalizing the neurons for operation in the complex domain. Training speed and reliability usually increase dramatically, and generalization quality is almost always superior to that obtained with strictly real-domain networks.

That fact is especially important in the context of this book. Many signal- and image-processing problems are best approached in the frequency domain, or in phase space. When this is done, the fundamental data on which the network operates is usually complex. The implication is that when the data is presented to a traditional real-domain neural network, half of the network's inputs are real numbers and half are imaginary. This is perfectly legal, and good results can usually be obtained with this approach. However, the intuition rightly balks when there is a mismatch between a problem and its solution tool. In such cases, it would be more virtuous to employ neurons that themselves operate in the complex domain. And far more than just intuition is pleased when the network's domain is matched to the problem's domain. Donald Birx and Stephen Pipenberg appear to be the first to publicize the superiority of complex-domain neural networks for complex-domain problems. This book will present many more problems for which complex-domain networks significantly outperform their real-domain counterparts. They almost always train faster and are more resistant to becoming trapped in local minima. But most importantly, they have a strong tendency to generalize better. Suppose that one trains a real-domain network and a complex-domain network (having approximately equal numbers of weights) on identical training data. When these two networks are tested using an independent data set, the complex-domain network will almost invariably have significantly lower error than the real-domain network. This is despite the fact that the real-domain network will invariably score better on the training set.

How does the reader go about deciding whether a real-domain or complex-domain MLFN is more appropriate to a particular problem? In many cases, it will be best if the user tries both models and chooses whichever gives the best performance. However, there are some general guidelines that can be followed to give strong hints in advance as to which will be superior. A careful reading of the examples throughout this text will help to develop a feeling for good criteria. This section will explore this topic on the most general level.

First and foremost, it must be understood that complex-domain neural networks are not a universal solution to problems. They are appropriate primarily when the data intrinsically occurs as *pairs* of numbers. This text makes frequent use of complex numbers, where each is comprised of a real part and an imaginary part. There may certainly be other possible pairs of measurements that would benefit from processing in the complex domain. But the point is that if the measured data does not have a fundamental pair structure, there is rarely any benefit to using complex-domain neural networks. Of course, when the data is complex, there can be tremendous gains realized by processing entirely in the complex domain. Only realize that the first question the user must ask is whether or not the data has the requisite pairwise organization. If not, there is little point in wasting time on complex-domain networks.

Next, it must be understood that for many applications, the whole point to using complex-domain data is to capture phase information. The implication is that phase is important. Later on, some examples will appear in which discarding the phase and working only with magnitude provides results that are superior to results obtained by working entirely in the complex domain. In those examples, it will be shown that the phase actually contributes little or no useful information. In such cases, the addition of irrelevant phase information results in overfitting the model to idiosyncrasies of the training set. If possible, the question of the relevance of the phase should be carefully considered.

The primary advantage of complex-domain neural networks relative to traditional real-domain networks is their usually superior generalization ability. The author has seen many cases in which the training-set error of a complex-domain network is inferior to that of a comparable real-domain network, while its validation-set error is tremendously superior. Therefore, the user who trains networks of both types must never compare their performance based on the training error. *Always compute the error of an independent test set and use that error measure to compare the networks.*

The experience of the author has led to some general conclusions concerning relative performance under various conditions. No definitive statements are possible, as the following guidelines are all based on empirical evidence. Additional guidelines, perhaps even accompanied by rigorous proofs, would be greatly welcomed by the neural network community. Some guidelines now follow:

- If the training data is clean while the test data is noisy, real-domain networks will be favored. Complex-domain networks tend to perform best when the noise level in the training set is comparable to that in the test set.

- If the training data is dense, thoroughly covering the problem domain, real-domain networks will be favored. Complex-domain networks are at their best when dealing with sparse training sets.

- If there are very few variables, real-domain networks will be favored. Complex-domain networks are much less susceptible to the overfitting that often results from a large number of inputs.

- If training time is limited, complex-domain networks are favored. In virtually every application pursued by the author, they reached a satisfactory level of learning at least twice as fast as a comparable real-domain network. In many cases, real-domain networks required ten, or even more, times as much learning time as their complex-domain counterparts.

The above criteria point to a broad generalization that can be made. If a problem is relatively simple, featuring few variables and an abundance of good, clean training data, with plenty of training time available, a real-domain network may be the best choice. But if the problem is more typical of those found in the real world, having many variables of questionable importance, and if the training data is difficult to come by, perhaps unavoidably contaminated with noise, a complex-domain network will probably outperform any real-domain network.

## Sizing the Network

One troubling issue facing those who must implement an MLFN solution to a problem is how to select an appropriate size for the network. How many layers should be used, and how many hidden neurons should be in each? Those questions are dealt with in many other texts, so not much space will be devoted to them here. However, a few basic guidelines will be presented. Interested readers are directed to [Masters, 1993] for an extensive treatment of these topics.

There are a few fundamental rules that, if followed, will practically guarantee the best possible results. These are the following:

- Use a training set that is thoroughly representative of the actual population. All possibilities should be taken into account, and all classes and subclasses should be present in approximately the same proportions as they are expected in use. This will make determining the best size for the network easier.

- The only exception to the above rule comes about if the designer has some advance knowledge that some cases will be easy to classify, while others will be difficult. It usually helps if the borderline cases are overrepresented in the training set. Easy cases can be more sparse.

- Use only one hidden layer. There are a few very rare situations in which two hidden layers may be preferable to one. More than two hidden layers are never theoretically needed, and the author has never seen a real-life problem in which more than two are needed.

- Use as few hidden neurons as possible. Start out with just two (that's right!) and train and test the network. Add neurons one at a time as needed. The author has seen almost no practical networks that required more than ten hidden neurons, and usually about three to six are optimal. If your application seems to need more, consider the possibility that the problem can be broken down into several simpler subproblems. *Never determine the optimal number by starting with a lot, then removing some.*

- Train the network for as long as possible. Overtraining will not occur if the training set is thorough and an excessive quantity of hidden neurons is avoided. Then test the trained network on a validation set. If performance on that independent set of data is poor, then either the training set was inadequate or too many hidden neurons were used. If reducing the number of hidden neurons results in inadequate performance on even the training set, add a neuron, merge the validation set into the training set and retrain. Then test it with a new validation set.

The flowchart shown in Figure 1-1 illustrates the design and training process used by the author.

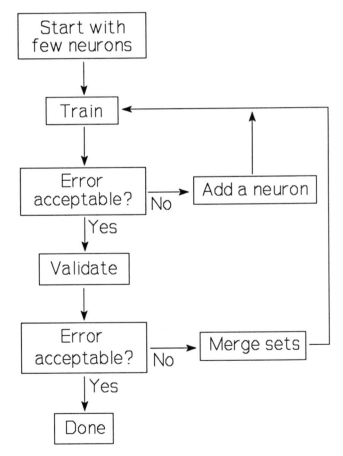

Fig. 1-1: Sizing and training a neural network.

# Integrating Neural Networks with Traditional Algorithms

There are an infinite number of ways that neural networks can be used to solve problems involving signal and image processing. This text focuses on using traditional algorithms as preprocessors for neural networks. For some very different approaches, see [Kosko, 1992]. The principal methodologies covered here are diagrammed in Figure 1-2.

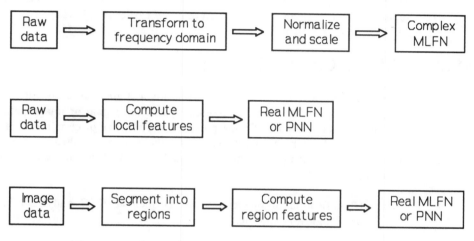

Fig. 1-2: The principal methodologies of this book.

Most of this text will follow the path shown at the top of the figure. The raw data, be it signal or image, is transformed into the frequency domain by means of the Fourier transform or some near relative, then normalized as needed. Time/frequency analysis by means of the Gabor transform is included here, as well as time/scale analysis by wavelets. In most cases, the transform will generate complex-domain data. That data will be processed by a complex-domain multiple-layer feedforward network. Actual applications will be given, and processing with real-domain networks will also be done to demonstrate the superiority of the fully complex approach.

Although Fourier techniques are a major focus of this book, other techniques are also presented. One application that is being actively pursued is developing the means to move a cursor around an image to select training areas, then teaching a neural network to classify similar images based on the training set. While Fourier methods can be valuable for this, it is also often the case that there is no pressing reason for operating in the frequency domain. A section of this text is devoted to the second path shown in Figure 1-2, in which local features that are not frequency-related are computed. Relatively simple and fast-to-compute measures of tone and texture can provide neural networks with effective inputs. Many of these measures can be easily adapted to time-series analysis as well.

A third approach to image processing involves segmentation *before* neural networks are called into play. This is shown at the bottom of Figure 1-2. Segmentation algorithms will not be discussed at all here, as they have received extensive treatment in many standard image-processing texts. However, we will present a variety of means of describing the shape of segmented regions. These shape descriptors will be defined in such a way that they are particularly suitable for use as neural network inputs.

Most of the non-Fourier techniques result in real-domain data. This means that they are best processed by either real-domain MLFNs or by probabilistic neural networks. Suggestions for making that choice were given earlier in this chapter.

## What Problems Can Be Solved?

The majority of the techniques presented in this text are geared toward pattern detection and classification. We want to recognize particular signal or image patterns. That in itself may be the end goal. Or the presence and absence of certain sets of patterns may be the means by which sample classification is accomplished. Some examples of applications for which the techniques of this book would be particularly valuable are the following:

- Classification of bioelectric signals such as EKG and EEG into normal or pathological conditions

- Classification of objects that cause SONAR returns

- Detection of signals having known form amidst noise

- Detection of stock market patterns that presage interesting moves

- Detection of mechanical faults in machinery based on visual or sonic information

- Detection of physical flaws in materials based on X-ray diffraction images

- Target acquisition and false alarm rejection using images obtained from military sensors

- Classification of lesions based on photomicrographs

- Parameter estimation for data generation models

Naturally, this list is far from complete. It is nothing more than a small sampling of typical applications of neural networks. These applications have been found to especially benefit from the integration of traditional approaches with neural networks.

# 2

# Neurons in the Complex Domain

There are many physical phenomena that inherently exist in the complex domain. The path of a point in two-dimensional phase space can be viewed as a tour of the complex plane. When the Fourier transform is applied to data, real or complex, the resultant mapping into the frequency domain brings it into the realm of complex numbers. Wavelets and many of their relatives often live in the complex domain. The list is endless.

It is good when the design of a neural network reflects the nature of the data that it is to process. Thus, when we are dealing with complex-valued data, we are led to construct our network from neurons that accept complex-valued inputs and produce complex-valued outputs. As will be seen in the applications presented later, it is not unusual for complex-domain neural networks to very significantly outperform their equivalent real-domain counterparts. This chapter will discuss multiple-layer feedforward networks which arise from neurons in the complex domain.

Note that no claim is made that complex-domain networks are theoretically superior to real-domain networks in terms of their ultimate capabilities. In fact, it is the experience of the author that for simple problems, and when unlimited training time is allowed, both types of network perform similarly. On the other hand, extensive experience indicates that complex-domain neural networks almost always train faster and more reliably than their real-domain counterparts when the data on which they operate is complex. They seem to be less likely to become trapped in local minima, and gradient descent descends more rapidly. Finally, and perhaps most importantly, complex-domain networks have a strong tendency to generalize better than real-domain networks having the same number of weights. The complex network may actually have a slightly higher training-set error than the corresponding real network, but its validation-set error will probably be lower. And that's what really counts.

## Review of Complex Numbers

This section contains a cursory review of what complex numbers are and how we perform basic arithmetic operations on them. Understanding this material is a prerequisite to understanding most of the remainder of this chapter. No attempt at strict mathematical rigor will be made. The presumed audience is comprised of persons having a basic grasp of introductory-level college mathematics, but needing a review of concepts important to this chapter.

Complex numbers may be thought of as pairs of real numbers, $(a, b)$, for which the operations of addition and multiplication are defined in a special way:

$$(a,b) + (c,d) = (a+c, b+d)$$
$$(a,b) \cdot (c,d) = (ac-bd, ad+bc)$$

$$(2\text{-}1)$$

Ambitious readers may wish to verify that the above operations satisfy the field axioms. They are associative and commutative, and the distributive law applies. The additive identity is $(0, 0)$, and the multiplicative identity is $(1, 0)$. Every nonzero complex number has a multiplicative inverse that will be shown as soon as a few other definitions are in place.

It is convenient to write $a$ as shorthand for the complex number $(a, 0)$. This allows us to consider the real numbers to be a subset of the complex numbers. We also define the special quantity $i = (0, 1)$. This means that we can write $(a, b)$ as $a + bi$, which is just what we shall do from now on.

Applying the rule for multiplication, we see that $i^2 = -1$. The fact that $i$ is the square root of $-1$ is probably the most famous feature of complex numbers. We are also in a position to define the reciprocal of a nonzero complex number.

$$\frac{1}{a+bi} = \frac{a-bi}{(a+bi)(a-bi)} = \frac{a}{a^2+b^2} + \left(\frac{-b}{a^2+b^2}\right)i \qquad (2\text{-}2)$$

One special definition with which we should be familiar is the *conjugate* of a complex number. This is computed by flipping the sign of the imaginary part. It is usually written by placing a bar above the variable:

$$\text{if } z = a+bi \quad \text{then} \quad \bar{z} = a-bi \qquad (2\text{-}3)$$

It is often profitable to think of complex numbers as points in the Cartesian plane. This facilitates representing them in polar coordinates. Look at Figure 2-1, which shows the single complex number $z = a + bi$ plotted as a point.

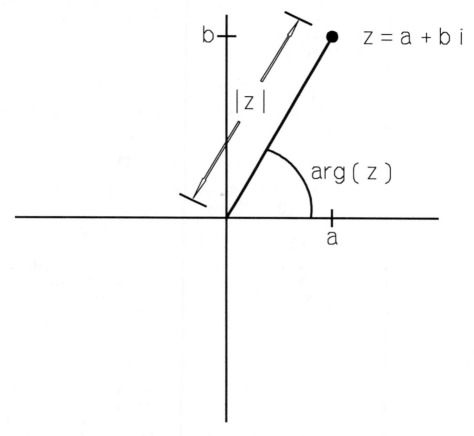

Fig. 2-1: A complex number on a plane.

The *absolute value* or *magnitude* of a complex number is defined as its length as shown in the figure. The *argument* of $z$, written $arg(z)$, is the counterclockwise angle that the vector from the origin to $z$ makes with the positive real axis. These definitions are formalized in Equation (2-4).

$$
\begin{aligned}
z &= a + bi \\
|z| &= \sqrt{a^2 + b^2} \\
a &= |z|\cos(arg(z)) \\
b &= |z|\sin(arg(z))
\end{aligned}
\qquad (2\text{-}4)
$$

The notion of a complex number being expressed in polar coordinates gives us a frequently useful alternative to the $a + bi$ notation. Observing that the real part of a unit length complex number is the cosine of its argument, and the imaginary part is the sine, we are inspired to write

$$\text{cis } \theta = \cos \theta + i \sin \theta \qquad (2\text{-}5)$$

Complex numbers that are not unit length would be written by multiplying by the length. For example:

$$3 + 4i \approx 5 \text{ cis } 53 \qquad (2\text{-}6)$$

It is interesting to note that the multiplication rule for complex numbers can be nicely written using polar coordinates.

$$(r_1 \text{ cis } \theta_1)(r_2 \text{ cis } \theta_2) = r_1 r_2 \, cis(\theta_1 + \theta_2) \qquad (2\text{-}7)$$

That alternative formulation for the multiplication rule falls into place if we explore a rather exotic but immensely useful mathematical property of complex numbers. We should all be familiar with exponentiation. We know that $e^0 = 1$, $e^1 = e$, $\log(e^x) = x$, and $e^a e^b = e^{a+b}$. But what happens if we raise $e$ to a complex power, rather than a plain old real power? We now state without justification the following fact:

$$e^{iz} = \cos z + i \sin z \qquad (2\text{-}8)$$

Although Equation (2-8) is true for any complex value of $z$, we will here be concerned with strictly real values of $z$. Readers should carefully ponder how that equation gives us an alternative form of polar representation, as that will be the method of choice in the next chapter when we discuss Fourier transforms. And how about that multiplication rule now?

Later in this chapter we will be discussing functions that map complex numbers to other complex numbers. In that context, we will also need to consider *derivatives* of these functions. Let us now briefly review derivatives in the real domain, hint at what fierce things derivatives are in the complex domain, and finally show how we will sidestep that perilous field by working with *partial derivatives*.

The reader may recall from basic calculus that the derivative of a function (from the reals to the reals) is the relative degree to which that function changes in response to a small change in its domain variable.

$$f'(x) = \lim_{h \to 0} \frac{f(x+h) - f(x)}{h} \qquad (2\text{-}9)$$

Implicit in that definition are that the limit exists, and that it is the same regardless of whether we approach $x$ from the left (negative $h$) or from the right (positive $h$).

The fact that we can approach $x$ from only the left or the right makes real-domain derivatives straightforward. Would that it were so simple in the complex domain. Here, there are an infinite number of directions from which we can approach a point. And the limit of the relative change in the function value must be the same for all of those directions! Little thought is needed to become convinced that this is a very stiff requirement. If we have some function that has a derivative throughout its domain, and if that derivative is continuous, then we say that our function is *analytic*. This term will come up later, so it was roughly defined here. But this is as deep as we need to dig.

We will often be dealing with functions whose domain is real-valued vectors and whose range is the reals. Such functions can have partial derivatives with respect to any element of the vector. (Higher-order derivatives will not be discussed.) This is the relative change effected in the function value in response to a small change in one of the components of the domain vector. For example:

$$\frac{\partial}{\partial x} f(x,y) = \lim_{h \to 0} \frac{f(x+h,y) - f(x,y)}{h} \qquad (2\text{-}10)$$

When we have a function from the complex domain to the complex domain, derivatives become much simpler if we separate the real and imaginary parts of the domain and range. Treat it as two functions, each from the complex domain to the real range. One of these functions defines the real part of the complex-valued function, and the other defines the imaginary part.

$$f(a+bi) = f_r(a+bi) + f_i(a+bi)\,i \qquad (2\text{-}11)$$

Then we can deal with four partial derivatives: the real part of the function with respect to the real part of the domain variable, the imaginary part of the function with respect to the real part of the domain variable, the real part of the function with respect to the imaginary part of the domain variable, and the imaginary part of the function with respect to the imaginary part of the domain variable. Functions from the complex domain to the complex domain for which those four partial derivatives exist and are continuous are vastly more common than functions that are analytic.

## Network Architecture in the Complex Domain

The title of this section may be a bit misleading in that it implies that network architecture in the complex domain is significantly different from that in the real domain. In reality, they are essentially identical. Instead of using ordinary addition and multiplication, we must use the special versions defined in Equation (2-1). We generally need fewer hidden neurons, as we have twice as many weights for each. And we need special consideration for the outputs when our network is a classifier. But beyond those minor considerations, complex-domain and real-domain neural networks look and act identically. Nevertheless, it is in our best interest to review the structure of a multiple-layer feedforward network and to set forth the formulas used to compute the activations of all neurons.

The basic network model that we will be examining in this chapter has one or more *input neurons* by which worldly data is presented to the network, and one or more *output neurons* through which the network returns its results to the world. It may (and usually will) have one or more *hidden neurons* that have no contact with the outside world. The hidden neurons are often the workhorses of the network, the means by which complicated patterns are processed in useful ways. The hidden neurons are organized into one or more layers. This chapter will deal almost exclusively with networks having one hidden layer, although extensions of results to more hidden layers will be discussed when appropriate. Figure 2-2 shows a small neural network.

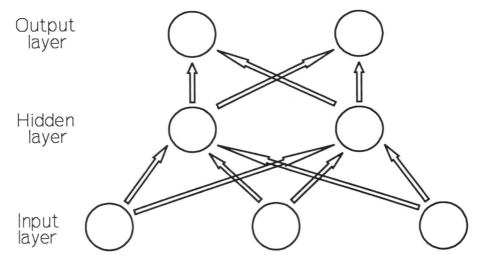

Output
layer

Hidden
layer

Input
layer

Fig. 2-2: A three-layer feedforward network.

Information passes in only one direction through the network. The outputs of the input neurons go to the inputs of the hidden neurons, and the outputs of the hidden neurons go to the inputs of the output neurons. Information does not pass backwards, from output toward input. There is no communication among neurons in a given layer. These are the properties that are reflected in the term *feedforward* when we speak of a multiple-layer feedforward network. Some models do allow information to skip layers. Input neurons can feed directly to output neurons without intervention of the hidden layer. These models will not be discussed here.

The input neurons are purely hypothetical constructs. They do not really exist. They do no computations and effect no actions. They simply have outputs, and those outputs are really the inputs to the network. When we present a pattern to the network, that pattern is sent directly to the hidden neurons (or to the output neurons if there is no

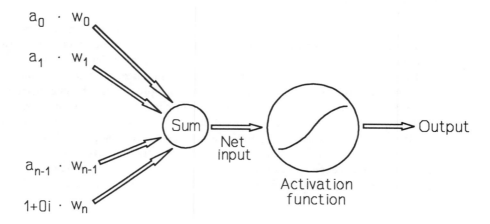

Fig. 2-3: Computing the activation of a neuron.

hidden layer). When we do this, we say that the input layer has outputs equal to the input presentation.

The output activation level of each neuron in the network is computed with the same two-step operation. First, its inputs from all neurons' outputs in the previous layer are individually multiplied by weights, and the sum of these weighted inputs is found. One additional hypothetical input, called the *bias*, is weighted and included in the sum. Its value is always equal to $1 + 0i$. This weighted sum of inputs is usually called the *net* input to the neuron. The *dot product* operation involved in its computation is normally the major eater of time in neural network learning and execution and should always be programmed with extra care.

The second step in computing a neuron's output is that the net input is acted on by an *activation function*. This function is described on page 18. The value of the activation function determines the output of the neuron. Traditionally, all neurons in a given layer employ the same activation function. Usually, all neurons in the entire network use the same function, although the output layer sometimes has its own special function. The computation of a single neuron's output activation given its inputs is shown pictorially in Figure 2-3, and is made mathematically explicit in Equation (2-12). Remember that all terms in that equation are complex numbers, and that their addition and multiplication are governed by Equation (2-1).

$$\text{out} = f(\textit{net}) = f\left(\sum_{j=0}^{n-1} a_j w_j + w_n\right) \qquad (2\text{-}12)$$

## Classification Networks

We will sometimes use complex-valued neural networks to perform classification tasks or to make binary decisions. In these cases it is silly to employ complex-valued outputs. Granted, we could conceivably use them in the binary decision. We could train one decision to produce $.3 - .6i$ and the other to produce $-.3 + .6i$, or some other bizarre values. But it would be far more efficient to simply discard the imaginary part and train for extreme values of the real part. In subsequent sections, many formulas and code fragments will appear. In every case, it will be seen that setting to zero the imaginary part of every term associated with the network's final output will produce tremendous simplification. It is highly recommended that programmers take this approach. The code shown in this chapter does not address this simplification, although it should be obvious how to modify it. However, the MLFN code supplied on disk does address this issue.

# The Activation Function

There is an infinite number of choices for activation functions. Early networks used simple threshold functions for each neuron's activation function. Such a neuron's output is equal to one if its net input reaches a predefined threshold. Otherwise, its output is zero. This function is neither continuous nor differentiable. It was soon found that performance could often be improved by modifying this function to make it continuous. The modified function still returns zero when the threshold is not reached. But when the net input reaches the threshold, the output is equal to the net input minus the threshold. This function is still not differentiable at the threshold, but its continuity constitutes an improvement over its predecessor.

Perhaps the greatest single leap in neural network technology was made when Rumelhart, Hinton, and Williams described in [Rumelhart *et al.*, 1986] how the gradient of the mean squared error function of a neural network could be computed. This paved the way for practical learning algorithms. Their method required that the activation functions be differentiable throughout their domain, though. This meant that a new generation of activation functions was needed.

It has become generally agreed that the most effective activation functions have a *sigmoid* (S-shaped) form. A sigmoid function is a function from the reals to the reals which is bounded, differentiable, and has a positive derivative everywhere. By convention, sigmoid activation functions are also assumed rapidly to approach their bounds asymptotically as their domain variable grows in absolute value, and to be roughly linear for values near the center of their domain. Years of experience with various sigmoid activation functions have led to the conclusion that the exact choice of the function has very little effect on the capabilities of the network. On the other hand, the shape of the function can have a profound effect on the speed with which the network learns. Papers are published regularly that propose and justify activation functions. The smoke has not yet cleared.

The longest-running contender is the logistic function, shown in Equation (2-13). It satisfies all of the basic criteria for a suitable activation function. In addition, its derivative can be computed quickly and easily.

$$f(x) = \frac{1}{(1 + e^{-x})}$$

$$f'(x) = f(x)(1 - f(x))$$

$$(2\text{-}13)$$

Practical experience indicates that the logistic function has one minor drawback in that it is positive everywhere. Connection weights, especially those for bias terms, tend to fall in more easily learned patterns when the activation function outputs can be both positive and negative and are centered at zero. This leads us to rescale the logistic function.

$$g(x) = 2f(2x) - 1 = \tanh(x) = \frac{e^x - e^{-x}}{e^x + e^{-x}} \qquad (2\text{-}14)$$

The *hyperbolic tangent* function shown in Equation (2-14) is one of the most popular activation functions at this time and is the author's favorite. Furthermore, [Kalman and Kwasny, 1992] make an eloquent argument in its favor. The logistic and hyperbolic tangent functions are graphed in Figures 2-4 and 2-5, respectively.

The discussion up to this point has been confined to activation functions in the real domain. It is now time to venture into the complex domain. Things are not so neat and tidy here. One of the most complicating factors is that derivatives are not straightforward extensions of their real-domain counterparts. A real point can be approached only from the left and right. A point in the complex domain can be approached from any direction. This makes even the very existence of derivatives in the complex domain a much more serious affair. We will sidestep this problem by avoiding direct use of complex derivatives. Rather, we will write all complex numbers as real and imaginary parts and refer to the partial derivatives of the real and imaginary parts of the function with respect to the real and imaginary parts of its domain variable. As will be seen later, this gives us far more latitude in choosing an activation function.

It is not generally possible to find complex extensions of real activation functions that are usable in the complex domain. For example, it has already been established that the hyperbolic tangent function is an excellent real-domain activation function. Now take a look at the real and imaginary parts of the *tanh* function in the complex domain. These are plotted (with vertical truncation) in Figures 2-6 and 2-7, respectively. Our intuition tells us in no uncertain terms that this does not look like a useful activation function! Moreover, the fact that it has violent discontinuities across its domain precludes its use if gradients are to be computed for learning algorithms. Something better is needed.

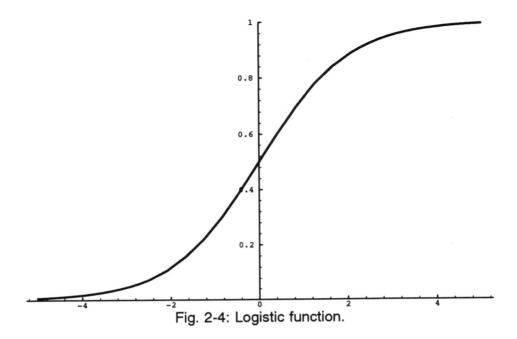

Fig. 2-4: Logistic function.

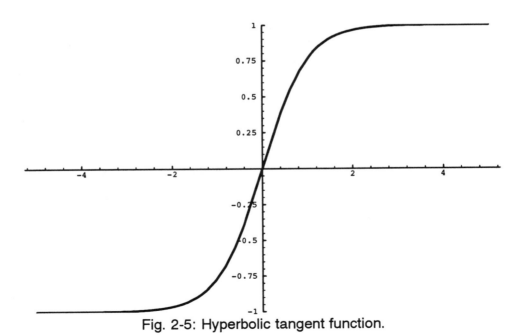

Fig. 2-5: Hyperbolic tangent function.

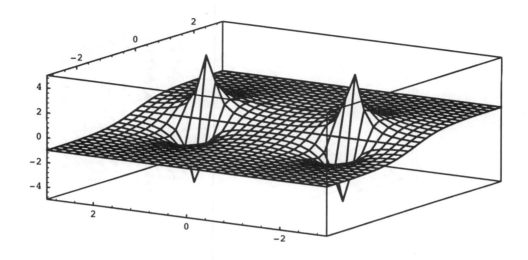

Fig. 2-6: Real part of complex tanh function.

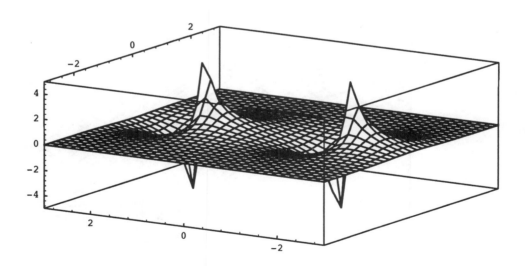

Fig. 2-7: Imaginary part of complex tanh function.

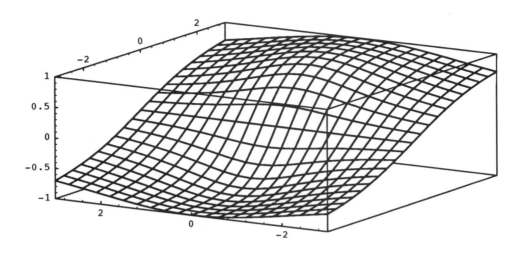

Fig. 2-8: Real part of complex squashing function.

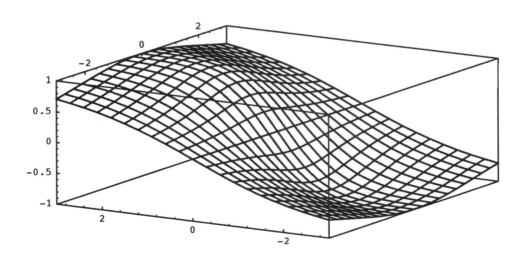

Fig. 2-9: Imaginary part of squashing function.

Consider the properties that a good activation function in the complex domain should have. These properties are intuitive extensions of the definition of a sigmoid function in the real domain. Thus, we would certainly like its real and imaginary parts to be continuously differentiable with respect to the real and imaginary parts of the domain variable. (It is unreasonable to expect the function to be analytic, though. This would impose a severe and unnecessary restriction on our choices.) We would like its magnitude to be bounded, and we would like it to approach those bounds rapidly as the magnitude of the domain variable increases. Finally, we would like the function to be approximately linear when the magnitude of the domain variable is small.

An immediately obvious possibility is to apply some squashing function to the real and imaginary parts separately. For example:

$$f(x+yi) = \tanh(x) + \tanh(y)\,i \qquad (2\text{-}15)$$

[Birx and Pipenberg, 1992] report good results using this technique.

There is another appealing property that functions like that shown in Equation (2-15) do not have, though. It would be nice if the activation function preserved the direction of the net input, squashing only its length. This is easily implemented. Suppose that $s(x)$ is a sigmoid function having $s(0) = 0$. The hyperbolic tangent is one such function. We will only be concerned with nonnegative values of its domain variable. Then we can squash the length of a complex number without changing its direction by means of the function shown in Equation (2-16).

$$f(x+yi) = px + pyi$$
$$p = \frac{s\left(\sqrt{x^2+y^2}\right)}{\sqrt{x^2+y^2}} \qquad (2\text{-}16)$$

The real and imaginary parts of that function for $s(x) = \tanh(x)$ are graphed in Figures 2-8 and 2-9, respectively. That activation function computes the output by simply multiplying the net input by a factor. The factor is the ratio by which the real-domain squashing function, $s(x)$, compresses the length of the net input. The activation function defined by Equation (2-16) when $s(x) = \tanh(1.5\,x)$ is the author's standard workhorse. The factor of 1.5 is suggested in [Kalman and Kwasny, 1992] but is certainly not vital.

## Output Activation Limits

Experienced users of real-domain neural networks know that it is futile to try to train a network to achieve its theoretical maximum or minimum output. For example, if the logistic function is used, an activation of 0.1 is generally considered to be *off* and 0.9 to be *on*. For the hyperbolic tangent function, these limits are typically –0.9 and 0.9, respectively. Trying to train the network to learn values outside this range will seriously degrade performance.

One must be careful when extrapolating these limits to complex-domain neurons. When, for example, Equation (2-16) is used for output activations, we must be concerned with the *length* of the output, not its real and imaginary parts taken individually. We must not try to train a neuron to respond with a value of 0.9 + 0.9*i*, for the length of that complex number is 1.27, an unobtainable level!

If we are training the outputs to learn real numbers, so that the imaginary part is guaranteed to be zero, we can impose the traditional limits of –0.9 to 0.9. In the fully general case, we are safer imposing limits of –0.64 to 0.64 or so. This limits the length to about 0.9.

## Derivatives of the Activation Function

In the next section, when we compute the error gradient, we will need the derivatives of the activation function. This section will show how to compute those derivatives for the model expressed by Equation (2-16).

Let us start by rewriting half of Equation (2-16) as a function of two real variables. It should be apparent that we can use Equation (2-17) to compute both the real and imaginary parts of the activation function according to whether we let *x* take the role of the real or imaginary part of the net input, respectively.

$$h(x,y) \; = \; \frac{x\, s\!\left(\sqrt{x^2+y^2}\right)}{\sqrt{x^2+y^2}} \tag{2-17}$$

To simplify the next few equations, define the length of the net input.

$$L \; = \; \sqrt{x^2+y^2} \tag{2-18}$$

We can now write the partial derivative of $h(x,y)$ with respect to $x$.

$$\frac{\partial h}{\partial x} \; = \; \frac{s(L)}{L} - \frac{x^2 s(L)}{L^3} + \frac{x^2 s'(L)}{L^2} \tag{2-19}$$

When Equation (2-16) is used to compute the real part of the activation function by letting $x$ play the role of the real part of the net input (with $y$ being the imaginary part), Equation (2-19) gives us the partial derivative of the real part of the output with respect to the real part of the input. Similarly, when $x$ is the imaginary part of the input and $y$ is the real part, Equation (2-19) is the partial derivative of the imaginary part of the output with respect to the imaginary part of the input.

To complete the picture, we need the partial derivative of $h(x,y)$ with respect to $y$.

$$\frac{\partial h}{\partial y} = \frac{xys'(L)}{L^2} - \frac{xys(L)}{L^3} \qquad (2\text{-}20)$$

That equation allows us to compute the partial derivative of the real part of the output with respect to the imaginary part of the input, and the partial derivative of the imaginary part of the output with respect to the real part of the input.

Code for computing all four partial derivatives is shown below. Several features of this code should be noted. First, as is frequently the case, it is most efficient to compute both the activation and its partial derivatives simultaneously, so that is what we do. Second, note the provision for a linear activation function. If this is an output neuron, we may want linearity. In this case, computations are trivial. The input and coefficient vectors contain complex numbers stored as pairs. The real part of each element is immediately followed by the imaginary part. The output is a single complex number and is stored the same way. Finally, observe that we call the subroutines act_func and actderiv to compute the real-domain squashing function (heretofore referred to as $s(x)$) and its derivative. These routines will be discussed shortly.

```
void partial (
   double *input ,              // Input vector from previous layer
   double *coefs ,              // Weight vector: real, imag, real, imag, ...
   double *output ,             // Computed output, real then imaginary
   int ninputs ,                // Number of inputs (not counting bias)
   double *deriv_rr ,           // Partial of real wrt real
   double *deriv_ri ,           // Partial of real wrt imaginary
   double *deriv_ir ,           // Partial of imaginary wrt real
   double *deriv_ii ,           // Partial of imaginary wrt imaginary
   int linear                   // Is this squashing function linear?
   )
{
   double rsum, isum, raw_length, squashed_length, ratio, d, l2, temp ;

   dotprodc ( ninputs , input , coefs , &rsum , &isum ) ;   // Find net input
   rsum += coefs[2*ninputs] ;                               // Include bias
   isum += coefs[2*ninputs+1] ;                             // which is 1+0i

   if (linear) {                                            // Trivial: linear squashing
      *output = rsum ;
      *(output+1) = isum ;
      *deriv_rr = *deriv_ii = 1.0 ;
      *deriv_ri = *deriv_ir = 0.0 ;
      return ;
      }
```

```
raw_length = sqrt ( rsum*rsum + isum*isum ) + 1.e-30 ;
squashed_length = act_func ( raw_length ) ;        // Squashing function
d = actderiv ( squashed_length ) ;    // Its derivative
ratio = squashed_length / raw_length ;             // Amount we squash length

*output = rsum * ratio ;                           // Compute output
*(output+1) = isum * ratio ;                       // And its imaginary part

l2 = raw_length * raw_length + 1.e-30 ;            // Work areas
temp = (d - ratio) / l2 ;                          // avoid recomputation

*deriv_rr = ratio  +  rsum * rsum * temp ;         // Equation (2-19)
*deriv_ii = ratio  +  isum * isum * temp ;         // Ditto
*deriv_ri = *deriv_ir = rsum * isum * temp ;       // Equation (2-20)
}
```

The following routine computes the dot product of two complex-valued vectors. It uses the traditional technique of unfolding the summation loop to reduce the number of branch decisions. An obvious benefit is that decrements of the loop counter k, along with decisions to see if it is zero, are reduced by a factor of four. A more subtle but often more valuable result is attained for processors that rely heavily on pipelining. By avoiding breaking the pipeline, operation is streamlined.

```
void dotprodc (
   int n ,                           // Length of vectors
   double *vec1 ,                    // One of the vectors to be dotted
   double *vec2 ,                    // The other vector
   double *re ,                      // Real part of output
   double *im )                      // and imaginary
{
   int k, m ;

   *re = *im = 0.0 ;                 // Will cumulate dot product here
   k = n / 4 ;                       // Divide vector into this many groups of 4
   m = n % 4 ;                       // This is the remainder of that division

   while (k--) {   // Do each group of 4
     *re += *(vec1  ) * *(vec2  ) - *(vec1+1) * *(vec2+1) ;
     *re += *(vec1+2) * *(vec2+2)  - *(vec1+3) * *(vec2+3) ;
     *re += *(vec1+4) * *(vec2+4)  - *(vec1+5) * *(vec2+5) ;
     *re += *(vec1+6) * *(vec2+6)  - *(vec1+7) * *(vec2+7) ;
     *im += *(vec1  ) * *(vec2+1) + *(vec1+1) * *(vec2  ) ;
     *im += *(vec1+2) * *(vec2+3) + *(vec1+3) * *(vec2+2) ;
     *im += *(vec1+4) * *(vec2+5) + *(vec1+5) * *(vec2+4) ;
```

```
      *im += *(vec1+6) * *(vec2+7)  +  *(vec1+7) * *(vec2+6) ;
      vec1 += 8 ;
      vec2 += 8 ;
      }

   while (m--) {    // Do the remainder
     *re += *vec1 * *vec2  -  *(vec1+1) * *(vec2+1) ;
     *im += *vec1 * *(vec2+1)  +  *(vec1+1) * *vec2 ;
     vec1 += 2 ;
     vec2 += 2 ;
     }
}
```

## The Squashing Function

This section will briefly discuss the real-domain squashing function, $s(x)$, needed by the activation function defined in Equation (2-16). Its derivative will also be considered, and code for act_func, actderiv and inverse_act given.

Recall that Equation (2-16) evaluates $s(x)$ only for nonnegative arguments. Therefore, our function need only resemble the right half of a sigmoid. It should be approximately linear for small values of $x$, with $s(0) = 0$. It should monotonically increase, rapidly approaching an asymptote as $x$ increases. Anything resembling Figure 2-10 will do.

The author's favorite squashing function is the one recommended in [Kalman and Kwasny, 1992]. Its derivative is easily computed also.

$$
\begin{aligned}
s(x) &= \tanh(1.5x) \\
s'(x) &= 1.5(1 - s^2(x))
\end{aligned}
\tag{2-21}
$$

When dealing with small problems and real-domain neurons, small but significant amounts of time can be saved by precomputing a table of activation function values. This is used as a lookup table whenever the activation function is to be computed. Code for doing this is provided in [Masters, 1993]. On the other hand, complex-domain neurons have so much computation associated with their partial derivatives and various other intricacies that the time saved by tabling the squashing function is probably not worth the effort involved. This makes their code trivially simple.

```
double act_func ( double x )
{
  return tanh ( 1.5 * x ) ;
}
```

```
double actderiv ( double f )
{
  return 1.5 * (1.0 - f * f) ;
}


double inverse_act ( double f )
{
  if (f < -0.999999)
    f = -0.999999 ;

  if (f > 0.999999)
    f = 0.999999 ;

  return 0.3333333 * log ( (1.0 + f) / (1.0 - f) ) ;
}
```

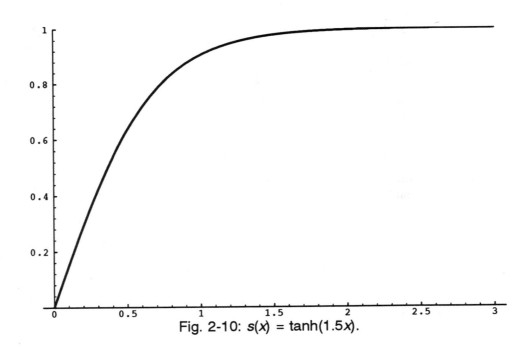

Fig. 2-10: $s(x) = \tanh(1.5x)$.

### Inverse of the Activation Function

If we use regression to compute output weights, as described on page 56, we need to be able to compute the inverse of the activation function. Equation (2-16) is trivially inverted. Code for doing so is shown here. Note that inverse_act is the inverse of the squashing function $s(x)$ and is supplied on page 28.

```
void inverse_act_cc ( double *out , double *net )
{
  double raw_length, squashed_length, f ;

  squashed_length = sqrt ( out[0] * out[0]  +  out[1] * out[1] ) + 1.e-30 ;
  raw_length = inverse_act ( squashed_length ) ;
  f = raw_length / squashed_length ;

  net[0] = f * out[0] ;
  net[1] = f * out[1] ;
}
```

## Computing the Gradient

Many standard neural network textbooks derive the equations for computing the gradient of the mean square error for real-domain networks. [Rumelhart *et al.*, 1986] is one of the most famous. However, due to the fact that there is an infinite number of directions from which a point in the complex plane can be approached, derivatives in the complex domain are not always a straightforward extension of their real-domain counterparts. It is, therefore, worthwhile to dedicate several tedious pages to explicit derivation of the gradient for complex-valued neural networks.

Several special notation conventions will be employed here. Their purpose is to minimize complexity in what could otherwise be an overwhelming exhibition. Any ambiguities will be easily resolved from context.

Our goal in the following is to compute the partial derivative of a three-layer network's error with respect to a single weight connecting a neuron in one layer to a neuron in another layer. Extensions to more hidden layers are easily done by recursion. The gradient will be found for a single input presentation. The gradient for an entire training epoch is the sum of the gradients for all presentations in that epoch.

Traditional real-domain derivations use subscripts on the weight, $w_{ij}$, to indicate the source and destination neurons. We will avoid this, as the subscripts introduce unnecessary complexity. Instead, we will use a subscript of "r" to indicate that we are referring to the real part of the weight and a subscript of "i" for the imaginary part. Thus, a single complex weight would be $w = (w_r + w_i i)$. It will be seen that subscripts indicating the neurons being connected by the weight are nearly always superfluous. They will be used only when absolutely necessary.

We now give definitions of the terms that will appear. The error, $E$, is a real number. All other terms are complex. Therefore, like the weight described above, they may have a subscript of "r" or "i" to indicate whether they refer to the real or imaginary part. The subscript is not shown in the definitions here.

| | |
|---|---|
| *in* | A single input to the network |
| *hnet* | The net input to a hidden neuron |
| *h* | The activation of that hidden neuron |
| *anet* | The net input to an output neuron |
| *a* | The activation of that output neuron |
| *t* | The target activation for that output neuron |
| *E* | The error for that output neuron |
| $E_{TOT}$ | The total error for all output neurons |
| *w* | A weight connecting one neuron to another |

$E$ refers to the error of the single output neuron under consideration when it does not have a subscript. Occasionally, though, it will appear as part of a summation across all output neurons. In these cases, it will be subscripted to indicate that it refers to a particular neuron. So, for example, we have the following:

$$E_{TOT} = \sum_k E_k \qquad (2\text{-}22)$$

Note that it is possible to define more complicated network errors than the sum of individual output neuron errors. That will be briefly discussed later.

Unless otherwise noted, all computations will be done with real numbers, using the definitions of complex arithmetic to operate on the real and imaginary parts of complex numbers separately. This is because most of the following equations will be programmed directly, so a strictly real presentation simplifies the translation from equations to code when the target language does not directly support efficient complex arithmetic.

The effect of a weight on the network error is the result of percolation through one or more *net* computations, one or more activation functions, and a final error function. We will apply tedious repetition of the generalized chain rule, shown in Equation (2-23), to break up that string of operations.

$$u = g_1(x, y)$$
$$v = g_2(x, y)$$
$$h = f(u, v)$$
$$\frac{\partial h}{\partial x} = \frac{\partial h}{\partial u}\frac{\partial u}{\partial x} + \frac{\partial h}{\partial v}\frac{\partial v}{\partial x}$$
$$\frac{\partial h}{\partial y} = \frac{\partial h}{\partial u}\frac{\partial u}{\partial y} + \frac{\partial h}{\partial v}\frac{\partial v}{\partial y} \qquad (2\text{-}23)$$

Let's start with the easy weights, those that connect the hidden layer to the output layer. In particular, we are interested in a weight, $w = (w_r + w_i i)$, connecting a hidden neuron to an output neuron. The activation of the hidden neuron is $h = (h_r + h_i i)$. The net input to the output neuron is *anet*, and the activation of that neuron is $a$. (The breakdown into real and imaginary parts of such terms no longer will be explicitly shown, as the reader certainly gets the idea by now.) Examine once again Equation (2-12), which shows how the output activation is computed. For now, also assume that the sum of squared errors is used. We will show on page 38 how other error measures can be incorporated into gradient calculations. The error of the single neuron under consideration, which we shall call output neuron $k$, is the squared length of the difference between its target and achieved activations.

$$E_k = |a-t|^2 = (a_r - t_r)^2 + (a_i - t_i)^2 \qquad (2\text{-}24)$$

We begin by using the chain rule to break down our desired partial derivative. Note that we need not be concerned about output neurons other than the one under consideration, as the weight $w$ affects only that one. In other words, the partial derivative of the total error with respect to $w$ is equal to the partial derivative of the error of that single output neuron with respect to $w$.

$$\frac{\partial E_{\text{TOT}}}{\partial w_r} = \frac{\partial E_{\text{TOT}}}{\partial anet_r}\frac{\partial anet_r}{\partial w_r} + \frac{\partial E_{\text{TOT}}}{\partial anet_i}\frac{\partial anet_i}{\partial w_r}$$

$$\frac{\partial E_{\text{TOT}}}{\partial w_i} = \frac{\partial E_{\text{TOT}}}{\partial anet_r}\frac{\partial anet_r}{\partial w_i} + \frac{\partial E_{\text{TOT}}}{\partial anet_i}\frac{\partial anet_i}{\partial w_i} \qquad (2\text{-}25)$$

The right-hand factors in Equation (2-25) are not difficult. Equation (2-12) shows that the net input to this output neuron depends on this weight only through the activation of the single hidden neuron from which this weight leads. Remember how the product of the weight with the hidden neuron activation is computed.

$$\begin{aligned} hw &= (h_r + h_i i)(w_r + w_i i) \\ &= (h_r w_r - h_i w_i) + (h_r w_i + h_i w_r)\, i \end{aligned} \qquad (2\text{-}26)$$

Thus, we immediately know the right-hand factors in Equation (2-25). They are shown in Equation (2-27).

The left-hand factors in Equation (2-25) require a little more work. The chain rule must be applied again, as shown in Equation (2-28). Note that these factors are labeled $\delta_{or}$ and $\delta_{oi}$ in that equation. This stands for the real and imaginary parts of the output delta. There are two reasons for this extra label. One is that we will refer back to these quantities later in this development. The other is that most traditional derivations of the gradient for real-domain neurons use $\delta$ as an intermediate term at this same point. This enables the reader to see the correspondence between the derivation in other standard texts and the derivation here.

$$\frac{\partial anet_r}{\partial w_r} = h_r \qquad \frac{\partial anet_r}{\partial w_i} = -h_i$$

$$\frac{\partial anet_i}{\partial w_r} = h_i \qquad \frac{\partial anet_i}{\partial w_i} = h_r \tag{2-27}$$

$$\delta_{or} = \frac{\partial E_{TOT}}{\partial anet_r} = \frac{\partial E_{TOT}}{\partial a_r}\frac{\partial a_r}{\partial anet_r} + \frac{\partial E_{TOT}}{\partial a_i}\frac{\partial a_i}{\partial anet_r}$$

$$\delta_{oi} = \frac{\partial E_{TOT}}{\partial anet_i} = \frac{\partial E_{TOT}}{\partial a_r}\frac{\partial a_r}{\partial anet_i} + \frac{\partial E_{TOT}}{\partial a_i}\frac{\partial a_i}{\partial anet_i} \tag{2-28}$$

The right-hand factors in Equation (2-28) are the partial derivatives of the output neuron's activation function. For optimal generality, computation of those derivatives should not be embedded in the gradient routine. A separate function should be written. This facilitates easily changing activation functions. Computation of partial derivatives of activation functions is discussed on page 24.

The left-hand factors in Equation (2-28) are dependent on the way we measure the network's error. For now, we are assuming the traditional sum of squared errors as defined in Equations (2-22) and (2-24). This gives us a relatively simple partial derivative.

$$\frac{\partial E_{TOT}}{\partial a_r} = 2(a_r - t_r)$$

$$\frac{\partial E_{TOT}}{\partial a_i} = 2(a_i - t_i) \tag{2-29}$$

Several things should be noted at this time. First, it is often reasonable to ignore the factor of two, as it is constant for all components of the gradient and hence does not affect the direction of the gradient. Similarly, the mean squared error, rather than the sum of squared errors, is often used. This, too, involves constant factors that can be ignored.

Finally, observe that because we are using the *sum* of individual output neuron errors as our error measure, Equation (2-29) does not involve output neurons other than the one under consideration. The partial derivative of a sum is the sum of the partial derivatives, and all other partial derivatives are zero due to the fact that the output neurons do not interconnect. This significant simplification is a good reason for using an error measure that involves only the *sum* of individual output neuron errors. Higher-order components can complicate things beyond the simplicity of Equation (2-29). For example, using an error defined as the sum of the logs of the individual neuron's squared errors is still simple; we just slightly modify Equation (2-29). But defining the error as the log of the sum of individual squared errors means that the effect of the weight *w* on the error depends on the errors of the other output neurons. Equation (2-29) must now include the activation of all output neurons. This topic is discussed on page 38.

With the preceding derivation tucked under our belts, it is time to examine the partial derivative of the error with respect to weights connecting the input layer to the hidden layer. It will now be necessary to peel more layers from the onion. It is the same onion, though, and the technique is similar to the one just described. As before, we start by invoking the chain rule. Remember that the weight, *w*, now refers to a connection from the input layer to the hidden layer.

$$\frac{\partial E_{TOT}}{\partial w_r} = \frac{\partial E_{TOT}}{\partial hnet_r}\frac{\partial hnet_r}{\partial w_r} + \frac{\partial E_{TOT}}{\partial hnet_i}\frac{\partial hnet_i}{\partial w_r}$$

$$\frac{\partial E_{TOT}}{\partial w_i} = \frac{\partial E_{TOT}}{\partial hnet_r}\frac{\partial hnet_r}{\partial w_i} + \frac{\partial E_{TOT}}{\partial hnet_i}\frac{\partial hnet_i}{\partial w_i} \tag{2-30}$$

Look back at Equations (2-26) and (2-27). We can find the right-hand factors in Equation (2-30) by analogy. They are shown in Equation (2-31).

$$\frac{\partial hnet_r}{\partial w_r} = in_r \qquad \frac{\partial hnet_r}{\partial w_i} = -in_i$$

$$\frac{\partial hnet_i}{\partial w_r} = in_i \qquad \frac{\partial hnet_i}{\partial w_i} = in_r \tag{2-31}$$

Apply the chain rule to the left-hand factors. Once again we will label those terms to correspond to the δ (delta) terms in other standard real-domain derivations.

The right-hand factors in Equation (2-32) are the partial derivatives of the hidden-neuron activation function. Those derivatives are discussed on page 24.

The left-hand factors in that equation are a bit more difficult than they were for the output neurons. This is because the network error, $E_{TOT}$, is affected by the hidden neuron's activation, *h*, through all of the output neurons. This is illustrated in Figure 2-11.

$$\delta_{hr} = \frac{\partial E_{TOT}}{\partial hnet_r} = \frac{\partial E_{TOT}}{\partial h_r}\frac{\partial h_r}{\partial hnet_r} + \frac{\partial E_{TOT}}{\partial h_i}\frac{\partial h_i}{\partial hnet_r}$$

$$\delta_{hi} = \frac{\partial E_{TOT}}{\partial hnet_i} = \frac{\partial E_{TOT}}{\partial h_r}\frac{\partial h_r}{\partial hnet_i} + \frac{\partial E_{TOT}}{\partial h_i}\frac{\partial h_i}{\partial hnet_i}$$

(2-32)

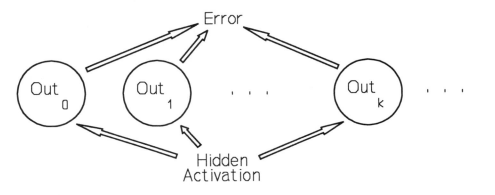

Fig. 2-11: Hidden neuron affecting error.

The chain rule as stated in Equation (2-23) is for just two functions. We now generalize it to as many functions as there are output neurons. Unfortunately, we must introduce a new subscript, $k$, for $anet_r$ and $anet_i$ in order to indicate which output neuron they represent. Note that the summations in the following equation have nothing to do with the fact that our network error is the sum of the individual output neuron errors. They are a direct consequence of the chain rule. Even if more complicated network error functions are used, such as those discussed on page 38, this equation still holds true.

$$\frac{\partial E_{TOT}}{\partial h_r} = \sum_k \left(\frac{\partial E_{TOT}}{\partial anet_{kr}}\frac{\partial anet_{kr}}{\partial h_r}\right) + \sum_k \left(\frac{\partial E_{TOT}}{\partial anet_{ki}}\frac{\partial anet_{ki}}{\partial h_r}\right)$$

$$\frac{\partial E_{TOT}}{\partial h_i} = \sum_k \left(\frac{\partial E_{TOT}}{\partial anet_{kr}}\frac{\partial anet_{kr}}{\partial h_i}\right) + \sum_k \left(\frac{\partial E_{TOT}}{\partial anet_{ki}}\frac{\partial anet_{ki}}{\partial h_i}\right)$$

(2-33)

Look back at Equation (2-26). We can now easily write the right-hand factors in each of the terms in Equation (2-33). This is done in Equation (2-34). Note that we must subscript the weights so that we know to which output neuron they lead.

$$\frac{\partial anet_{kr}}{\partial h_r} = w_{kr} \; , \qquad \frac{\partial anet_{kr}}{\partial h_i} = -w_{ki}$$

$$\frac{\partial anet_{ki}}{\partial h_r} = w_{ki} \; , \qquad \frac{\partial anet_{ki}}{\partial h_i} = w_{kr}$$

$$(2\text{-}34)$$

The left-hand factors in the summations in Equation (2-33) should look a little familiar. They are the output deltas already computed in Equation (2-28). There is one for each output. Responsible programmers will compute them only once.

The derivation of the gradient terms is complete. Let us now lay out a typical course of action for their computation. Note that it differs slightly from the usual course for real-domain neurons. This is because it is usually more efficient to compute the expensive partial derivatives of the various activation functions as each neuron's activation is computed, saving them for later use in computing the gradient. Memory is cheaper than time.

1. Compute and save the hidden neuron activations (Equation (2-12)) and their partial derivatives (page 24).
2. Compute and save the output neuron activations and their partial derivatives in the same way as for the hidden neurons.
3. Compute the network error and its partial derivative with respect to output activations using Equation (2-29) or whatever is appropriate for the error measure.
4. Compute delta for each output neuron using Equation (2-28).
5. Compute the output-layer gradient using Equation (2-25).
6. For each hidden neuron:
   a. Cumulate the product of all output deltas times the weight connecting that output to the hidden neuron being done (Equation (2-33)).
   b. Compute this hidden neuron's delta using Equation (2-32). If there will be another hidden layer, save delta.
   c. Compute this hidden neuron's gradient using Equation (2-30).

Some code fragments will now be shown to illustrate the preceding description. A complete listing of a gradient subroutine can be found on the accompanying disk (GRADIENT.CPP).

```
/*
    Partial derivatives of each neuron's activation function will be computed as the activation is
    computed, and they will be stored here.
*/

    double *dar10 ;   // Partial of real attained output wrt real net
    double *dai10 ;   // Partial of imaginary attained output wrt real net
```

```
          double *dar01 ;   // Partial of real attained output wrt imaginary net
          double *dai01 ;   // Partial of imaginary attained output wrt imaginary net
          double *dhr10 ;   // Partial of real hidden wrt real net
          double *dhi10 ;   // Partial of imaginary hidden wrt real net
          double *dhr01 ;   // Partial of real hidden wrt imaginary net
          double *dhi01 ;   // Partial of imaginary hidden wrt imaginary net


  /*
          Compute and save activations and partial derivatives of activations.
          The subroutine "partial" can be found on page 25.
  /*

          for (i=0 ; i<nhid ; i++)
            partial ( inputs , hid_coefs[i] , hid_act+2*i , ninputs ,
                    dhr10+i , dhr01+i , dhi10+i , dhi01+i ) ;

          for (i=0 ; i<nout ; i++)
            partial ( hid_act , out_coefs[i] , out+2*i , nhid ,
                    dar10+i , dar01+i , dai10+i , dai01+i ) ;


  /*
          Save difference between attained and target activations, which is the partial derivative of the
          error with respect to output activations (except for a constant factor).
          Cumulate error.
          The output and target activations are stored real, imaginary, real, imaginary, ...
          Thus there are 2*nout components in each.
          By using (target - output) rather than (output - target) as in Equation (2-29), we end up
          computing the negative of the gradient, which is what we want for descent.
  */

          for (i=0 ; i<2*nout ; i++) {
            d = target[i] - out[i] ;
            error += d * d ;
            diffs[i] = d ;
            }


  /*
          For each output neuron, compute and save delta.  Then cumulate the gradient into the
          epoch sum.  We store the bias components last in the array.  Remember that the bias is fed
          from a hypothetical neuron having an activation of 1+0i.
  */

          for (i=0 ; i<nout ; i++) {           // Do every output neuron
            rdiff = diffs[2*i] ;               // Target minus attained output
```

```
        idiff = diffs[2*i+1] ;                    // And its imaginary part
        rdelta = rdiff * dar10[i]  +  idiff * dai10[i] ; // Equation (2-28)
        idelta = rdiff * dar01[i]  +  idiff * dai01[i] ;
        delta_r[i] = rdelta ;                     // Save for hidden gradient later
        delta_i[i] = idelta ;
        for (j=0 ; j<nhid ; j++) {                // Equation (2-25), real then imaginary
           *grad++ += rdelta * hid_act[2*j] + idelta * hid_act[2*j+1] ;
           *grad++ += -rdelta * hid_act[2*j+1] + idelta * hid_act[2*j] ;
           }
        *grad++ += rdelta ;                       // Do the bias gradient.
        *grad++ += idelta ;                       // Activity = 1+0i
        }


/*
     For each hidden neuron:
        1) Sum weight times output delta for all output neurons.
        2) Compute delta for this hidden neuron.  No need to save it.
        3) Cumulate gradient into epoch sum.
*/

   for (i=0 ; i<nhid ; i++) {                 // For every hidden neuron
     rsum = isum = 0.0 ;                       // Sum weight times delta for all outputs
     for (k=0 ; k<nout ; k++) {               // Equation (2-33)
        rsum += delta_r[k] * out_coefs[k][2*i] +
             delta_i[k] * out_coefs[k][2*i+1] ;
        isum += -delta_r[k] * out_coefs[k][2*i+1] +
             delta_i[k] * out_coefs[k][2*i] ;
        }

     rdelta = rsum * dhr10[i]  +  isum * dhi10[i] ;      // Equation (2-32)
     idelta = rsum * dhr01[i]  +  isum * dhi01[i] ;      // No need to save it

     for (j=0 ; j<ninputs ; j++) {            // Equation (2-30), real then imaginary
        *grad++ += rdelta * input[2*j] + idelta * input[2*j+1];
        *grad++ += -rdelta * input[2*j+1] + idelta* input[2*j];
        }
     *grad++ += rdelta ;                       // Bias gradient
     *grad++ += idelta ;                       // Input is 1+0i
     } // For every hidden neuron
```

## Alternative Error Functions

Nearly all learning algorithms for multiple-layer feedforward networks operate by minimizing the mean squared error of the outputs. There are several reasons for this choice. It is fast and easy to compute. Its derivatives with respect to weights are easy to compute. Its intuitive meaning is readily grasped. It is mathematically optimal for certain statistical tests. And last but certainly not least, its behavior under directed descent seems to be at least as good as that of most common competitors. These and perhaps other reasons make it an excellent general-purpose error measure for guiding learning.

Mean squared error is not by any means the only possibility, though. This section will examine a variety of other candidates. Many of them are useless in practical applications. They are offered for pedagogical purposes only. The reader who works through the derivation and programming of them all will be qualified to design his or her own custom error function. On the other hand, several of them are immensely useful. These will be noted individually.

Several terms will regularly appear in the subsequent presentations. Let us define them now, so that tedious repetition may be avoided. We have $n$ output neurons. Their attained activation levels are $a_0, a_1, ..., a_{n-1}$. Their target activations are $t_0, t_1, ..., t_{n-1}$. Each of these can be expressed in terms of their real and imaginary parts. For example, $a_0 = a_{0r} + a_{0i}i$. When it is clear that we are talking about a particular output neuron, the identifying subscript may be eliminated: $a = a_r + a_i i$. This will usually be done when writing the equation for a derivative, in which case the use of an extraneous subscript would add to the clutter. Finally, the network error, taking into account all outputs, is $E$. (We called this $E_{TOT}$ in the previous section.)

It is important to understand that only one small part of the derivation of the gradient in the previous section is affected by changing the error function. That is the calculation of the partial derivative of the error with respect to the output activation. Equation (2-29) is for the sum of squared errors. *Only this one equation needs to be revised.* The entire remainder of the derivation is independent of the error function used.

When programming a gradient-computation algorithm, the normal order of operation is to compute all neuron activations, then compute the network error and the partial derivatives of that error with respect to output activations, then compute the gradient. We are concerned here with the second step. The most general approach is to employ a subroutine dedicated to the derivative calculation and call it for each output neuron. That routine will need to be provided with the entire vector of output activations, as some error measures require them all simultaneously to compute any one derivative. But it will need to be callable for individual outputs, as it should not be expected to know how to find the target activations. A typical calling sequence is now shown.

```
err = 0.0 ;    // Will cumulate network error here
for (i=0 ; i<nout ; i++) {
  // Compute target for this output here...
  errderiv( nout, i, out_activations, target_real, target_imag, &err, derivs);
  }
```

As each error measure is discussed in subsequent sections, code fragments for computing the error and its partial derivatives will be given. These fragments all share a common calling parameter list. They also share a few preliminary computations. This common beginning is shown now so that it need not be repeated for each error type.

```
void errderiv (
    int nout ,                      // Number of output neurons
    int iout ,                      // Of which we are doing this one (org 0)
    double *outs ,                  // All output activations (real, imaginary, ...)
    double target_r ,               // Target for this one
    double target_i ,               // And imaginary part
    double *err ,                   // Input of error so far, output cumulated
    double *deriv                   // Output: deriv[iout] = d(err) wrt outs[iout]
    )
{
    int i ;
    double dr, di, dsq, p, denom ;

    dr = outs[2*iout] - target_r ;    // Difference of real parts
    di = outs[2*iout+1] - target_i ;  // And imaginary parts
    dsq = dr * dr  +  di * di ;       // Squared difference
```

The last call parameter, deriv, is an array 2*nout long. Its first element is the returned derivative of the error with respect to the real part of the first output neuron's activation. The second element is for the imaginary part, and so on.

One implementation detail should be mentioned. Careful study of the gradient derivation in the previous section will reveal that constant factors in the network error partial derivatives are exactly reflected in the final gradient. In other words, if one multiplies the real and imaginary derivatives of the error by a constant, the computed gradient will be multiplied by that same constant. This means that we can increase program efficiency by factoring as many constants as possible out of the training epoch. For example, the derivatives for the mean squared error include multiplying the error difference by two and dividing it by $n$. (See Equation (12-5).) It is inefficient to do this for every output in every training presentation. Simply return the error difference as the derivative. Then, after the gradient for the entire epoch is summed, multiply the final gradient by two and divide by $n$. The result will be the same, but much less computation will have been done. For purposes of clarity, though, the sample code given will include any necessary constant factors. This strict correctness avoids confusing the reader. Programmers may wish to use the examples exactly as given until correct operation is verified, then deal with common factors for the final version.

Figures 2-12 through 2-23 at the end of this section (starting on page 50) show typical instances of the error function and its negative derivative for many of the following examples. It would not hurt to examine these plots as the error measures are described. These plots are described in more detail on the page preceding the first plot.

## Mean Squared Error

It is good to start by reviewing the standard choice for the error function. The previous section assumed that we were minimizing the sum of squared errors. This, of course, is equivalent to minimizing the mean squared error, as they are related by a constant factor, the number of output neurons. It is usually preferable to work with the mean squared error. It is more intuitive in that it is a "per output" quantity. Also, this immunity to variations in the number of output neurons can make its behavior under minimization more consistent than a simple sum. Therefore, all network errors will be normalized per the number of outputs. The mean squared error and its gradients are computed as follows:

$$E = \frac{1}{n} \sum_k \left[ (a_{kr} - t_{kr})^2 + (a_{ki} - t_{ki})^2 \right] \tag{2-35}$$

$$\frac{\partial E}{\partial a_r} = \frac{2}{n}(a_r - t_r)$$

$$\frac{\partial E}{\partial a_i} = \frac{2}{n}(a_i - t_i) \tag{2-36}$$

A code fragment for computing this error and its partial derivatives is shown below. Remember that this code is intended to follow the common code fragment given at the beginning of this section.

```
*err += dsq / nout ;
deriv[2*iout]   = 2.0 * dr / nout ;
deriv[2*iout+1] = 2.0 * di / nout ;
```

The plots on page 50 show why this error measure is such a reliable performer. The error has a smooth bowl shape, and the magnitude of the derivative is proportional to the error. All in all, this is a relatively pleasant function to minimize. When in doubt about a sensible choice for the error measure to minimize, the mean squared error is usually a safe bet. It has a long-established record of satisfactory performance and is difficult to criticize.

## Mean Absolute Error

A frequently useful alternative to the mean *squared* error is the mean *absolute* error. With this error measure, all output errors are treated equally. This is in contrast to the former, in which the squaring operation causes the learning algorithm to concentrate more on reducing large errors, at the expense of ignoring small errors. If your application would benefit from reducing small errors just as much as large errors, even though that means that some errors may be left relatively large, then consider using the mean absolute error. A classic application in which this may be the case is when the network is used to compute negative feedback in a control situation. It is known that minimizing the absolute error often reduces unwanted oscillations more effectively compared to minimizing the squared error. The mean absolute error and its derivatives are computed as shown in the following equations and code fragment.

$$E = \frac{1}{n} \sum_k \sqrt{(a_{kr} - t_{kr})^2 + (a_{ki} - t_{ki})^2} \qquad (2\text{-}37)$$

$$\frac{\partial E}{\partial a_r} = \frac{a_r - t_r}{n \sqrt{(a_r - t_r)^2 + (a_i - t_i)^2}}$$

$$\frac{\partial E}{\partial a_i} = \frac{a_i - t_i}{n \sqrt{(a_r - t_r)^2 + (a_i - t_i)^2}} \qquad (2\text{-}38)$$

```
if (dsq == 0.0)
   deriv[2*iout] = deriv[2*iout+1] = 0.0 ;
else {
   p = sqrt ( dsq ) ;  // Absolute value of error distance
   *err += p / nout ;
   deriv[2*iout] = dr / (p * nout) ;
   deriv[2*iout+1] = di / (p * nout) ;
   }
```

The plots on page 51 show that this error function is satisfactory, but probably more difficult to minimize than the mean squared error. Its principal problem is that its derivative is not proportional to the magnitude of the error. It has constant magnitude, changing sign instantly at the optimal error value. This contributes to some instability in learning, but usually not to a serious extent.

## Kalman-Kwasny Error

[Kalman and Kwasny, 1991 and 1992] assert that when a neural network is trained using traditional gradient descent on the mean squared error, the algorithm can become trapped in one of many special *exterior* local minima and saddle points in the error function. These areas are located where the attained activation is extreme. The more output neurons that exist, the more this becomes a problem. They propose an error function that discourages gravitating toward such areas. It is effected by dividing each output neuron's squared error by a quantity that shrinks as extreme activations are approached. This causes the network error to blow up as an attained activation approaches the negative of its (extreme) target. This error function is especially useful for classification and binary decisions. It may not be so desirable when intermediate activation levels are to be learned. The Kalman-Kwasny error and its derivatives are computed as shown in the following equations and code fragment.

$$E = \frac{1}{n} \sum_k \frac{(a_{kr} - t_{kr})^2 + (a_{ki} - t_{ki})^2}{1 - a_{kr}^2 - a_{ki}^2} \tag{2-39}$$

$$p = 1 - a_r^2 - a_i^2$$

$$q = (a_r - t_r)^2 + (a_i - t_i)^2$$

$$\frac{\partial E}{\partial a_r} = \frac{2}{np} \left( \frac{a_r q}{p} + a_r - t_r \right)$$

$$\frac{\partial E}{\partial a_i} = \frac{2}{np} \left( \frac{a_i q}{p} + a_i - t_i \right) \tag{2-40}$$

```
denom = 1.0 - outs[2*iout] * outs[2*iout]
        - outs[2*iout+1] * outs[2*iout+1] ;
if (denom < 1.e-10)  // SHOULD never be negative
   denom = 1.e-10 ;
*err += dsq / (denom * nout) ;
deriv[2*iout]   = 2.0 / (denom * nout) * (outs[2*iout]   * dsq / denom + dr) ;
deriv[2*iout+1] = 2.0 / (denom * nout) * (outs[2*iout+1] * dsq / denom + di) ;
```

Note that denom in the above code should never be negative. This is because we assume that we are using an output activation function that is bounded at +/– 1. Kalman and Kwasny recommend the hyperbolic tangent function. Also note that their paper applies to real-domain neurons. In practice, we would usually assume that the imaginary parts of the above equations are all zero if we are classifying. The author has taken the liberty of extending their results to the complex domain.

The plots on page 52 reveal the secret of this method's avoidance of false minima at extreme values. The error and its derivative explode as the activation approaches its theoretical extremes. Unfortunately, moderate errors are tolerated comparatively well. Performance of this method is erratic, and it is not recommended if there are few classes. It often performs well when there are many classes.

## Cross Entropy

Research into class-probability estimators for classifiers (including neural networks) has led to a novel new objective function. Since this error measure is appropriate only for classification models, which in this text always have real-valued outputs, it will be formulated here only for the strictly real case.

The traditional definition of cross entropy for a single output neuron is shown in Equation (2-41). That definition assumes that the target, $t$, and attained activation, $a$, are in the range 0–1.

$$\text{Xent} = t \log\left(\frac{t}{a}\right) + (1-t) \log\left(\frac{1-t}{1-a}\right) \tag{2-41}$$

In this text we work with symmetric outputs. The default output activation function is the hyperbolic tangent, which ranges from –1 to 1. Classification training is done using –0.9 and 0.9 as targets. Therefore, we need to define a simple function to map our values to 0–1. This is easily done with Equation (2-42).

$$m(x) = 0.5\,(x+1) \tag{2-42}$$

We can now write our cross-entropy error function as Equation (2-43).

$$E = \frac{1}{n} \sum_k \left[ m(t_k) \log\left(\frac{m(t_k)}{m(a_k)}\right) + (1-m(t_k)) \log\left(\frac{1-m(t_k)}{1-m(a_k)}\right) \right] \tag{2-43}$$

The derivative of that error function is easily expressed in terms of both the original and mapped activations.

$$\frac{\partial E}{\partial a} = \frac{1}{2n}\left(\frac{1-m(t)}{1-m(a)} - \frac{1+t}{1+a}\right) \tag{2-44}$$

Code for computing the cross-entropy error and its derivative is shown below. Note that unlike other code in this section, this is dealing with strictly real values for both the output activation and target. The derivative is similarly real. This code, then, does not directly correspond to the other code fragments in this section. However, the meaning should be obvious.

```
xx = outs[iout] ;
if (xx < -.99999999)                    // Avoid dividing by zero later
   xx = -.99999999 ;
if (xx > .99999999)
   xx = .99999999 ;
t = 0.5 * (target + 1.0) ;              // Map our range to 0-1
x = 0.5 * (xx + 1.0) ;
*err += (t * log ( t/x ) + (1.0 - t) * log ((1.0-t) / (1.0-x))) / nout ;
deriv[iout] = -0.5 / nout * ((1.0 - t) / (1.0 - x)  -  (1.0 + target) / (1.0 + xx )) ;
```

The plots on page 53 show that this error measure is a compromise between mean squared error and Kalman-Kwasny error. It and its derivative blow up at extreme activations, like the Kalman-Kwasny error. But it responds fairly well to intermediate errors, nearly as well as the mean squared error. This error measure is highly recommended for classification problems, but tends to be too tolerant of moderate errors for other uses.

## Maximum Squared Error

This is the first error function we have encountered that is not computed as the mean of the errors of the individual output neurons. For this function, the network's error is equal to the squared error of the output neuron having maximum error. This means that for each training presentation, the gradient will be computed such that only the output having maximum error will be affected. Numerical problems happen when, as is inevitable, outputs have nearly equal errors that are maximum. The gradient becomes ill-defined and convergence suffers. This error function is not generally recommended. It is presented as a demonstration only.

$$E = \max_{k} \left[ (a_{kr} - t_{kr})^2 + (a_{kl} - t_{kl})^2 \right] \tag{2-45}$$

$$\frac{\partial E}{\partial a_r} = \begin{cases} 2(a_r - t_r) & \text{if this neuron has max error} \\ 0 & \text{otherwise} \end{cases} \tag{2-46}$$

$$\frac{\partial E}{\partial a_l} = \begin{cases} 2(a_l - t_l) & \text{if this neuron has max error} \\ 0 & \text{otherwise} \end{cases} \tag{2-47}$$

Note in the following code that we preserve important quantities as the derivative routine is called, then compute the final results when the last output neuron is done.

```
if (dsq > *err) {
  *err = dsq ;
  ibest = iout ;                  // Keep track of max error neuron
  }
deriv[2*iout]   = dr ;            // Save all derivatives for now
deriv[2*iout+1] = di ;
if (iout == nout-1)              // After all outputs are done
  for (i=0 ; i<nout ; i++) {     // Fix the derivatives
    if (i == ibest) {
      deriv[2*i]   *= 2.0 ;
      deriv[2*i+1] *= 2.0 ;
      }
    else
      deriv[2*i] = deriv[2*i+1] = 0.0 ;
    }
  }
```

## High Powers of the Error

Interesting (though not often useful) learning behavior can be obtained by using the mean of high powers of the output error. We are all familiar with the mean squared error. What about the sixteenth power of the error, instead of just the square? This measure of error will obviously focus on large errors, while almost totally ignoring small errors. Even moderate errors will nearly vanish in importance when learning is done. Superficial-ly, this seems to resemble the maximum error discussed in the previous section. However, there is a subtle but vitally important difference. Using high powers of the error has a global effect, permeating the entire training epoch more than most other error types. When the maximum output error measure is used, every training presentation will most likely make a noticeable contribution to the epoch gradient. That is because each presentation has *some* output neuron that has maximum error, and the gradient will be generated so as to reduce the error of that neuron in that presentation. But the situation is very different when we raise the error to a high power. Only those training presentations that have one or more outputs that are seriously in error will significantly contribute to the gradient. Presentations whose output responses are all at least moderately good will be practically ignored. This tends to reduce the maximum error *across the training epoch*, something not seen to such a great degree in other error measures. Of course, it always happens to some degree. Seriously erroneous presenta-tions will make a larger contribution to the epoch gradient than will better presentations. But raising the error to a high power accentuates this effect greatly.

The price paid is that numerical stability suffers when all errors are reduced to moderate levels. Typically, when learning commences, the worst errors are rapidly cut down. At this point, though, the network error is already reduced to what are numerically small values. Gradients become ill-defined, and only small changes in error occur in response to relatively large weight changes. This error function should be used with caution.

$$E = \frac{1}{n} \sum_k \left[ (a_{kr} - t_{kr})^2 + (a_{ki} - t_{ki})^2 \right]^8 \qquad (2\text{-}48)$$

$$\frac{\partial E}{\partial a_r} = \frac{16}{n} (a_r - t_r) [(a_r - t_r)^2 + (a_i - t_i)^2]^7$$

$$\frac{\partial E}{\partial a_i} = \frac{16}{n} (a_i - t_i) [(a_r - t_r)^2 + (a_i - t_i)^2]^7 \qquad (2\text{-}49)$$

```
p = dsq ;
dsq = dsq * dsq ;
dsq = dsq * dsq ;
dsq = dsq * dsq ;
*err += dsq / nout ;
deriv[2*iout]   = 16.0 * dr * dsq / (p * nout) ;
deriv[2*iout+1] = 16.0 * di * dsq / (p * nout) ;
```

The plots on page 54 clearly indicate why this is such a poor error measure. There is a terrible transition region of error tolerance. Larger errors are blitzed by the gigantic derivative. But moderate errors, eminently serious, are given essentially the same attention as tiny errors.


## Mean Log Error

This error measure is presented as an example of how easy it is to design an abhorrent error function. Despite the fact that this function is almost never appropriate, it is tutorial to explore its properties, to deduce the reason for its poor general behavior, and to try to imagine a problem for which it would be appropriate.

The primrose path of misguided intuition starts with the premise that we may not want our error function to be skewed by one or more extremely large errors. Sure, they should be reduced if possible. But let's worry about the smaller errors too. The traditional mean square error favors the reduction of large errors. The mean absolute error distributes attention more equitably. We may want to go even further, though. The obvious candidate is to take the log of the error. We compute the average of each output's log error. This way, very large output errors will affect the network's error only slightly more than moderate errors affect it.

An obstacle is immediately placed in our path. What if an output neuron has zero error? The log of zero is not defined. The traditional solution to this dilemma is to add a tiny constant before taking the log. We shall take this approach. Thus, our error function and its derivatives are defined as shown in Equations (12-19) and (12-20), respectively.

$$ E = \frac{1}{n} \sum_k \log \left( (a_{kr} - t_{kr})^2 + (a_{ki} - t_{ki})^2 + \epsilon \right) \tag{2-50} $$

$$ \frac{\partial E}{\partial a_r} = \frac{2(a_r - t_r)}{n \left( (a_r - t_r)^2 + (a_i - t_i)^2 + \epsilon \right)} $$

$$ \frac{\partial E}{\partial a_i} = \frac{2(a_i - t_i)}{n \left( (a_r - t_r)^2 + (a_i - t_i)^2 + \epsilon \right)} \tag{2-51} $$

What is wrong with that function? Think about what happens if one of the output neurons has nearly zero error. The log of the error will be a very negative number — so negative, in fact, that it will dominate both the network error and the gradient. Errors in other outputs will be virtually ignored. The learning algorithm will focus the majority of its effort to further reducing the already minuscule error of that one neuron, for that is how the most additional reduction in the network error can be obtained. Other neurons, many of which may have significant error, are left out in the cold. Numerical convergence of the learning algorithm will be obtained when the error of one neuron in one training sample has been reduced to essentially zero.

How can this situation be improved? Examination of the equations for the error and gradient clearly indicate that the tiny offset in the denominator provides a safety net. If we make this constant large enough, tiny errors will not exert the overwhelming influence that they would otherwise have. But then the basic shape of the log function is compromised. This will in general be an improvement, but why bother anyway? It is hard to imagine a situation in which our main goal is to find a network that has a tiny error for one neuron in one presentation, and let the rest of the world be ignored. The principal reason for including this error measure is to demonstrate to the reader that careful thought is necessary if a custom error measure is to be designed.

In case the reader wants to implement this unusual function, suitable code is now given.

```
*err += log ( dsq  + epsilon ) / nout ;
deriv[2*iout]   = 2.0 * dr / ((dsq + epsilon) * nout) ;
deriv[2*iout+1] = 2.0 * di / ((dsq + epsilon) * nout) ;
```

The plots on page 55 show the novel shape of this error measure and its derivative. Epsilon is set to 0.0001 for these plots, and the error is positively offset to improve readability of the graph. The narrow channel surrounding the minimum obviously impedes learning algorithms. The fact that the derivative becomes large only when we are already near the minimum, and is small when the error is large, is just the opposite of what we want. Don't even think about using this error measure.

## Log Mean Error

The collection of alternative error functions will conclude with one that demonstrates how to handle functions with interactions between outputs. With the exception of the maximum error function presented earlier, all of the functions seen so far involved the *sum* of errors for individual outputs. This made the derivative of that error with respect to a particular output activation dependent only on the activation of the one output involved. Now we will examine a function whose derivative with respect to any output activation depends on all output activations. Like the mean log error, this function would be useful in only very special situations. However, it demonstrates in a clear and simple manner just how such interactions could be handled in a practical program. This error and its derivatives are defined as follows:

$$E = \log\left(\epsilon + \frac{1}{n}\sum_k \left((a_{kr}-t_{kr})^2 + (a_{ki}-t_{ki})^2\right)\right) \qquad (2\text{-}52)$$

$$\frac{\partial E}{\partial a_r} = \frac{2(a_r-t_r)}{\sum_k \left((a_{kr}-t_{kr})^2 + (a_{ki}-t_{ki})^2\right) + n\epsilon}$$

$$(2\text{-}53)$$

$$\frac{\partial E}{\partial a_i} = \frac{2(a_i-t_i)}{\sum_k \left((a_{kr}-t_{kr})^2 + (a_{ki}-t_{ki})^2\right) + n\epsilon}$$

It is apparent that a serious problem of the mean log error has been resolved by interchanging the order of finding the mean and taking the log. The network error and gradient are no longer held hostage by one particularly good neuron. All output errors are averaged before the log is taken. We still have the problem of one good training case dominating, though. As was the case for the mean log error, using a large constant offset helps, but does not completely solve the problem unless it is made so large that the log function is entirely subverted.

Code for computing the log mean error and its derivatives is now given. Note how we save temporary values as this code fragment is called for successive outputs, then compute the final values when the last output is done.

```
*err += dsq ;
deriv[2*iout]   = dr ;
deriv[2*iout+1] = di ;
if (iout == nout-1) {  // Complete things if this is last output
  denom = *err + epsilon * nout ;
  for (i=0 ; i<nout ; i++) {
    deriv[2*i]   *= 2.0 / denom ;
    deriv[2*i+1] *= 2.0 / denom ;
    }
  *err = log ( *err / nout + epsilon ) ;
  }
```

## Summary

This section has presented a variety of alternatives to the traditional mean squared error for evaluating network performance to guide learning. Of these, only three are considered by the author to have real practical value. One is the mean absolute error. This measure equalizes the importance of all errors, regardless of their magnitude. Another is the Kalman-Kwasny function. It appears to have some utility for classification problems, or other situations in which the network must learn output values that are always near their extreme activation levels. Last, but certainly not least, is cross entropy. This is sometimes *superior* to mean squared error in classification problems.

The remainder of the error functions comprise a variety of simple models that may serve as the foundation for more complex custom-built functions. The maximum error and log mean error are particularly unusual approaches in that the error and its derivatives are dependent on all outputs, rather than just one at a time. The others are intended to show the reader how seemingly useful error functions can have unexpected effects on learning, and to encourage the designer of custom functions to think carefully about candidate functions.

This section concludes with a look at some of the error functions just described. To keep things simple, we assume a classification model that has real outputs and targets. The $x$-axis in each graph is the attained output, and the implicit target is 0.9, corresponding to full activation. The upper graph on each page is the error of that single neuron as a function of its activation. Since the target is nearly full activation, the error is at a minimum near the right side of the graph. The lower graph on the page is the negative derivative of the error. Careful study of the shape of these curves can contribute to understanding the properties of the various error measures.

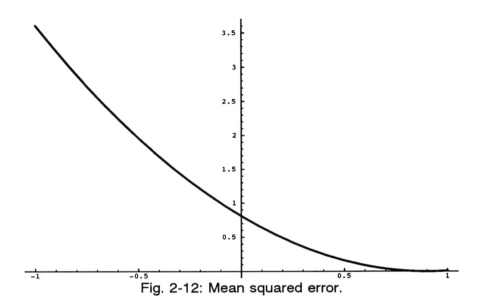

Fig. 2-12: Mean squared error.

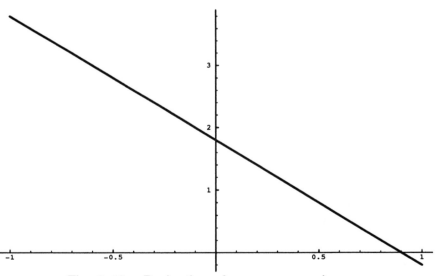

Fig. 2-13: −Derivative of mean squared error.

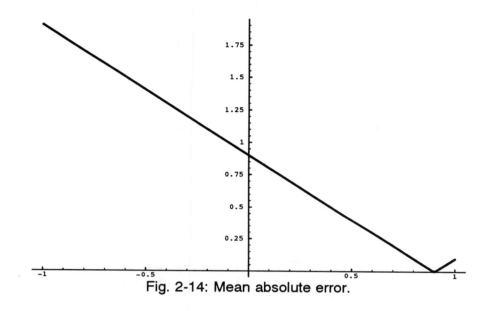

Fig. 2-14: Mean absolute error.

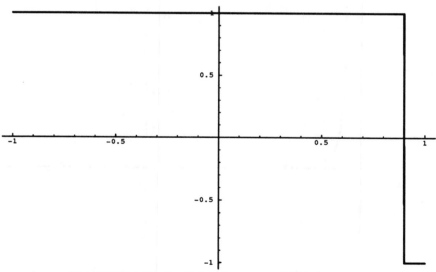

Fig. 2-15: −Derivative of mean absolute error.

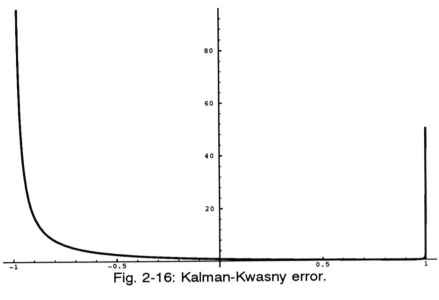

Fig. 2-16: Kalman-Kwasny error.

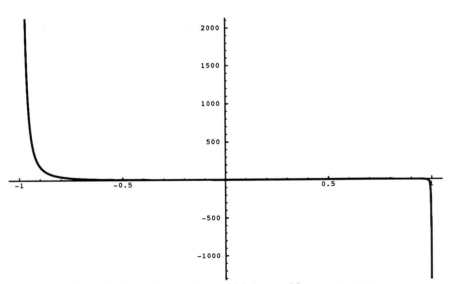

Fig. 2-17: −Derivative of Kalman-Kwasny error.

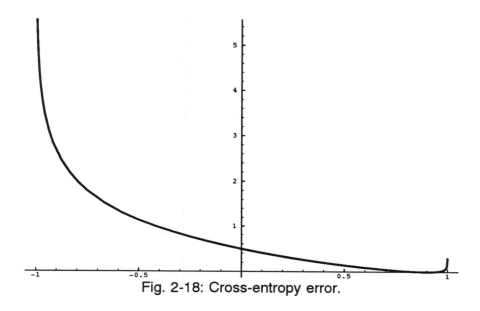

Fig. 2-18: Cross-entropy error.

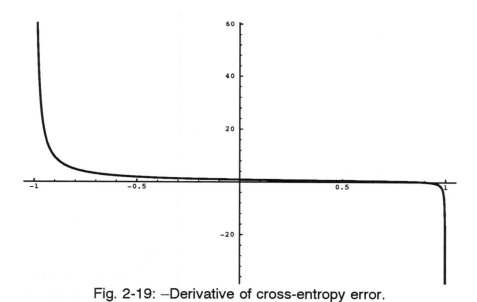

Fig. 2-19: −Derivative of cross-entropy error.

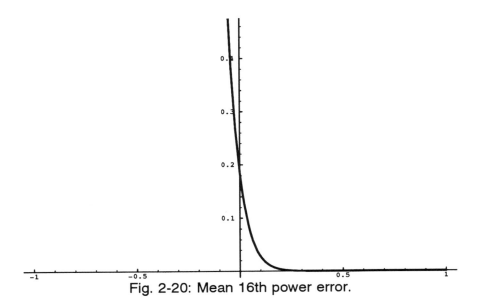

Fig. 2-20: Mean 16th power error.

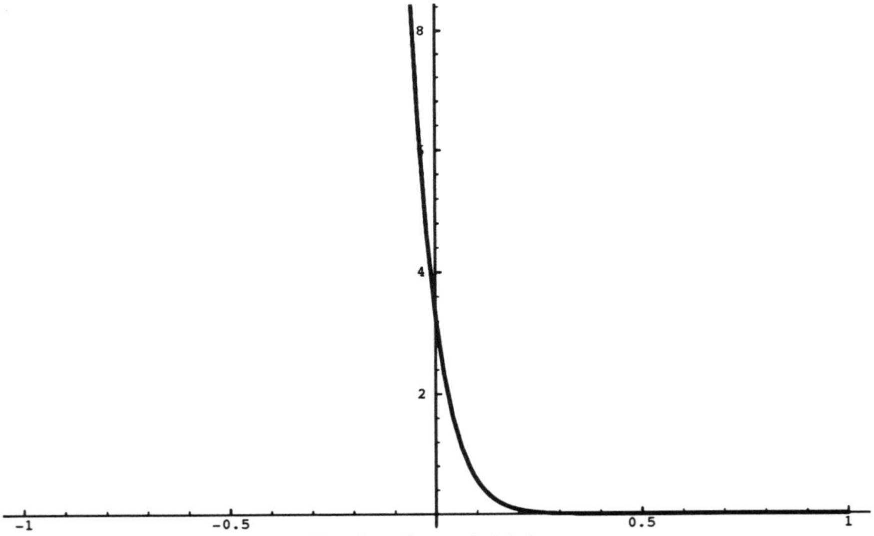

Fig. 2-21: −Derivative of 16th power error.

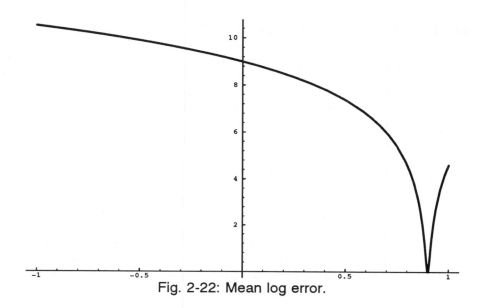

Fig. 2-22: Mean log error.

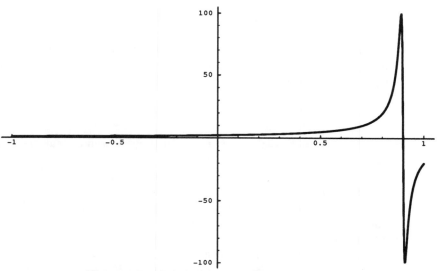

Fig. 2-23: −Derivative of mean log error.

# Regression in the Complex Domain

There is an algorithm for explicitly computing the weights connecting the final (and usually only) hidden layer to the output layer. This algorithm assumes that all other weights in the network are fixed. It then uses a direct technique to compute the output weights. When the activation functions of the output neurons are linear, these weights minimize the mean squared error of the outputs. If the activation functions are nonlinear, the error is approximately minimized.

The primary use for this technique is as an adjunct to stochastic minimization. It is fine to apply methods based on randomness to hidden-layer weights. But once these are set, there is no point in randomly choosing output weights when optimal or near-optimal weights can be explicitly computed!

The reader must be cautioned against using the method of this section injudiciously, though. If output activations are nonlinear, regression is guaranteed to provide results that are inferior to those which could be obtained from methods that directly minimize the network error. Even if the outputs are linear, it is doubtful that any computation time could be saved by applying descent methods to only the hidden-layer weights, relying on regression to find the output weights. This is because there are usually far more hidden weights than output weights, so not enough would be gained to offset the added burden of repeated regression.

## Review of Real-Domain Regression

Linear regression is the process by which linear combinations of the values of one or more *independent* variables are used to predict the value of a *dependent* variable. For example, we may have

$$\hat{y} = ax_1 + bx_2 + c \qquad (2\text{-}54)$$

The constants a, b, and c may be estimated by obtaining many samples of $x_1$, $x_2$, and $y$, and then applying to this data the correct equations.

Let us start with some notation. The sample contains $n$ cases. The independent variables are stored as a matrix $A$. This matrix has $n$ rows (one for each case), and it has one column for each independent variable. It also has one extra column that contains the value 1 throughout. That last column is the regression constant, and it corresponds to the bias in neural computation. The number of columns is $m$. Since the last column is constant, we have $m$-1 independent variables. The dependent variable, $Y$, is stored in a column vector of $n$ elements, one for each case. Finally, the regression coefficients to be estimated, ß, are also stored in a column vector. There are $m$–1 independent variables, plus the constant, so this vector contains $m$ elements. The general regression equation can be written in matrix form as shown in Equation (2-55).

$$A \; \beta \; = \; Y \tag{2-55}$$

This equation has an exact solution if and only if $Y$ lies in the subspace spanned by the columns of $A$. Unfortunately, this solution is not guaranteed to be unique. That fact can cause serious difficulties that must be dealt with. Furthermore, random sampling errors do in practice guarantee that $Y$ will virtually never lie in the subspace spanned by the columns of $A$. So we must find a solution ß that approximately solves Equation (2-55). Nearly any statistics text will show that Equation (2-56) minimizes the mean squared error in the predicted values of $Y$.

$$\beta \; = \; (A'A)^{-1} \, A' \, Y \tag{2-56}$$

Unfortunately, that traditional old workhorse of an equation will do us little good. In neural network applications, the matrix that must be inverted will often be either totally singular, and hence not invertible, or so ill-conditioned that it might as well be singular. We will need a technique called *singular value decomposition*. The mathematics of that algorithm is beyond the scope of this text. [Press *et al.*, 1992] gives a good summary. [Forsythe *et al.*, 1977] is even more detailed. A good intuitive discussion of the technique, including both the basic mathematics and the motivation for that mathematics, is in [Masters, 1993]. Finally, complete subroutines for performing singular value decomposition and its accompanying back-substitution are included on the program disk. The remainder of this section will assume that the algorithm is available and will show how it can be applied to neural network weight computation.

## Regression in Real-Domain Neural Networks

The net input applied to an output-layer neuron is a linear combination of the activations of the neurons in the last hidden layer. If we consider all cases, this relationship for a single output neuron can be expressed using Equation (2-55). We know the desired output of that output neuron, the target, for each case. Thus, we can apply the inverse of its activation function to compute the desired net input to that output neuron. That is the dependent variable, $Y$, in Equation (2-55). The final hidden layer's activations are the independent variables, $A$, in the regression equation. The weight vector connecting the last hidden layer to the output neuron is the ß vector in Equation (2-55). This weight vector will be optimal in that it minimizes the mean squared error of the *input* to that output neuron. Of course, what we really want to minimize is the mean squared error at the *output* of that neuron, so this is no panacea if the output activation functions are nonlinear. It can be immensely useful, though, as the approximation is usually excellent. And it is perfect if the outputs are linear.

Computation of the output-layer weights is a three-step process. First, the $A$ matrix is computed as the casewise activations of the hidden layer just prior to the output layer. Simultaneously, the inverse activation function of the targets are computed and stored in a separate $Y$ vector for each output neuron. Second, the singular value decompo-

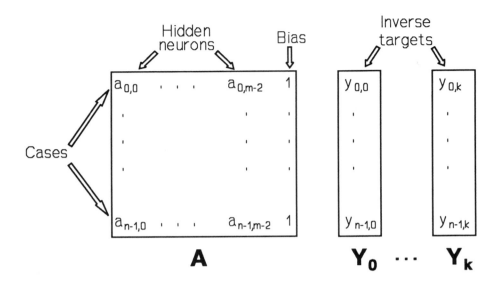

Fig. 2-24: Structure of the A matrix and Y vectors.

sition of $A$ is found. Last, the weights for each output neuron are computed by back-substitution. These steps will be discussed in more detail. Examination of Figure 2-24, which shows the internal structure of the $A$ matrix and the $Y$ vector for each output neuron, will aid understanding.

We start by processing each case in the training set. For each case, compute the activation of each neuron in the last hidden layer. Each case defines a row of the $A$ matrix, and each hidden neuron defines a column. The matrix will also have one additional column of 1's for the bias.

As we process each training case, we also compute the inverse activation function of each output neuron's target value. Each output neuron's desired net input computed that way defines a different dependent variable $Y$, and each must be placed in a separate column vector. In other words, we will be solving several regression problems. They will all share the same matrix of independent variables. But they will have a separate column vector of dependent variables for each output neuron. We use a standard subroutine to find the singular value decomposition of $A$. Finally, we use another standard algorithm, back-substitution, to compute ß for each $Y$. Both of these subroutines are supplied on the accompanying disk, and the listing of the MLFN program on that disk demonstrates their use.

In summary, understand that if the output activation function is nonlinear, the weights that are computed this way do not minimize the mean squared error of the *outputs*, which is what we really want. What they minimize is the mean squared error of the *desired net input* to the output neurons. In practice, that is 99.999% of the battle.

## Extension to the Complex Domain

The method of the preceding section is easily extended to the case of complex-valued dependent and independent variables. All we do is treat the real and imaginary parts separately. Each case in the training set now generates two rows in the $A$ matrix and two rows in each output neuron's $Y$ vector. One of these rows is for predicting the real part of the target, and the other is for predicting the imaginary part. For a given case, these two equations would be written (without subscripts identifying the case number) as follows:

$$a_{0,r} w_{0,r} - a_{0,i} w_{0,i} + \cdots + w_{m-1,r} = y_r$$
$$a_{0,i} w_{0,r} + a_{0,r} w_{0,i} + \cdots + w_{m-1,i} = y_i$$

(2-57)

We now see how the $A$ matrix and $Y$ vectors are constructed. Figure 2-25 shows the analogue of Figure 2-24 with a few slight differences. The real-domain version showed multiple cases, with a subscript on the elements to indicate the case number. Since we now need a subscript to indicate whether an element is the real or imaginary part, we show the two rows corresponding to a single case and omit the subscript identifying the case. The numerical subscripts used in this figure refer to hidden neurons. Also, as before, we have $m$-1 hidden neurons subscripted from 0 through $m$-2. But now we have two columns for each hidden neuron, reflecting the real and imaginary parts of the activation of each corresponding weight, and two columns for the bias of $1 + 0i$. Each of the $m$ computed weights will have a real part and an imaginary part, corresponding to the even and the odd columns in the $A$ matrix. Finally, only one $Y$ vector is shown. There would, of course, be one for each output neuron. Complete code for implementing regression in the complex domain can be found on the accompanying disk.

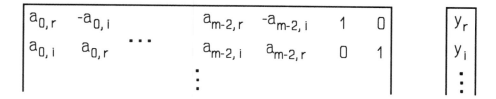

Fig. 2-25: Regression in the complex domain.

# 3

## Data Preparation

## for Neural Networks

Neural networks are extremely tolerant of many aberrations that would render other procedures worthless. They do not assume linearity of effects. They tolerate all sorts of interactions between variables. Correlation among inputs, including total redundancy, is accepted without question. Random noise contamination is usually well tolerated no matter what its statistical distribution. Wild outliers generally have little ill effect on the correct operation of neural networks. However, most neural network models do impose a few very reasonable restrictions on the data that they are asked to process. This chapter will discuss some of the most important considerations involved in preparing data for presentation to a neural network.

## Deterministic Components

When we analyze a time series so that we can predict future values, classify its characteristics, or deduce other useful facts, we are usually most interested in subtle patterns in the data. Gross properties of the series are of little interest. In fact, they can often obscure the interesting components of the data. For example, we may want to examine long-term retail sales figures of a certain product, looking for consumption patterns. The fact that its sales volume peaks every year in December tells us nothing interesting if we already know that the product is a popular Christmas gift. When we visually examine a multiyear chart of sales, those annual peaks will surely distract us, perhaps causing us to miss an important pattern that occurs at a level much lower than the seasonal pattern. Neural networks can be distracted in exactly the same way. They can eventually learn to ignore unimportant variation, even if it is large compared to the important variation. However, it is always in our best interest to make the network's job as easy as possible. Therefore, when we ask a neural network to process a time series, we must do two things:

1.  Determine whether or not deterministic components such as trend or seasonal variation are present.

2.  Remove these components if they are undesirable. We may want to remove them in such a way that they can be easily added back in later. This would be the case, for example, if we were using the network to predict future values of the series.

It should be noted that although most deterministic components are nearly always unwelcome, seasonal components may sometimes contribute important information to the network. This does not happen often, but the user should always be aware that it can happen. The usual reason why seasonal variation might be valuable is if the smaller, more subtle variation has different meanings depending on the seasonal state of the data. For example, suppose that we are studying an industrial manufacturing process. Raw products are injected into a reaction vessel. Temperature builds, reaches a peak, then drops rapidly as the synthesis becomes complete. The product is expelled, then the cycle repeats with a new batch of raw products being injected. The process is monitored by means of a temperature sensor. The temperature defines a time series having a strong periodic component. Suppose that we also know that the degree of buildup of undesirable byproducts on the walls of the reaction vessel can be detected by looking for certain patterns of temperature variation at particular stages of the reaction. In a situation like this, it may be better if we do not remove the quasi-periodic component, as that may be an important indicator of the stage of the reaction. Even if we synchronize the sampling with the injection, which we should do, it may be that the significance of the subtle patterns depends primarily on the current reaction state, as defined by the current temperature, rather than on simple elapsed time.

That example should not be interpreted as generally endorsing retention of seasonal variation. It is an exception to the rule, presented for education only. In the vast majority of situations, it is nearly mandatory that large, slow variation be eliminated. This is because neural networks, particularly the MLFNs that are central to this text, are inherently nonlinear. Their reasoning ability depends to a great extent on this non-linearity. Hidden neurons must often pass between extremes of saturation in response to different input patterns. If the important subtle variation is immersed in large variation that makes no useful contribution, the subtle variation will not be able to manipulate the hidden neurons effectively. The network's ability to solve complicated logic problems will be adversely impacted.

How do we know if our time series contains potentially troublesome trend, seasonal variation, or other components? It is often easy. We just plot the data and look at it. We would not want to put statisticians out of a job. But the fact of the matter is that for neural network purposes, statistical decisions regarding these components is often overkill. There are many useful tests for detecting the presence of statistically significant trend and seasonal variation. However, detecting their presence is not our goal. The key question is, "Are they present *enough to be troublesome*?" Given that we will probably have a lot of data to examine, it is not unlikely that these components will often be found at high statistical confidence levels. But remember that detecting *some* deterministic component is very different from detecting one that is *troublesome*. We are almost always better off simply trusting our intuition for this decision.

The main problem with using statistics to detect a deterministic component, then removing it, is that it is too easy to actually *create* one where none belongs. Then we are really in trouble! For example, suppose that we are sampling a phenomenon that contains a strong periodic component. We know about it and intend to remove it. As long as we are at it, we intend to eliminate any linear trend that may be present, even though we do not have an explicit reason to suspect the presence of a linear trend. Many popular and otherwise effective detrending algorithms, including the one that will be presented soon, can falsely find a trend in periodic data that is sampled over relatively few periods. A sine wave that is detrended with a least squares fit will come out looking like Figure 3-1 if the sample is not balanced relative to the top and bottom cycle positions.

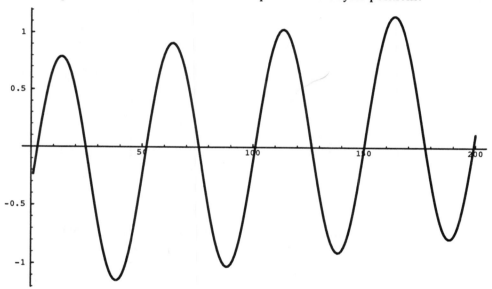

Fig. 3-1: Detrending can be dangerous.

In summary, think about the physical process that generated the data and examine a plot of the data. If you have reason to believe that the process would generate a trend or seasonal variation, or if you clearly see evidence of them or of any other deterministic component in the plot, then by all means remove them. But if you have no reason to expect their presence and see no sign of them in a plot, then do not try to remove them based on a statistical test. You may be doing more harm than good.

## Removing Linear Trends

Suppose that a linear trend is definitely present in the data. Unlike slow or seasonal variation, which may occasionally be useful, linear trends are virtually always undesirable. The most common way to eliminate a linear trend is to do a simple least-squares fit of a line to the series. We have a variable $t_i$ that is the time corresponding to sample point $i$. This variable usually has origin zero, although the actual origin is not important as long as a consistent standard is maintained. The values of $t_i$ in the subroutine presented shortly are 0, 1, ..., $n - 1$, where $n$ is the number of points in the series. The value of the signal at each point is $s_i$. We model the trend component of the series using Equation (3-1).

$$s = \text{slope} * t + \text{intercept} \qquad (3\text{-}1)$$

The slope and intercept are computed in such a way that the sum of squared errors between the modeled and the actual values of the series is minimized. The method of accomplishing this is well known. First, compute the mean of each variable using Equation (3-2).

$$\bar{t} = \frac{1}{n} \sum_{i=0}^{n-1} t_i = \frac{n-1}{2}$$

$$\bar{s} = \frac{1}{n} \sum_{i=0}^{n-1} s_i \qquad (3\text{-}2)$$

Then compute the sum of squares (SS) of the time variable about its mean. Also compute the cross-product of the time variable with the signal as shown in Equation (3-3). The slope and intercept are computed with Equation (3-4).

$$SS_t = \sum_{i=0}^{n-1} (t_i - \bar{t})^2$$

$$SS_{ts} = \sum_{i=0}^{n-1} (t_i - \bar{t})(s_i - \bar{s}) \qquad (3\text{-}3)$$

$$\text{slope} = \frac{SS_{ts}}{SS_t} \qquad (3\text{-}4)$$

$$\text{intercept} = \bar{s} - \text{slope} * \bar{t}$$

Once the slope and intercept have been found, the series is easily detrended. For each point, compute the trend component, then subtract that component from the series as shown in Equation (3-5).

$$\tilde{s}_i = s_i - (\text{slope} * t_i + \text{intercept}) \qquad (3\text{-}5)$$

If we are using a neural network to predict future values of the series, we will need to reapply the trend to the predicted series. This is done using Equation (3-6).

$$s_i = \tilde{s}_i + (\text{slope} * t_i + \text{intercept}) \qquad (3\text{-}6)$$

Detrending and retrending is best implemented using a class structure. This allows the user to create a detrending/retrending object based on a training model. This object can be reused on other samples. Suitable code now follows.

```
class Detrender {

public:
   Detrender ( int n , double *model ) ;
   void detrend ( int n , double *input , double *output ) ;
   void retrend ( int n , double *input , double *output ) ;

private:
   double slope, intercept ;
} ;


/*
   Constructor computes and saves slope and intercept
*/

Detrender::Detrender (
   int n ,                          // Length of model signal
   double *model                    // Use this signal as model
   )
{
   int i ;
   double temp, sig, *sigptr, tmean, sigmean, tss, tsig ;

   if (n < 2) {                     // Trivial, but should make provision
      slope = 0.0 ;
      intercept = model[0] ;
      return ;
      }
```

```
/*
   Compute the mean of the time variable and the signal
*/

   tmean = 0.5 * (double) (n-1) ;
   sigmean = 0.0 ;
   i = n ;
   sigptr = model ;
   while (i--)
     sigmean += *(sigptr++) ;
   sigmean /= (double) n ;


/*
      Compute the sum of squares and cross-product of the time variable and the signal
*/

   tss = tsig = 0.0 ;                   // Time sum squares, crossproduct
   i = n ;                              // This many points
   sigptr = model + n ;                 // Will work backwards
   while (i--) {                        // From last point to first
     temp = (double) i - tmean ;        // Time offset from its mean
     sig = *(--sigptr) - sigmean ;      // Ditto for signal
     tss += temp * temp ;               // Time sum of squares
     tsig += temp * sig ;               // Cross product
     }

   slope = tsig / tss ;
   intercept = sigmean - slope * tmean ;
}



/*
   Routines for detrending and retrending
*/

void Detrender::detrend ( int n , double *input , double *output )
{
  while (n--)
     output[n] = input[n] -  slope * n - intercept ;
}
```

```
void Detrender::retrend ( int n , double *input , double *output )
{
  while (n--)
    output[n] = input[n] + slope * n + intercept ;
}
```

## Removing Seasonal Components and Slow Variation

The correct removal of seasonal components and other relatively slow variation is a surprisingly complicated task. A detailed treatment is outside the scope of this text. A good, practical discussion can be found in [Kendall and Stuart, vol. 3, 1976]. A quite different approach is taken in [Box and Jenkins, 1976]. Brutal theory is given in [Brillinger, 1975]. A vast number of standard signal-processing texts discuss digital filters in detail. Finally, [Masters, 1993] discusses this problem in the context of neural networks. In this section we will touch on only a few of the most important concepts.

One reason that this problem is so difficult is that *trend and seasonal variation are inexorably bound up with one another*. When one attempts to remove a seasonal component from a time series, then either all trend must have been removed first, or the two must be removed simultaneously with an appropriate algorithm. A simple example will reveal why this is so important. Suppose that we are dealing with monthly sales figures of some product. We notice that there is a strong seasonal component across the years that we are studying. Perhaps the product is sun screen or a popular Christmas gift. We want to isolate that seasonal component and subtract it from the raw series. Suppose also that there is an upward trend caused by steadily increasing popularity of the product. This trend will cause the difference between months to be different from what it would be due to the seasonal component alone. If the sampled time period starts in January of one year and ends in December several years later, the sales figures for December relative to January would be higher than what would be caused by seasonal effect alone. The trend throughout each year would be falsely interpreted as a seasonal effect. Removing this sawtooth-shaped function would result in rubbish.

Because seasonal component analysis is extremely sensitive to slow variation in the series, the linear detrending method previously described may not be adequate if seasonal trend is also to be removed. The slowly varying trends that remain after linear detrending confound the seasonal estimates. A common and moderately effective way of handling this problem is to apply a simple low-pass filter to the series. This filter passes only variation that is slower than the seasonal trend. A primitive implementation of such a filter is a centered moving average. If the period of the seasonal component, $n$, is odd, sum every $n$ points and divide by $n$. If $n$ is even, sum $n - 1$ contiguous points, then add in 0.5 times the sum of one more point on each end, then divide by $n$. For example, if we were expecting a period of 4 points, the filter coefficients would be {0.5, 1, 1, 1, 0.5}, with the weighted sum being divided by four. The result of applying this filter is a series that contains only variation slower than the period of the seasonal component (plus some annoying stuff described later). If we subtract this filtered series from the original series, we are left with three components: seasonal variation, higher-frequency variation that is presumably most important, and random noise. The troublesome slow trends, linear and otherwise, have been removed. It is now easy to average the samples for each position in the season (months or whatever) in order to safely estimate the seasonal component. That component is then subtracted from the series.

The method just described has some virtue. It is quick, easy, and intuitively solid. But that is where its virtue ends. We can do much better. Its main problem is the primitive nature of the low-pass filter. A moving average (MA) filter has side lobes that are large and that extend into quite high frequencies. A plot of the relative amplitude frequency response of a 12-month (13-point) moving average filter is shown in Figure 3-2.

The filtered signal that supposedly contains only low-frequency components actually contains a lot of components above the seasonal frequency. The seasonal component is well removed, as can be seen by examining the plot at a frequency of $1/12 = 0.083$. But look at what gets through at higher frequencies! When the filtered signal is subtracted from the original, it is possible that important information is being removed along with the slow trends. There are much better ways. Unfortunately, they are all beyond the scope of this text. Digital filters are widely discussed in the literature. On the other hand, the simple method presented here has been used successfully for many years in many situations. It could well be satisfactory for your application.

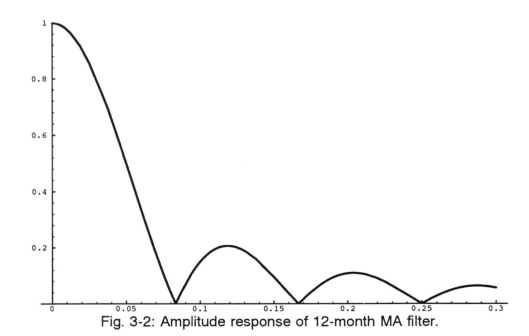

Fig. 3-2: Amplitude response of 12-month MA filter.

## Removing Other Deterministic Components

Linear trend and seasonal variation are the two most common troublesome deterministic components. Slow drift, perhaps caused by sensor problems, is also an occasional problem. That is easily removed with an appropriate filter. But there is another type of deterministic component that sometimes occurs and whose removal is beneficial. This is data dependency that is easily modeled if the model is known. These can be difficult to see in a plot of the data. However, if the presence of a deterministic model component is suspected, we should certainly try to compensate for its effects.

For example, the ARIMA model is an extremely popular model used in time series analysis and prediction. It is described in the classic text [Box and Jenkins, 1976]. Suppose that we have a times series for which future values are to be predicted. Traditionalists would perhaps fit an ARIMA model and use that model for the predictions. Modernists would take the alternative approach of training a neural network to make the prediction. But we postmodernists can beat them both. The correct method is to use the two approaches together. Start by fitting an ARIMA model. Use that model to remove the deterministic components from the series, so that all that remains is what is supposed to be noise (as far as ARIMA is concerned). Train a neural network to predict that noise. Then, to predict a future value of the original series, use the trained network to predict the future noise. Finally, apply the previously fit ARIMA model, using the predicted noise as the noise term in the ARIMA model. The results will often be surprising in that what was thought to be noise was definitely not!

# Differencing

In many applications, good results can be obtained by working with the differences between successive samples rather than with the samples themselves. Economic series are a case in point. Day-to-day values of stock market averages convey one sort of information. The *change* from one day to the next conveys an entirely different sort of information. They may be mathematically equivalent in that one begets the other. But in terms of the picture that can be quickly drawn from them, they are quite different. Often, intuition can be a valuable guide to choosing between raw or differenced data. Simply ask which is more relevant: the current value of the variable, or the amount that it has recently changed.

There is a major advantage to differencing a time series before processing it. This operation is a potent high-pass filter. A linear trend is converted to a constant offset, which is easily removed. Slowly varying components are dramatically reduced in amplitude. Even strong seasonal variation is often so attenuated that it can be safely ignored. As we have seen, that is no small accomplishment.

Recall that sometimes it is better to retain seasonal information. When we difference the data, we have removed the current status information that may be of use to the network in some applications. On the other hand, the seasonal variation may be so large that it swamps out variation that may be vital to a correct decision. The solution is simple. Give the network *both* the raw and the differenced data. It can then find the significant patterns easily. This is an extremely powerful technique, and it should always be given consideration. Redundancy of this sort is not only easily handled by neural networks, but it often even helps their performance.

# Scaling

For most neural network models, it is absolutely vital that the data presented to these models be scaled appropriately. In most cases, this is a trivially easy job. A few network models, such as the Kohonen, can sometimes be more demanding. This section will discuss some of the most important general considerations.

## Weight Normalization

Many neural networks, including the MLFN featured in this text, have no limitation imposed on their inputs. They can theoretically handle the distance to the moon measured in millimeters and the thickness of a piece of paper measured in lightyears. But if a wide range of input magnitude is possible, a similarly wide range of connection weights should also be expected. This obviously makes it difficult to learn optimal weights. It would not be feasible to search the range of possibilities, looking for promising localities in weight space. And even if the learning algorithm did manage to find its way to such a locality, a ridiculously large distance might still have to be traversed to find the best weights at a local minimum. The task of training the network is greatly simplified if it can be expected that the optimal weights will probably lie within a small, known range. The best way to insure this is to make the data presented to the network do likewise. The author always tries to scale the data in such a way that most of the cases lie within plus or minus three, with few or no cases exceeding ten in absolute value. This insures that the weights will have a strong tendency to do the same.

## Balancing Relative Importance

Another reason for imposing the scaling discussed earlier is to encourage equitable distribution of importance. Input variables that have a large absolute magnitude will tend to exert more influence on the network's response than those having small magnitude. The training process can eventually compensate for this by adjusting the weights connecting the inputs to the network. But that compensation takes time and sometimes may fail to occur at all. It is better to start out with no prejudices and then let the network choose its own course.

For output variables, the situation is even more critical. If we are using the network to estimate more than one parameter, or to predict more than one future value of a time series, it is vital that we pay close attention to the relative scaling of those output values. It must be stated that *multiple prediction is never encouraged.* It is nearly always better to train separate neural networks for each output variable. But sometimes that is not possible. We may want a single autoassociative network, or we may want to assume that meaningful hidden layer patterns will be similar for all variables. When multiple outputs absolutely must be learned, then special care is needed.

Many models, including the MLFN, learn by minimizing a measure of output error, typically the mean squared error. Outputs having larger magnitude will contribute more to that error. Thus, the training algorithm can gain more by reducing the error of large magnitude outputs than by reducing that of smaller magnitude outputs. The learned weights will favor the larger magnitude output variables. Sometimes we want an inequitable distribution of performance, so we deliberately scale the outputs differently. But if we do not want learning bias, we must take care to scale the output variables so that their ranges are comparable.

## Centering

For many models, especially the MLFN, learning is expedited if the input variables are roughly centered about zero. This is not absolutely necessary. The bias term used by each neuron in the first layer after the input layer can compensate for a skewed input range. However, it is slightly easier on the training algorithm if the role of the bias can be strictly limited to logic, without it being used to compensate for unbalanced inputs as well.

## Output Activation Limits

The preceding discussion has focused almost exclusively on input variables. Some network models impose strict limitations on the outputs that they can learn. In particular, MLFN models that use a nonlinear output activation function require special care. It is obvious that there is no point in trying to teach a network to respond with values that are not in the range of the activation function. On the other hand, it may not be so obvious that it is also risky to ask a network to produce values that are near the extremes of its output activation function. For example, the good old standard logistic function has a theoretical range of from zero to one. But traditional practice is to limit the range of target output values to the reduced range of 0.1 to 0.9 or so. Asking for more extreme outputs requires the input to the output neurons to be larger than is practical. Similarly, the highly regarded hyperbolic tangent function has a theoretical range of minus one to one. The actual learnable range is more like –0.9 to 0.9.

An MLFN that has linear output neurons does not impose this limitation. The outputs can theoretically attain any value. However, the scaling already recommended for inputs will benefit outputs as well. Keeping the target values in a moderate, centered range will simplify the life of the training algorithm by keeping output weights in a similar range.

## Special Input Limitations

There are a few neural network models that have special needs. One of the most famous of these is the Kohonen network. Most members of the Kohonen family demand that the sum of squares of its inputs be constant for all cases. This is because these models treat the inputs as vectors, and the learning algorithm assumes that all input vectors lie on the surface of a sphere. Sometimes this is no problem. For example, if all inputs are binary, they can be coded as plus or minus one. In some applications, though, this is a severe restriction. If the inputs to the network are the height and weight of experimental subjects, requiring every case to have the same sum of squared inputs is a daunting challenge! Luckily, there is a solution that can often be used. It involves including an extra, synthetic input. The value of this input is adjusted so that the sum of squares of all inputs, including the synthetic input, is constant. This technique is often called a *Z-axis transform*. Readers wanting more details on this transform are referred to [Masters, 1993].

## Transformations

A neural network will have a hard time learning training patterns if the majority of the cases lies in one region of the input data space while a few cases lie far away. And it is a harsh reality that this situation arises frequently. Acoustical data is famous for exhibiting this problem. Sounds that seem to be only moderately different in volume may actually be orders of magnitude apart in power. When this happens, it behooves us to apply a compressing transform to the data so that wild samples are dragged back into the neighborhood of the majority. When data arises from multiplicative processes, as often happens with economic series, the logarithmic transform is often appropriate. That function is also good for acoustical data. A weaker compression is obtained with the square root or cube root. Transformations like these will often be needed. Do not neglect to consider this possibility.

Much of the data treated in this book lies in the complex domain. This data needs special treatment. We should not scale or transform complex data by operating on the real and imaginary parts separately. We should always convert the data to polar form, transform the magnitude only, then convert back. Note that the log function is often inappropriate for this task, as it can take on negative values. Either an offset can be used to prevent negatives, or an alternative like a root can be used.

## Code for Scaling

This section presents a class for centering, scaling, and optionally transforming a time series. Both strictly real and full complex versions are supplied. The optional transformation is the cube root, as that is a good general-purpose compressing transform. However, the program is written in such a way that any transformation function can be easily substituted.

The class constructor is called using a training sample as a model. The sample itself is not modified by the constructor. The scale member function can then be called as often as desired. It will apply the same transformation each time. The unscale member function can be used to undo the scaling operation.

The scaling takes place in three steps. The first step, which is optional, is application of the transformation function. For complex data the transformation will affect only the magnitude of each case, not its argument (angle). In the second step, the data is centered to have a median of zero. The last step is to scale the data so that its interquartile range is 3. In the case of complex data, this is approximately equivalent to having a median magnitude of 3. The class declaration and constructor now follow.

```
class Scaler {

public:
   Scaler ( int n , double *r_model , double *i_model , int transform ) ;
   void scale ( int n , double *r_input , double *i_input ,
           double *r_output , double *i_output ) ;
   void unscale ( int n , double *r_input , double *i_input ,
           double *r_output , double *i_output ) ;

private:
   int is_complex, tfunc ;
   double r_offset, i_offset, factor ;
} ;

/*
   Constructor computes and saves offset and scaling factor
*/

Scaler::Scaler (
   int n ,                          // Length of model signal
   double *r_model ,                // Use this signal as model
   double *i_model ,                // It may be complex (This is NULL if real)
   int transform                    // Transform also?
   )
{
```

```
int i ;
double mag, trans, *work, *r_work, *i_work, rtemp, itemp ;

is_complex = (i_model != NULL) ;
tfunc = transform ;

if (n < 2) {                              // Make provision for degenerate case
  r_offset = r_model[0] ;
  if (is_complex)
    i_offset = i_model[0] ;
  factor = 1.0 ;
  return ;
  }

if (is_complex  &&  transform) {
  work = (double *) malloc ( 3 * n * sizeof(double) ) ;
  r_work = work + n ;
  i_work = r_work + n ;
  }
else {
  work = (double *) malloc ( n * sizeof(double) ) ;
  r_work = r_model ;
  i_work = i_model ;
  }

if (work == NULL) {
  factor = -1.0 ;   // Flag error
  return ;
  }

if (is_complex) {

  if (transform) {                      // Transform series
    for (i=0 ; i<n ; i++) {
      mag = sqrt ( r_model[i] * r_model[i] + i_model[i] * i_model[i] ) ;
      if (mag > 0.0) {
        trans = cube_root ( mag ) ;
        r_work[i] = r_model[i] * trans / mag ;
        i_work[i] = i_model[i] * trans / mag ;
        }
      else
        r_work[i] = i_work[i] = 0.0 ;
      }
    }
```

```
      for (i=0 ; i<n ; i++)                  // r_offset is real median
        work[i] = r_work[i] ;
      qsort ( 0 , n-1 , work ) ;
      r_offset = work[n/2] ;

      for (i=0 ; i<n ; i++)                  // i_offset is imaginary median
        work[i] = i_work[i] ;
      qsort ( 0 , n-1 , work ) ;
      i_offset = work[n/2] ;

      for (i=0 ; i<n ; i++) {                // Complex scale from median mag
        rtemp = r_work[i] - r_offset ;
        itemp = i_work[i] - i_offset ;
        work[i] = rtemp * rtemp + itemp * itemp ;
        }
      qsort ( 0 , n-1 , work ) ;

      if (work[n/2] > 0.0)
        factor = 3.0 / sqrt ( work[n/2] ) ;
      else
        factor = 1.e30 ;                     // Arbitrary, as this is degenerate
      }

   else {                                    // Pure real series
     if (transform) {
       for (i=0 ; i<n ; i++)
         work[i] = cube_root ( r_model[i] ) ;
       }
     else {
       for (i=0 ; i<n ; i++)
         work[i] = r_model[i] ;
       }
     qsort ( 0 , n-1 , work ) ;
     r_offset = work[n/2] ;
     factor = 3.0 / (work[3*n/4] - work[n/4]) ;
     }

   free ( work ) ;
}
```

The constructor looks at the address of the imaginary series input. If it is NULL, then the series is assumed to be strictly real. A private flag is set accordingly. We also flag the user's desire to transform. It is good programming practice to bail out a careless user by checking for degenerate calling parameters.

At a minimum we will need a work vector for sorting, as the input data is not to be touched. In addition, the most complicated situation, a complex series needing transformation, will need two extra work vectors to hold the transformed inputs. Memory is allocated as needed. Note that if the extra work vectors are not allocated, their pointers are set to point to the input data. This is important, as those pointers will be used later.

The most complicated case is handled first. If complex data is to be transformed and scaled, the transformation is done immediately. The magnitude is computed and transformed. The original case's magnitude is then set equal to the transformed magnitude without changing the argument. The transformed case is placed in the complex work vector.

The medians of the real and imaginary parts are then computed individually. Note that if the data was not transformed, the work pointers were set to point to the input data.

The squared magnitude of each centered case is computed and stored in the work vector. The scaling factor is then based on the median magnitude.

The case of real data is much simpler. The only difference is that we use the actual interquartile range to compute the scaling factor.

The scale and unscale routines are very straightforward. The former optionally transforms the data exactly as was done in the constructor. It then subtracts the offset and multiplies by the scaling factor. The unscale routine reverses these operations. Their code now follows.

```
void Scaler::scale ( int n , double *r_input , double *i_input ,
              double *r_output , double *i_output )
{
  double mag, trans, rtemp, itemp ;

  if (factor < 0.0)                     // Check memory error flag
    return ;

  if (is_complex) {

    if (tfunc) {
      while (n--) {
        mag = sqrt ( r_input[n] * r_input[n] + i_input[n] * i_input[n] ) ;
        if (mag > 0.0) {
          trans = cube_root ( mag ) ;
          rtemp = r_input[n] * trans / mag ;
          itemp = i_input[n] * trans / mag ;
          }
        else
          rtemp = itemp = 0.0 ;
```

```
            r_output[n] = factor * (rtemp - r_offset) ;
            i_output[n] = factor * (itemp - i_offset) ;
            }
         }

      else {
         while (n--) {
            r_output[n] = factor * (r_input[n] - r_offset) ;
            i_output[n] = factor * (i_input[n] - i_offset) ;
            }
         }
      }
   else {
      if (tfunc) {
         while (n--)
            r_output[n] = factor * (cube_root ( r_input[n] ) - r_offset) ;
         }
      else {
         while (n--)
            r_output[n] = factor * (r_input[n] - r_offset) ;
         }
      }
}

void Scaler::unscale ( int n , double *r_input , double *i_input ,
                  double *r_output , double *i_output )
{
   double fac, mag, trans, rtemp, itemp ;

   if (factor <= 0.0)
      return ;

   fac = 1.0 / factor ;

   if (is_complex) {

      if (tfunc) {
         while (n--) {
            rtemp = r_input[n] * fac + r_offset ;
            itemp = i_input[n] * fac + i_offset ;
            mag = sqrt ( rtemp * rtemp + itemp * itemp ) ;
            if (mag > 0.0) {
               trans = cube ( mag ) ;
```

```
              r_output[n] = rtemp * trans / mag ;
              i_output[n] = itemp * trans / mag ;
              }
            else
              r_output[n] = i_output[n] = 0.0 ;
            }
          }

      else {
        while (n--) {
          r_output[n] = r_input[n] * fac + r_offset ;
          i_output[n] = i_input[n] * fac + i_offset ;
          }
        }
      }
  else {
    if (tfunc) {
      while (n--)
        r_output[n] = cube ( r_input[n] * fac + r_offset ) ;
      }
    else {
      while (n--)
        r_output[n] = r_input[n] * fac + r_offset ;
      }
    }
}
```

The above code refers to several other subroutines. One is the optional transformation, here the cube root. Care is needed in defining this routine. Suitable code for it and for its inverse is now given.

```
double cube ( double x )
{
  return x * x * x ;
}


double cube_root ( double x )
{
  if (x == 0.0)
    return 0.0 ;
  else if (x < 0.0)
    return -pow ( -x , 1.0 / 3.0 ) ;
```

```
      else
        return pow ( x , 1.0 / 3.0 ) ;
    }
```

Most readers will have access to a quicksort routine. However, for those who may not, here is a simple version. It is nearly as fast as versions that are more sophisticated. Its only problem is that it uses explicit recursion. Readers who include this subroutine in their program must make sure that sufficient runtime stack space is allocated to allow for whatever is needed to accommodate their data.

```
void qsort ( int first , int last , double data[] )
{
    int bottom, top ;
    double ftemp, split ;

    split = data[(first+last)/2] ;
    bottom = first ;
    top = last ;

    do {
        while ( split > data[bottom] )
            ++bottom ;
        while ( split < data[top] )
            --top ;
        if (bottom == top) {
            ++bottom ;
            --top ;
            }
        else if (bottom < top) {
            ftemp = data[bottom] ;
            data[bottom++] = data[top] ;
            data[top--] = ftemp ;
            }
        } while ( bottom <= top ) ;

    if (first < top)
        qsort ( first , top , data ) ;
    if (bottom < last)
        qsort ( bottom , last , data ) ;
}
```

It should be pointed out that the primary use for the quicksort routine in the Scaler class is to compute the median of arrays. There are algorithms dedicated to that single task that are considerably faster. In fact, [Press *et al.*, 1992] gives excellent treatment to a family of routines for computing not only the median, but the other quartiles that are needed also. A possibly greater advantage of the algorithms given there is that there is no need for a large runtime stack to support recursion. Many readers will want to ignore the quicksort given above and implement one of the superior choices. However, this quicksort is nothing to be ashamed of. It serves as an excellent general workhorse.

This section will be concluded with a brief discussion of a topic that is sure to pique the curiosity of readers throughout much of the rest of this book. The Scaler class used two separate arrays to store the real and imaginary parts of the series. That tradition will be maintained in nearly every piece of code that appears in the remainder of this text. Most programming languages implement complex numbers as a single array in which the real part of the first element is followed by its imaginary part, then the real and imaginary parts of the second element, and so forth. Why does the author depart from that convention? There are at least three reasons. The first reason is that many, if not most, common algorithms that operate in the complex domain execute slightly faster if programmed this way when compiled with a modern optimizing compiler. The difference is small but significant in some situations. The second reason is that subscripting is invariably more clear. Instead of referring to x[2*i+1], we can refer to xi[i]. This is nice. The third reason is that in most cases it is simply more convenient to work with two separate arrays. This last reason is difficult to quantify, as subjective opinion is involved. Suffice it to say that the author prefers to work that way.

# 4

# Frequency-Domain Techniques

Recall from the review of complex numbers on page 12 that complex numbers are inherently *pairs* of real numbers. Beyond that fact, there is nothing magical about them. The key is that they work together in pairs. In particular, their arithmetic is special, being governed by Equation (2-1). The significance of this observation is that we should always consider complex-valued neural networks when our problem involves pairs of numbers. This means that we need not limit ourselves to applications in which the data itself is inherently in the complex domain. Rather, we should take the broader view, including all applications in which the data comes to us in pairs of numbers that are somehow bound together. This includes trajectories of a point in a plane and many other types of data in which a pair of variables is parameterized by a third variable that does not directly interest us. It can even include common bivariate statistical measurements, such as the height and weight of experimental subjects. But the most common naturally pairwise data is that which arises from a Fourier transformation. Therefore, this chapter will concentrate on that area. It would be a pity, though, if such concentration led the reader to believe that Fourier-transformed data is the only application that can benefit from complex-valued neural networks.

## Review of the Fourier Transform

This section will briefly review the salient features of the Fourier transform, especially as it applies to data preparation for neural networks. Many of its most common and useful applications, such as filter design and power spectrum analysis, will be glossed over or totally ignored. Mathematical theory, unless essential to the topic, will be dispensed with. The focus here will be only on what is needed for a modest family of neural network applications. It is assumed that the reader possesses a basic familiarity with the material. Those to whom the Fourier transform is completely new should read this section in conjunction with a text that provides a solid introduction to Fourier techniques.

Suppose we have some physical process that gives rise to a variable that varies as time passes. We may write that variable as a function of time using the notation $h(t)$. Much of the foundation theory treats $t$ as a continuous variable, so we will pay due homage to that treatment. On the other hand, virtually all practical problems involve samples of $h$ at discrete time intervals, so we will emphasize that aspect. Also, most (but, as we will see, not all) problems assume that $h$ takes on real values. That will be emphasized at first, but the general complex case will be taken up later.

When our physical process is considered as a function of time, we say that we are in the *time domain*. Under reasonable conditions, which will be mercifully dispensed with here, $h(t)$ may also be considered to be the sum of sine and cosine functions that extend to infinity. The process may then be represented by a function of frequency, $H(f)$. This remarkable result lets us study the process in the *frequency domain*. The relationship between these two representations is given (in the continuous case) by the *Fourier transform* equations.

$$H(f) = \int_{-\infty}^{\infty} h(t)\, e^{2\pi ift}\, dt$$

$$h(t) = \int_{-\infty}^{\infty} H(f)\, e^{-2\pi ift}\, df \qquad (4\text{-}1)$$

Recall from Equation (2-8) on page 14 that $e$ raised to the power of an imaginary quantity is equivalent to the cosine of the quantity plus $i$ times the sine of the quantity. In particular, the transform can also be written as shown in Equation (4-2). That form clarifies the wave nature of the Fourier transform. But it is not often used, partly because it is longer to write, and partly because many operations are more easily expressed in exponential form.

$$H(f) = \int_{-\infty}^{\infty} h(t)\cos(2\pi ft)\, dt\ +$$

$$i \int_{-\infty}^{\infty} h(t)\sin(2\pi ft)\, dt \qquad (4\text{-}2)$$

Note that some (primarily electronics engineering) authorities reverse the signs in the complex exponent of the Fourier transform. A negative sign is used when computing the forward transform, with the inverse transform having a positive exponent. As long as the signs are opposite, the relationship holds true. The author prefers the version stated here, which is most often used in mathematics and physics.

Equation (4-1) is fine for theoretical work, but it does us little good when we are faced with values of $h(t)$ experimentally measured at discrete time intervals. In that case, we need the *discrete Fourier transform* (*DFT*), shown in Equation (4-3).

$$H(f) = \sum_{t=0}^{n-1} h(t)e^{i\frac{2\pi tf}{n}}$$

$$h(t) = \frac{1}{n} \sum_{f=0}^{n-1} H(f)e^{-i\frac{2\pi tf}{n}}$$

(4-3)

Several things should be noted about Equation (4-3). First, we have $n$ measured complex values for both the time-domain series and the frequency-domain series. This should please the intuition, as it implies that information is being neither created nor destroyed. Also, the forward and reverse transformations are nearly identical. They differ only in the sign in the exponent and in the division by the sample size for the inverse transform. This tells us that we can use one subroutine to do both operations. To use a forward-transform routine to do the reverse transform, simply take the complex conjugate, do a forward transform, take the complex conjugate again, then divide by $n$. (Verification of that fact is a nice exercise for the reader.) Finally, be warned that, as was the case for Equation (4-1), some authorities switch the signs of the exponents in the two equations. Division by $n$ in the reverse transform is fairly universal, though.

It is absolutely vital that the reader sees Equation (4-3) as computing the dot product of our data vector with sine waves and cosine waves at each of the frequencies under consideration. For example, look at the forward transform at a frequency of $f = 1$. As the index $t$ moves from 0 through $n - 1$, accessing each of our measured values, the fraction $2\pi t/n$ sweeps up from 0 to nearly $2\pi$, exactly one full cycle. If $h$ is strictly real, the real component of $H(1)$ is the dot product of the time-domain sample with a single cosine wave. The imaginary component is the dot product with a sine wave. For complex $h$, the math gets a little trickier, but it is still dot products. It should be obvious that when $f = 2$, the waves will sweep out exactly two complete cycles, and so forth. Thus, $H(f)$ can be viewed as the strength of the component of the time-domain series which repeats $f$ times during the total time interval stretching from $t = 0$ to $t = n - 1$. If this is not clear, ponder it until it becomes clear, as it is a crucial concept.

## The Fast Fourier Transform

It should be obvious that something on the order of $n^2$ complex multiplications and additions are required to naively compute the Fourier transform using Equation (4-3). Each value of $f$ in $H(f)$ requires $n$ terms in the summation, and there are $n$ values of $f$. When $n$ is large, as it often is, the size of $n^2$ becomes a serious problem. This stifled widespread use of Fourier techniques during the early days of computing. Then, a few clever individuals discovered that the computing time could be halved by splitting the series into two halves, transforming each half separately (in a quarter of the time apiece), then rapidly combining the two halves to get the transform of the original series. After a little more time passed, some *really* clever people discovered that if $n$ is a power of two, the splitting can be applied recursively! This brings the time into the realm of $n \log_2 n$, a huge improvement over $n^2$.

   *Fast Fourier transform* (FFT) programs are widely available in both source and object form, so we will not waste trees by providing detailed algorithm descriptions or code for any of the highly sophisticated versions that exist. However, for the sake of those readers who may have trouble procuring any version at all, we will list a short but effective subroutine. This is a hybrid derived from [Press *et al.*, 1992], [IEEE Digital Signal Processing Committee, 1979], and some of the author's own work. Remember that $n$ must be a power of two. This routine does not verify it!

```
void bit_reverse ( int n , double *xr , double *xi ) ;      // Forward
void butterflies ( int n , double *xr , double *xi ) ;      // Declare

void fft (
   int n ,                          // Length of vectors
   double *xr ,                     // Real components
   double *xi                       // Imaginary components
   )
{
   bit_reverse ( n , xr , xi ) ;
   butterflies ( n , xr , xi ) ;
}

static void bit_reverse ( int n , double *xr , double *xi )
{
   int i, bitrev, k, half_n ;
   double temp ;

   bitrev = 0 ;     // Will count in bit-reversed order
   half_n = n / 2 ;

   for (i=1 ; i<n ; i++) {   // Do every element

/*
   Increment bit-reversed counter
*/

      ++bitrev ;
      k = half_n ;
      while (bitrev > k) {
        bitrev -= k ;
        k >>= 1 ;
        }
```

```
          bitrev += k-1 ;

/*
    Swap straight-counter element with bit-reversed element (just once!)
*/

      if (i < bitrev) {
        temp = xr[i] ;
        xr[i] = xr[bitrev] ;
        xr[bitrev] = temp ;
        temp = xi[i] ;
        xi[i] = xi[bitrev] ;
        xi[bitrev] = temp ;
        }
      }
}

static void butterflies ( int n , double *xr , double *xi )
{
   int i, m, mmax, step, other ;
   double theta, tr, ti, wr, wi, wkr, wki ;

   for (mmax=1 ; mmax<n ; mmax=step) {
     step = 2 * mmax ;

     wr = 1.0 ;
     wi = 0.0 ;
     theta = PI / (double) mmax ;
     tr = sin ( 0.5 * theta ) ;
     wkr = -2.0 * tr * tr ;
     wki = sin ( theta ) ;

     for (m=0 ; m<mmax ; m++) {
       for (i=m ; i<n ; i+=step) {
         other = i + mmax ;
         tr = wr * xr[other]  -  wi * xi[other] ;
         ti = wr * xi[other]  +  wi * xr[other] ;
         xr[other] = xr[i] - tr ;
         xi[other] = xi[i] - ti ;
         xr[i] += tr ;
         xi[i] += ti ;
         }
       tr = wr ;
       wr += tr * wkr  -  wi * wki ;
```

```
wi += tr * wki  +  wi * wkr ;
   }
 }
}
```

There are at least two improvements that can be made to the above algorithm. Most notable is that it demands that $n$ be a power of two. More modern versions factor $n$ into as many small prime factors as possible and handle each separately. Factors that are not small are processed with the straightforward, abominably slow direct method. This generality makes the user's life easier by allowing any $n$, but it can trap the unwary into thinking that they are using a sophisticated algorithm when an unwise choice of $n$ is dragging it back to $n^2$ operation. Be warned.

The other improvement is that the very best modern FFT's use highly optimized kernels of size four and perhaps even eight. This can speed operation to a small but significant degree. But do not despair if these programs are unobtainable. The code given above is quite respectable and should suffice for all but the most demanding applications.

## Transforming Real Data

The following table shows a strictly real time series along with its Fourier transform. The series contains eight points. Each point in the series and its transform is represented by two numbers: the real part and then the imaginary part.

| $h(t)$ | | $H(f)$ | |
|---|---|---|---|
| 1.00 | 0.00 | −1.00 | 0.00 |
| 3.00 | 0.00 | 1.29 | 5.54 |
| −2.00 | 0.00 | 6.00 | 3.00 |
| 2.00 | 0.00 | 2.71 | 1.54 |
| −1.00 | 0.00 | −11.00 | 0.00 |
| 1.00 | 0.00 | 2.71 | −1.54 |
| −4.00 | 0.00 | 6.00 | −3.00 |
| −1.00 | 0.00 | 1.29 | −5.54 |

Do you notice anything interesting about $H(f)$? With the exception of the first line, the real part is symmetric about the middle entry. And the imaginary part is nearly symmetric, save for the sign reversal. Recall the definition of the complex conjugate. We may write this mathematically as Equation (4-4).

$$H(i) = \bar{H}(n-i), \qquad 0 < i < n \qquad (4\text{-}4)$$

Ambitious readers will certainly want to prove this symmetry using Equation (4-3) along with basic trigonometric identities. The proof is not difficult. While at it, also prove that the imaginary parts of the first and middle transform components are always zero. (Remember that all of this is true only when $h(t)$ is real.)

All readers who want to transform real series should take note of one more fact about the transform. The above table shows that there are five unique complex numbers in the transform. The other three can be obtained by symmetry. In general, there will be $n/2 + 1$ complex numbers required to represent the Fourier transform of a real series. Since each complex number can also be considered to be two real numbers, the real and imaginary parts, we have $n + 2$ numbers altogether. But two of these are always zero (the imaginary parts of the first and middle terms). This means that we actually have exactly $n$ numbers in the transform. And we started out with $n$ real numbers. Isn't mathematics beautiful?

At this point there should be a little voice whispering in our heads, saying that it is silly to transform a real series of length $n$ by treating it as a complex series of length $n$, setting all of the imaginary terms to zero. As is usually the case throughout life, that little voice is absolutely correct. It can be shown that a far better way is to generate a complex series of length $n/2$. Place the even-indexed components of the original series in the real parts of this new complex series, and place the odd-indexed components in the imaginary parts. After transforming this half-length fully complex series, there is a simple algorithm for computing the transform of the original real series. A nice derivation of the algorithm can be found in [Press *et al.*, 1992]. Code for doing this is now given.

```
void real_fft (
   int n ,                       // Length of each vector (Total is 2n)
   double *xr ,                  // In: 0,2,4,... Out:Real parts
   double *xi                    // In: 1,3,5,... Out: Imaginary parts
   )
{
   int i, j ;
   double theta, wr, wi, wkr, wki, t, h1r, h1i, h2r, h2i ;

   fft ( n , xr , xi ) ;

/*
   Use the guaranteed zero xi[0] to actually return xr[n]
*/

   t = xr[0] ;
   xr[0] = t + xi[0] ;
   xi[0] = t - xi[0] ;

/*
   Now do the remainder through n-1
*/
```

```
theta = PI / (double) n ;
t = sin ( 0.5 * theta ) ;

wr = 1.0 + (wkr = -2.0 * t * t) ;
wi = wki = sin ( theta ) ;

for (i=1 ; i<n/2 ; i++) {
  j = n - i ;

  h1r =  0.5 * (xr[i] + xr[j]) ;
  h1i =  0.5 * (xi[i] - xi[j]) ;
  h2r =  0.5 * (xi[i] + xi[j]) ;
  h2i = -0.5 * (xr[i] - xr[j]) ;

  xr[i] =  wr * h2r - wi * h2i + h1r ;
  xi[i] =  wr * h2i + wi * h2r + h1i ;
  xr[j] = -wr * h2r + wi * h2i + h1r ;
  xi[j] =  wr * h2i + wi * h2r - h1i ;

  t = wr ;
  wr += t * wkr - wi * wki ;
  wi += t * wki + wi * wkr ;
  }
}
```

Several aspects of the above program must be noted. First, the n in the parameter list is actually *half* of the length of the original series, *n*. It is the length of the packed complex series. The other important feature concerns that annoying extra component of the transform. Recall that the Fourier transform actually contains $n/2 + 1$ complex numbers, but xr and xi are each only $n/2 = $ n long. What to do? The problem is solved when we remember that the last (extra) element of the transform is strictly real, as is the first element. So we can be clever and place the real part of $H(n/2)$ in the imaginary part of $H(0)$. That is just what the previous subroutine does. It places what would logically belong in xr[n], an array element that will not in general have memory allocated for it, into xi[0], which would otherwise be guaranteed zero.

## Aliasing and the Nyquist Limit

Since we are sampling a physical process $h(t)$ at discrete time intervals, it is natural to ask how often we must register our measurements. We want to keep the storage and processing requirements to a minimum by sampling as slowly as possible. But certainly there must be a minimum rate if we are to capture the information we want. This section will discuss the sampling rate.

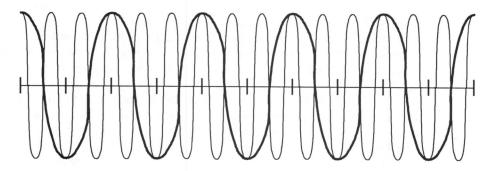

Fig. 4-1: Aliasing due to slow sampling.

It turns out that excessively slow sampling imposes a far harsher penalty than just *missing* information. We can actually end up with badly contaminated data. Examine Figure 4-1, which shows a high-frequency wave being sampled too slowly.

The tick marks in that figure show the sample collection times. The rapid variation of the true wave is totally missed. Worse, the data that is collected will look like it has been generated by a much slower wave, also shown. This effect, in which a high-frequency component takes on the character of a low-frequency component that does not truly exist, is called *aliasing*. It is obviously a problem worthy of careful attention.

The highest frequency we can correctly handle has a period equal to two sample intervals. If the time between adjacent samples is $s$ seconds, the highest frequency in $h(t)$ that can be tolerated is $1 / (2s)$ cycles per second. This frequency is often called the *Nyquist frequency*. Looked at in the frequency domain, $H(n/2)$ is the highest frequency that can be resolved.

How can we protect ourselves from aliasing? There are three things that should be kept in mind. First, we should hopefully possess some knowledge of the physical process that is producing the variable. If we know the highest frequency that it can generate, we know that we must sample at least twice that rate. Second, realize that we may have to resort to some sort of low-pass filtering before sampling. This may be an analogue device whose output is sampled. Or we may sample at an extremely high rate, guaranteed to be satisfactory, apply a digital filter to that rapid sample, then decimate the filter's output for our final sample. Finally, it never hurts to examine a plot of the power spectrum, discussed in the next section. If the power in $H(f)$ resolutely drops to zero by the time we get to $H(n/2)$, we have some assurance (though no absolute guarantee!) that all is well.

If $H(n/2)$ is the highest frequency component of the Fourier transform, what about the rest of them, on up to $H(n-1)$? We already know that if $h(t)$ is real, they are redundant. Nonetheless, they are there for us to contemplate. And if $h(t)$ is complex, they have unique meaningful values. But what is their meaning? Here we must resort to some loose talk, as a strictly correct answer requires far more mathematics than this text can support.

The easiest (really) answer is that they correspond to *negative* frequencies. A negative frequency is encountered when time runs backwards. If we are dealing with a time series, this would correspond to reversing the order of the series. In an application later in this chapter, we will be dealing with a point following a rotating trajectory. Positive frequencies refer to rotation in one direction, negative frequencies refer to the opposite direction.

The frequency is reflected around the center point, $H(n/2)$.   $H(0)$ is the zero-frequency (DC to engineers) component. $H(1)$ is the component whose frequency is one positive cycle per the total time interval. $H(2)$ is two positive cycles, and so on, up to $H(n/2-1)$ being $n/2 - 1$ positive cycles. $H(n/2)$ is $n/2$ cycles, being neither positive nor negative. (Do you see why it has no sign? Think about the symmetry of a cosine wave.) $H(n/2+1)$ is $n/2 - 1$ negative cycles. This continues to lower and lower frequencies, until we reach the last term in the transform. $H(n-1)$ is a frequency of one negative cycle per the total time.   Figure 4-2 illustrates this traditional method of storing the DFT in computer memory.

The author has a favorite way of visualizing the above relationship. Think of a clockwise rotating pointer viewed with a stroboscope. Each time the strobe flashes we take a sample. When the object is rotating slowly, we see it rotate. For example, the strobe may flash when it is pointing toward one o'clock, then again at two o'clock, and so forth. Its clockwise motion is obvious. As it speeds up, we can continue to see it rotating clockwise for a while, though it gets more difficult. When it reaches such a speed that the first flash is at noon, the second is at five o'clock, the next at ten o'clock, and so forth, we are just barely able to perceive its motion as being clockwise. Just a little faster and we are at the Nyquist frequency. The strobe flashes at noon and six o'clock. We can no longer say that it is moving forward or backward, as it is simply alternating between two positions. Now go even faster. The strobe flashes at noon, seven o'clock, two o'clock, and so forth. It has apparently reversed direction! When we eventually go so fast that it makes it to eleven o'clock by the first flash, then around to ten o'clock by the second flash, it will appear to be rotating slowly counterclockwise. This is not at all unlike what happens in the Fourier transform.

## The Power Spectrum

No discussion of the Fourier transform would be complete without at least a token mention of power spectra. Although the applications presented later in this chapter do not directly use power spectrum information, this topic is close enough to what we will do that it should be included here. And some reader's applications may well be served by presenting power spectrum information to a neural network.

Let us jump right in with the discrete version of *Parseval's theorem*. This famous theorem, shown in Equation (4-5), says that the total power in the time-domain series is closely related to the total power in the frequency-domain series.

Most of us, especially electronics engineers, like to think of power as mean squared amplitude. If we divide both sides of Equation (4-5) by $n$, we get mean squared

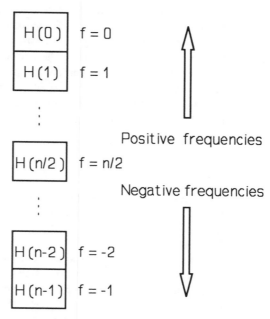

Fig. 4-2: Layout of all frequencies.

amplitude on the left side. Remember how positive and negative frequencies are laid out in the discrete Fourier transform. We can now break down the total power into power in each of $n/2 + 1$ frequency bands. Equation (4-6) is probably the most common way of computing these powers. It is called the *periodogram*. The sum of the powers in all of these bands is equal to the power in the original series.

$$\sum_{t=0}^{n-1} |h(t)|^2 = \frac{1}{n} \sum_{f=0}^{n-1} |H(f)|^2 \qquad (4\text{-}5)$$

$$P(0) = \frac{|H(0)|^2}{n^2}$$

$$P(i) = \frac{|H(i)|^2 + |H(n-i)|^2}{n^2} \qquad i = 1, 2, ..., \frac{n}{2} - 1$$

$$P(n/2) = \frac{|H(n/2)|^2}{n^2} \qquad (4\text{-}6)$$

Note that Equation (4-6) sums the positive and negative frequency powers for the $n/2 - 1$ cases in which they both exist. If the time-domain series is real, those components will be equal due to symmetry, so we can just multiply one of them by two.

Some applications benefit from using the periodogram as input to the network, rather than the actual Fourier transform. But such applications are not common, as the periodogram discards phase information that is often valuable. Readers are strongly encouraged to work with the full complex transform, perhaps employing a complex-domain neural network to process this data.

## Data Windows

There is one last annoying detail to be dealt with before this survey of the Fourier transform is complete. It is ugly, arbitrary, and cumbersome, but unfortunately very necessary. We speak of the process of applying a smooth window function to our time-domain sample.

The root of the problem that necessitates data windowing is that we have in all likelihood violated an as-yet-unmentioned assumption of the discrete Fourier transform. The DFT assumes that our time-domain sample is periodic and that we have captured an integral number of its periods. In other words, the end of the sampled series implicitly wraps around to its beginning. The entire infinite sequence of $h$'s could be obtained by endlessly repeating our finite sample. One implication of this scenario is that the exact frequencies defined by the DFT of the sample are the only frequencies that can exist in the sample. Ponder that.

Although such a situation could conceivably arise, it is highly unlikely in practice. If one attempts to wrap the last sample point around to the first, which is what the DFT implicitly does, the transition is rarely smooth! The implications are complicated, and we will only cover what is absolutely necessary.

Another way to look at the situation is to see that we have an infinite series containing components at frequencies that do not lie on the exact values defined by the sampled interval. We have taken this infinite series and multiplied it by a rectangular window function that is equal to one during the time of our sampling and equal to zero elsewhere. When we apply the DFT to our sample, we do not get the Fourier transform of the infinite series. Rather, the Fourier transform that we get is the *convolution* of the transform of the infinite series with the transform of the rectangular window. The net result is *leakage* of frequency components across relatively large distances. Readers who do not have the background to understand that statement should not worry. The net effect that will now be discussed is easier to comprehend.

What we ideally would like is for each $H(f)$ to represent the contents of a frequency bin centered at $f$ and extending halfway above and below to the adjacent bins. For example, the squared length of $H(7)$ should be proportional to the total power of all components having a frequency of from 6.5 to 7.5 cycles per total time. Unfortunately, what we want is not what we get. Because we have just a windowed section of an infinite series, which is a terrible violation of the wraparound assumption, components at frequencies significantly distant from *f* make it into *H(f)*.

We can do nothing about the wraparound violation. But we can try to lessen its evil effects. The cause of much of the leakage is the sharp corners of the implicit rectangular data window. It suddenly goes from zero to one, stays at one during our sampling time, then just as suddenly drops back to zero. The solution is to window the series with a smoother function. This function should equal zero (or nearly zero) at the start and end of the sampling interval, equal one in the middle, and make the transition smoothly. The collected sample would be multiplied, case by case, by the values of the window function. The Fourier transform would be done after this multiplication. Observe that the implicit wraparound of the DFT is now smooth, as the end of the series blends into the beginning, both being nearly zero. The Fourier transform of such a smooth function is far better behaved than the transform of a rectangle. Thus, when the transform of the infinite series, which is what we want but cannot get, is convolved with the transform of the smooth window, we end up with a transform having significantly less leakage.

There is a truly amazing variety of contenders for the window function, each having its own adherents. The author's favorite is the Welch window, defined in Equation (4-7).

$$ w_i = 1 - \left(\frac{i - 0.5(n-1)}{0.5(n+1)}\right)^2 , \qquad i = 0, ..., n-1 \qquad (4\text{-}7) $$

Inexperienced practitioners invariably balk at applying such a function to their precious data, for information is certainly being discarded. Their wariness is somewhat justified, as the computed Fourier coefficients have slightly higher error variance as a result of windowing. But it is a price well worth paying. Look at Figure 4-3 which shows the amplitude leakage associated with the rectangular window implied by using no explicit window. The leakage at a distance of 1.5 is around 20 percent. That means, for example, that the computed Fourier coefficient at a frequency of 7 cycles per total time will have contributions of 20 percent of components at frequencies of 5.5 and 8.5, not to mention many others! Worse still is the fact that the amplitude of the leakage lobes decreases abominably slowly as the distance from the center frequency increases. (They actually go on to infinity. Only the graph stops.) Do not be deceived. This is without doubt an intolerable situation.

Compare that dismal picture with the leakage function of the Welch window, shown in Figure 4-4. There is still some leakage to be endured. And another price we pay for the Welch window is that the important center lobe, which determines the primary sensitivity of the coefficients, has been widened. We gained quite a bit on the first side lobe, though. It has less than half the height of the first lobe of the rectangular window. But most important of all is the speed with which the side lobes vanish as they depart from the center frequency. This is vital if our computed Fourier coefficients are to have real meaning.

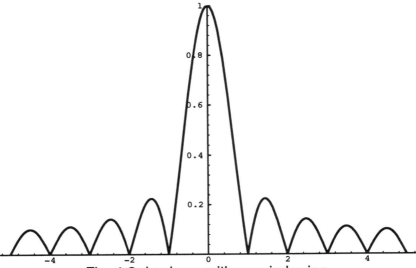

Fig. 4-3: Leakage with no windowing.

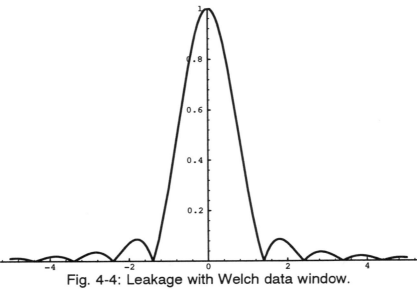

Fig. 4-4: Leakage with Welch data window.

This sort of data windowing can have another benefit. We may be unfortunate in that the ideal sample size is not a power of two, but the only DFT program we have available to us is restricted to powers of two. In this case we apply the data window to the ideal size sample, then append zeros to the end of the windowed sample to bring it up to a power of two in length. This is an entirely legitimate operation and is often imperative.

## Signal Detection in the Frequency Domain

Most traditional signal-detection techniques involving neural networks operate in the time domain. The network is trained with either actual samples of the pure signal or with noisy examples. Time-domain treatment is not to be disparaged. It can perform quite satisfactorily. On the other hand, seeking an elusive signal amidst noise is often more fruitful when the search is conducted in the frequency domain using complex-valued neural networks. This section will discuss two common variations on the signal-detection problem. Also, empirical results with a contrived example will be presented.

This section will emphasize real-valued signals in the time domain. Very little generalization is needed to handle complex time-domain signals. However, that topic will be the focus of the next section, so it will be deemphasized here.

The general signal-detection problem can be divided into two cases depending on whether or not we are able to collect training samples having known phase. For example, the pattern we are seeking may consist of two ascending sawtooth ramps followed by a negative pulse. We expect this pattern to repeat itself at least a few times when it is present. Our ultimate goal is to be able to collect a noisy sample and determine whether or not the signal pattern is present amidst the noise. It may be that we are able to set up our training-set collection apparatus, then repeatedly press a switch. Every time we press the switch we turn on both the sampling device and the signal generator. In this way, we know that every sample in the training set commences with the first of the two sawtooth ramps, or whatever the pattern may be. This is a valuable bit of knowledge, and it is worthwhile taking this approach if possible. Not only does it strengthen the power of the signal detector, but it allows the detector to more accurately locate the start and stop of the signal in a continuous sampling situation.

Sometimes we do not have the luxury of knowing the phase of the training patterns. Somebody just shows us a continuous signal and tells us to go ahead and collect samples as we see fit. Our samples may start anywhere in the pattern, and we have no way of knowing where we are. We now have two choices. We may train the network with the multiple phases. This approach usually works fairly well and is commonly employed. A generally better approach is to perform some sort of phase-independent normalization on the data, then train and test the network with the normalized data. When the neural network is working with frequency-domain data, there is a particularly simple method for doing this normalization. This method will be discussed later. For now, let us consider the easier case.

## Detection with Known Phase

Training and testing a complex-domain neural network with frequency-domain data is fairly straightforward. One simply uses a DFT program, such as the previously listed fft, to convert the windowed problem data to the frequency domain. The transformed data is then presented to the network. However, one must pay attention to several details.

In any signal-detection problem, real or complex, time-domain or frequency-domain, the data should be normalized in amplitude. There are two reasons for doing this. One is that we should always help the training algorithm by insuring that the learned weights will have reasonable values. By keeping the input data in a small range, perhaps with absolute value only rarely exceeding three or so, and roughly centered around zero, we can keep the weights in this same vicinity. If network inputs are typically around 0.0001, or perhaps around 1,000,000, the optimal weights will be so extreme that they will be learned only with great difficulty, if at all. The other reason for normalizing the input data is that we *must* train the network with data that is comparable to the data with which it will ultimately be used. If we present the trained network with a sample whose amplitude is significantly different from the amplitude of the training data, it cannot be expected to perform well.

If some characteristic of the process that generates the data *guarantees* that its amplitude will be consistent, then we are best off using a constant scaling factor rather than normalizing amplitude in a data-dependent way. For example, suppose that the data will always have a standard deviation of about 10. It would be reasonable to divide the data by 10 for training and final use. This way, if the presence of the signal to be detected affects the amplitude even slightly, we will capture that information.

On the other hand, it is much more common in practice that the amplitude varies from minute to minute, or at least from training time to testing time. In this case we must use dynamic scaling. The most common approach is to center each training and testing sample by subtracting its mean or median, then normalizing it to unit variance by dividing by its standard deviation. This is a good general-purpose technique. If the data distribution has heavy tails (many outliers), then the interquartile range may be better than the standard deviation for amplitude scaling.

Remember that the magnitude of the Fourier coefficients is affected by the sample size. Larger samples will produce larger coefficients. We may therefore need to multiply these coefficients by a constant that depends on the sample size. The goal is to keep all network inputs to single-digit numbers.

There is one common signal-processing normalization method that should be mentioned here. The author has not had good success with it with real-life signal detection, but it is entirely possible that some problems may be served well by it. The technique is to divide all Fourier coefficients by the magnitude (absolute value) of the coefficient corresponding to the fundamental frequency. This is the lowest, and typically strongest, frequency in the data. If the sample contains $p$ complete periods of the pattern being sought, the factor will be the absolute value of $H(p)$. This has the effect of expressing every harmonic as a fraction of the fundamental amplitude. If we do this, we will probably want to multiply the result by a constant equal to about 10 or more in order to present the network with more easily learned values. In practice, consistently finding a stable estimate of the power of the fundamental frequency is rarely easy, if it is possible at all. But for those situations in which it is feasible, this action may be a good component of a total normalization scheme.

There is one Fourier coefficient that we should nearly always discard. This is $H(0)$. Remember that its imaginary part is always zero (for a real input series). Its real part is the DC component (the sum) of the input signal. If we have centered the original time-domain data by subtracting its mean, $H(0)$ will always be zero. Thus, it contributes no information. The only exception is the situation, rare in practice, in which we do not center the signal due to the fact that its DC offset is physically meaningful. It may be that the with-signal and without-signal samples can be expected to have slightly different DC offsets on average. In this unusual case, we should preserve that information.

Another advantage of doing signal detection in the frequency domain is that we can nearly always significantly reduce the number of inputs to the network compared to the time domain. Perhaps, for example, our sample contains (by design) exactly $p$ periods of the pattern being sought, where $p$ is typically some small integer. This means that the fundamental frequency is $p$. Furthermore, all harmonics will be integer multiples of $p$. In other words, the only Fourier coefficients that have real meaning are $H(kp)$ for positive integers $k$. The upshot is that if we present all of the coefficients to the network, we are burdening it with far more data than is strictly necessary. All that we really need are $H(p)$, $H(2p)$, $H(3p)$, and so on. Except for data window leakage, the values of all of the Fourier coefficients that are not multiples of the fundamental frequency are determined solely by the system noise, not by the presence of the signal. Therefore, we can (and definitely should) reduce the number of inputs by a factor of $p$.

An example should clarify these issues. Suppose that we sample a real-valued variable once per millisecond and that the signal pattern being sought repeats itself every 25 milliseconds. Let each complete sample consist of 100 points, so we have $p = 4$ periods. The computed DFT will consist of 51 unique complex numbers, two of which are pure real. We would present the network with only 12 complex numbers. These are $H(4)$, $H(8)$,..., $H(48)$.

The above example assumed that we have a DFT program capable of transforming 100 points, which is not a power of two. We may not be so fortunate. Or, worse, the period of the signal may not break down into small factors. If this happens, our first tactic should be to attempt to change the sampling rate to get a power of two! It is definitely worth trying to do that. The next alternative is to procure a DFT that can handle 100 points. If that is absolutely impossible, we are stuck with a brute-force approach. We must apply the data window to each sample, then append zeros to each sample to bring it to a transformable size. We now have a mountain of data to present to the network. Very bad luck.

## Normalization for Unknown Phase

This section extends the previous section to the case in which we do not know the phase of the training samples. We could always simply go ahead and train the network using all possible phases. This often works reasonably well. But experience indicates that superior results can almost always be obtained by transforming the data into a new data set that is unaffected by the signal's phase. When the data is in the frequency domain, this is very easily done for many problems.

We shall start by exploring the effect that a shift in the time-domain signal phase has on the signal's Fourier transform. Equation (4-3) told us how to compute the DFT of a time series, real or complex. This is an excellent example of why it is useful to write a transform involving sine and cosine terms in complex exponential form. The mathematics becomes almost trivial. We will now rewrite that equation, this time incorporating a time shift of $t_0$. Assume that we have some series $h(t)$ whose DFT is $H(f)$. We need $H_{t0}(f)$, the DFT of the time-domain series obtained by shifting $h(t)$ by $t_0$ samples. Remember the implicit assumption of the DFT that our sample is periodic, and that we have an integer number of periods of it.

$$
\begin{aligned}
H_{t0}(f) &= \sum_{t=0}^{n-1} h(t-t_0) e^{i\frac{2\pi t f}{n}} \\
&= \sum_{t=0}^{n-1} h(t-t_0) e^{i\frac{2\pi (t-t_0)f}{n}} e^{i\frac{2\pi t_0 f}{n}} \\
&= H(f) e^{i\frac{2\pi t_0 f}{n}}
\end{aligned}
\tag{4-8}
$$

Equation (4-8) may look fierce to some, but it really tells us a very simple thing. Remember how we multiply complex numbers expressed in polar notation. Forgetful readers should see Equation (2-7). Now look at the factor that multiplies $H(f)$ in Equation (4-8). Its magnitude is equal to one, so the magnitude of the Fourier coefficients is not changed by time-shifting the time-domain series. All that happens is that the phase angle of each coefficient changes. By how much? That depends on which coefficient we are talking about, for the angle has a factor of $f$ in it. The remaining factor is the fraction $t_0/n$ of full revolution, $2\pi$.

For example, suppose that we have $n = 16$ points in our time-domain sample, and that we know its DFT, $H(f)$. Let us determine the DFT of the sample that would be obtained by starting the sampling two time slots later. For convenience, we will work in degrees rather than radians. The fraction of full revolution is 2 shifts / 16 points = 1/8. So we see that $H(1)$ will be shifted by 360 / 8 = 45 degrees. $H(2)$ is shifted by twice that basic amount, or 90 degrees. $H(3)$ moves by three times that, or 135 degrees, and so on.

It should not be surprising that $H(0)$ is not affected by a time-shift, as it is the DC component computed as the sum of the time-domain values. And it should be kept firmly in mind that the magnitude of the coefficients is not affected by a time-domain shift.

Alert readers may worry about the fact that $H(k)$ for values of $k$ greater than $n/2$, the Nyquist frequency, actually represent negative frequencies. In that case, what do we use for $f$ in the above operations? Do we use the array subscript, $k$, or do we use the corresponding negative frequency, $k - n$? The answer is that if we follow the derivation in Equation (4-8), which assumed a time shift equal to a multiple of exactly one discrete sample point, it doesn't matter. They are equivalent. It is left as a simple exercise for the reader to verify this. (Hint: What is the angle difference between the two subscripts?) However, we soon will need to be able to shift by any continuous amount. In this case, $f$ must not exceed the Nyquist frequency. We must use negative frequencies rather than values beyond the Nyquist limit. A fully general derivation of this fact is beyond the scope of this text.

Now that we know the effect of a phase shift in the time domain, we are prepared to find a means for normalizing the frequency-domain values against such a phase shift. Our goal is to find a set of frequency-domain variables that is representative of the data but that is unaffected by changes in the time-domain phase. The most common way of doing this is to simulate the effect of a time-domain shift sufficient to cause the phase angle of the fundamental frequency component to become zero. It should be apparent that the resulting phase-shifted Fourier transform will be the same regardless of the phase of the time-domain series. (The derivation in Equation (4-8) was for discrete time-domain shifts. We state here, without formal proof, that it applies for any shift as long as the absolute value of $f$ does not exceed the Nyquist frequency.)

In many applications the fundamental frequency is 1, so the required phase shift is equal to the phase angle of $H(1)$. In the more general case, the fundamental frequency is $p$, the number of complete periods embodied in each sample. Therefore, the required phase shift is $\arg(H(p)) / p$. Computation of the phase-normalized DFT is explicitly stated in Equation (4-9).

$$\theta = \frac{arg\,(H(p))}{p}$$

$$\hat{H}(f) = |H(f)|\; cis(\,arg\,(H(f)) - f\theta)$$

(4-9)

In plain language, the phase normalization is done as follows. First, compute the phase angle of $H(p)$, then divide it by $p$ to get the time-domain phase shift. Now pass through the entire set of Fourier coefficients. For each $H(f)$, express it in polar coordinates. Subtract $f$ times the time-domain shift from the phase of $H(f)$, leaving its magnitude unchanged. That's all there is to it. A code fragment for performing this operation is now given. Note that the normalization is done on only the fundamental frequency and its integer harmonics, as discussed at the end of the previous section. Also, the processing loop extends through $n/2$, as is the mathematical situation for real-valued input series. However, if the efficient transformation method discussed on page 86 is used to compute the DFT, xr[n/2] and xi[n/2] may not be valid addresses. Take appropriate steps. Also, note that this only processes up to the Nyquist limit, so we do not need to worry about negative frequencies that are redundant for real data. For a more general version, see the listing of normalize_phase on page 126.

```
mag = sqrt ( xr[p] * xr[p]  +  xi[p] * xi[p] ) ;
if (mag > 0.0)
   phase = atan2 ( xi[p] , xr[p] ) ;
else
   phase = 0.0 ;
theta = phase / p ;

for (i=p ; i<=n/2 ; i+=p) { // Beware if x[n/2] not there!
   mag = sqrt ( xr[i] * xr[i]  +  xi[i] * xi[i] ) ;
   if (mag > 0.0)
      phase = atan2 ( xi[i] , xr[i] ) ;
   else
      phase = 0.0 ;
   phase -= theta * i ;
   xr[i] = mag * cos(phase) ;
   xi[i] = mag * sin(phase) ;
   }
```

Alert readers may be troubled to realize that there are actually $p$ different values of theta, the time-domain phase shift, that will zero the phase angle of $H(p)$. Most other applications of this type of normalization zero the phase of $H(1)$, so there is no ambiguity. But here it is apparent that not only will the computed theta do the job, but also all angles obtained by adding integer multiple of $360 / p$ degrees to theta. This is because when the additional time-domain shift is multiplied by $p$, that addition becomes 360 degrees, which has no effect on the phase of $H(p)$. But it can certainly affect the others. So do we have a problem? Is this transformation ill-defined in that it is not unique? Not at all. Remember that we are presenting to the network only the integer harmonics of the fundamental. We are keeping only $H(kp)$ for positive integer values of $k$. The addition of angles equal to $360 / p$ will result in extra phase shifts of multiples of 360 degrees. No problem. The Fourier coefficients that would be affected by the ambiguity have been discarded.

This entire discussion has assumed that the total time interval encompassed by the sample is an integer multiple of the period of the pattern being sought. We can often contrive this by adjusting the sampling rate and/or interval length. It is definitely worth trying very hard to do so. But sometimes we simply cannot do it. Perhaps we do not know enough about the underlying process to be able to specify a period. Or there may be several possible signals being sought, each having a different period. This makes our job much more complicated. First of all, we are now forced to keep all Fourier coefficients, as we no longer have a clearly defined fundamental frequency and harmonics.

Skilled practitioners may be able to eliminate some or many of the coefficients, but general techniques for accomplishing this are beyond the scope of this text. Worse, we no longer have a clearly defined fundamental frequency to use for phase normalization. Maybe we can pick a small frequency, as small as possible but not less than the true fundamental frequency, and arbitrarily choose it for the coefficient to rotate to zero. Or perhaps we may choose, for each individual case, the coefficient having greatest magnitude and zero its phase. Ambiguity does become an issue here, so we need to choose a secondary normalization criterion also. The *positive real energy* criterion discussed on page 123 is a candidate. But this secondary criterion can be very problem dependent, so it will not be discussed here. The issues involved are messy, so expertise beyond what can be covered in a neural network text is called for.

## Example of Signal Detection

This section presents the results of an experiment designed to demonstrate the previously described techniques. Several waveforms are generated and mixed with noise. A traditional, real-domain neural network is trained and tested on time-domain signals. Then the Fourier transforms of the Welch-windowed signals are computed. This transformed data is used to train and test both real-domain and complex-domain neural networks. The results show that complex-domain neural networks operating on frequency-domain data dramatically outperform real-domain networks operating on time-domain data. The performance of the networks for time-domain and frequency-domain data is similar when real-domain networks are used.

One of the basis waveforms used in this test is an ascending sawtooth ramp which can be seen in Figure 4-5. The other is a cosine wave, shown in Figure 4-6. The noise is uniform random numbers. We work with four different combinations of these basis waves and define two classes. Each of the two classes contains two of the combinations. One class is composed of samples of pure noise and is also composed of samples of the ramp and the cosine superimposed, contaminated with noise. Figure 4-7 shows the appearance of the ramp plus cosine, without any noise contamination. The other class consists of samples of the ramp plus noise, as well as samples of the cosine plus noise. In other words, one class consists of cases containing *both* basis waves as well as cases containing *neither* basis wave. The other class consists of cases containing exactly one of the two basis waves. This problem structure means that our network will not only have to solve a signal detection problem, but it will also have to decode an XOR (exclusive or) pattern.

Pure waves are never presented to the network for training or testing. Input samples are always contaminated with noise. Typical examples are shown in Figures 4-8 through 4-10. Every case that is presented to the network is standardized to have zero mean and unit standard deviation. Each case has 32 time samples. Training is done with 500 cases from each of the four categories. After the network is trained, it is tested with 500 independent cases from each category. The results that will be shown are the number of the 2000 validation-set cases that were misclassified by the trained network.

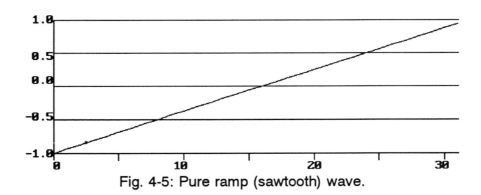

Fig. 4-5: Pure ramp (sawtooth) wave.

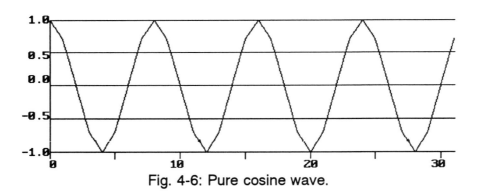

Fig. 4-6: Pure cosine wave.

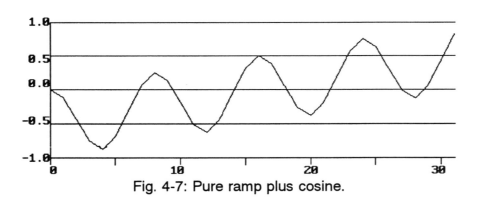

Fig. 4-7: Pure ramp plus cosine.

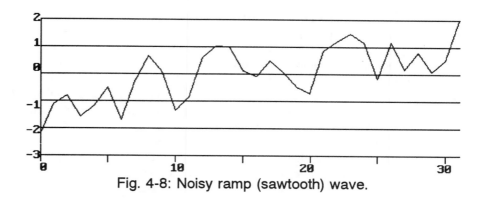

Fig. 4-8: Noisy ramp (sawtooth) wave.

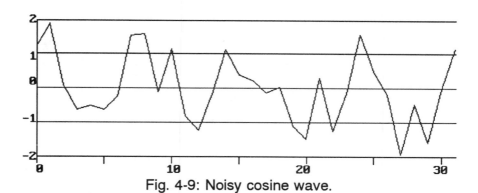

Fig. 4-9: Noisy cosine wave.

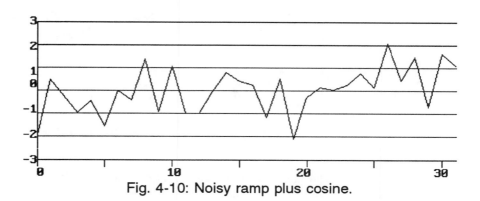

Fig. 4-10: Noisy ramp plus cosine.

Three different networks for each of the three tests (time-domain, Fourier real, Fourier complex) were trained and tested. The smallest had no hidden layer at all. As expected (due to the XOR nature of the problem), classification ability was about 50–50. An intermediate network size used two hidden neurons for the real networks and one hidden neuron for the complex network. This provides equivalence in terms of the number of weights. The time-domain model misclassified 550 of the 2000 validation cases. The frequency-domain model using a real network performed about the same, missing 517 cases. But when that same frequency-domain data was used with a complex network having one hidden neuron, misclassification dropped to 196 cases. The final networks had four hidden neurons for the real models and two hidden neurons for the complex. Misclassification in the time domain was 46 cases. In the frequency domain with a real network, it was 82. Once again, frequency-domain data fed to a complex-domain network was the winner, with 30 misclassifications. Figure 4-11 shows this in a graph.

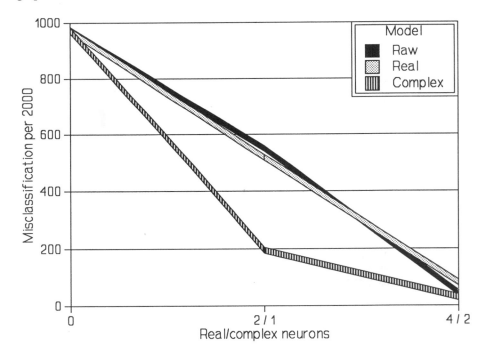

Fig. 4-11: Misclassification for the nine tests.

The slight inferiority of the frequency-domain data relative to the time-domain data for equal-size real-domain networks is often seen. This is especially true when few periods are sampled. (Just one period was sampled here.) The author has no rigorous explanation, but does have a hypothesis. When the Welch window is applied to prepare for the Fourier transform, some information is unavoidably discarded. Therefore, the frequency-domain data has an inherent disadvantage in that it contains less information than the original time-domain data. But even that is overcome when the frequency-domain data is presented to a complex network!

# Complex Data in the Time Domain

The previous section was directed primarily at real-domain time series. There is no reason that the time-domain series could not consist of complex numbers. All of the methods described in that section would apply just as well, as long as it is remembered that there is no longer conjugate symmetry about $H(n/2)$. However, the techniques described there are techniques that are usually most appropriate for real data. Complex numbers in the time domain usually arise from quite different physical processes, and different methods are needed to handle them. This section will focus on the special considerations needed for complex data in the time domain.

### Sources of Complex Data

We should start with some examples of how we might come to be faced with a complex-valued time series. One area receiving much attention lately concerns the path of a point in phase space. For example, van der Pol's Equation, (4-10), arises in the study of oscillation and chaotic dynamics.

$$
\begin{aligned}
x'(t) &= y \\
y'(t) &= (1 - x^2)\, y - x
\end{aligned}
\tag{4-10}
$$

This pair of differential equations describes the path of a point $(x, y)$ as a function of time, $t$. If we view $x$ as the real part and $y$ as the imaginary part, we can think of this as a complex-valued time series. A periodic solution to this equation is shown in Figures 4-12 through 4-14, respectively.

Data of this type, and from any of a multitude of similar differential equations, often occurs in nature (including when nature is electron noise in a vacuum tube). We may have a problem in which we need to detect the presence of this sort of pattern amidst noise. For example, [Birx and Pipenberg, 1992] construct a chaotic oscillator, then use the signal in question to perturb it. The phase-plane response of the oscillator is used as the input of a complex-valued neural network. They find that very low levels of the proper signal, buried under Gaussian noise, are able to excite the oscillator sufficiently for the network to respond.

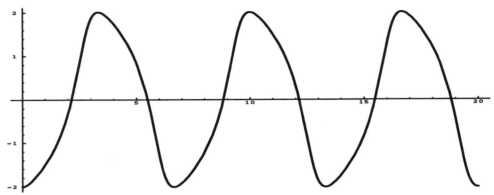

Fig. 4-12: x(t) in the van der Pol equation.

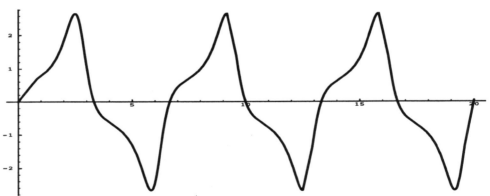

Fig. 4-13: y(t) in the van der Pol equation.

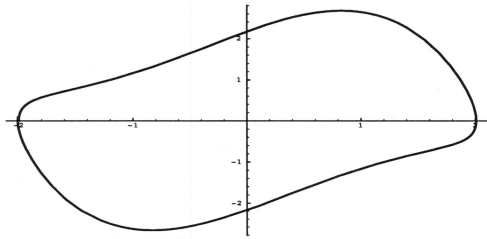

Fig. 4-14: Plot of the van der Pol equation.

      Another application that produces a complex-domain time series is shape classification based on perimeter. We may have objects coming down an assembly line, and a video camera taking an aerial view. The problem is to determine the identity or quality of each object based on its outline. When an object's perimeter is traced at a constant velocity and regular samples are taken, the $(x, y)$ coordinates of each snapped point along the perimeter can be viewed as a complex-valued time series.

      One area that is starting to receive attention is vibration analysis of mechanical components. Failure of helicopter gearboxes is notoriously difficult to predict. It is said that highly experienced mechanics can place their hand on the gearbox casing when it is running and tell from the feel of the vibration whether or not maintenance is needed. But the knowledge required to automate such decisions can be nearly impossible to quantify. One approach has been to analyze the sound made by the gearbox. Unfortunately, the wide variety of sounds possible under normal conditions makes this a very difficult problem. One promising alternative approach is to fasten motion sensors to the apparatus to follow the path of the cyclic vibration. This two-dimensional measurement supplies more information than a one-dimensional sound sample. In most cases, the path of the spot can be treated as periodic, and complex-domain Fourier analysis is fitting.

## Time-Domain or Frequency-Domain?

The first decision that must be made is whether we want to present complex-domain time series data directly to the neural network or apply a Fourier transformation to it first. The most important factor in making that decision is whether or not the time series is best characterized by periodic components. The path obtained by following the perimeter of an object is clearly periodic, so we would be inclined to present frequency-domain data to the network. On the other hand, a segment of a strange attractor will, by definition, not be periodic. In this case we may want to stay in the time domain.

The decision is not always clear-cut. Remember that a Fourier transform is just a linear transformation. The time-domain and frequency-domain versions of the data present essentially the same information to the network. They are nothing more than different ways of looking at the same thing. In fact, if no data window is applied, a linear transformation of the input weights exists such that networks that process time-domain and Fourier-domain data can be made equivalent! In case of doubt, it would not be a waste of time to try both approaches. It has been the author's experience, though, that if periodicity is at all inherent in the data, the frequency domain is usually better.

There is one more argument in favor of using frequency-domain data if possible. It will be seen on page 109 that, in general, relatively few Fourier coefficients are needed to capture the vast majority of the information in the data. It may be that the time-domain sample has hundreds of points. If, as is common, we find that only a dozen Fourier coefficients account for virtually all of the power in the data, it means that only this many network inputs are needed, as opposed to hundreds. The savings are often substantial.

## Do We Apply a Data Window?

When data windows were discussed on page 92, it was emphasized that we *must* use one. That is certainly true in the context of signal detection in a noisy environment, and in a lot of other cases as well. But for some of the applications discussed in this section, we may not want to use a data window.

The key to making this decision is the wraparound assumption of the DFT. Remember that when we apply the discrete Fourier transformation to time-series data, we are implicitly assuming that it is periodic and that the end of the series wraps right back to the beginning. If we do not know for sure that we have an integer number of periods of *all* components of the series, or if random noise destroys the notion of periodicity, we must use a window to avoid disastrous leakage. This will usually be the situation. But one notable circumstance in which we would *not* want a data window is when we are algorithmically tracing a fixed perimeter. Granted, there may be noise inherent in the data, perhaps due to sensor problems or other physical sources. Nevertheless, the perimeter, noise and all, is fixed. If we were to run the trace program over and over, we would get exactly the same series every time. There can be no leakage problem because there are no nonharmonic components to leak.

The distinction can be quite subtle. In the case of perimeter tracing, any noise is (in the vast majority of situations) embedded in the perimeter *before* tracing. Therefore, we trace exactly once and do not apply a data window. Contrast this with the less common situation of tracing a perimeter with a mechanical device that contributes noise of its own. Now we cannot be sure that multiple circumnavigations will give precisely the same results, so the wraparound assumption of the DFT is violated. In this case we would want to use an integer number of passes around the circumference (to average sample errors) and apply a data window. In practice, when we are not explicitly tracing a perimeter or doing something similar that physically guarantees perfect periodicity, we will want to sample multiple periods and apply a data window. That is by far the most common circumstance.

## How Many Fourier Coefficients Are Needed?

It is frightfully easy to overburden the neural network by presenting it with far more information than it really needs. In most situations, whether we are processing a real-valued time series as emphasized on page 95, or working entirely in the complex domain as discussed in the last few pages, the amount of power contained in frequency-domain components of experimental data decreases rapidly as the frequency increases. This tends to be especially true for the latter applications. Look back at Figure 4-14 which shows the path of a point in the complex plane when that path is governed by a common differential equation. The scaled amplitude spectrum of that path is shown in Figure 4-15. (The amplitude spectrum is the square root of the power spectrum as defined on page 90.)

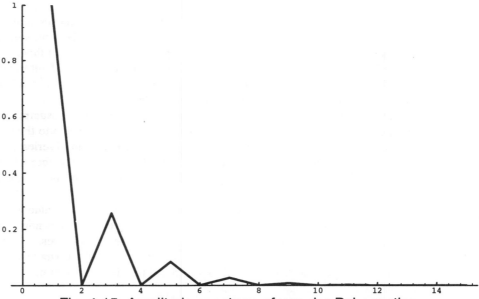

Fig. 4-15: Amplitude spectrum of van der Pol equation.

Because of symmetry, no even harmonics exist. The important thing to notice is how fast the coefficients diminish as the frequency increases. After 13 cycles per revolution, they do not even show up on the graph. What this means to us is that there is no point in presenting the neural network with meaningless data. High frequency components will be so engulfed by noise that they will lose all significance.

The significance of harmonics can be appreciated most easily by studying their contributions to a simple geometric shape. Figure 4-16 shows a T-shaped object. If one traces around its perimeter once and computes the Fourier transform of that path, the amplitude spectrum is shown in Figure 4-17.

The sharp corners of the T object generate far more high frequency components than were found in the smooth van der Pol path. Even so, it can be seen that the Fourier coefficients drop off quite rapidly.

It is interesting to explore the contributions of the various coefficients. Figures 4-18 through 4-23 show the object as represented by varying numbers of frequency components. These shapes were created by sampling the T perimeter at a large number of equally spaced points, computing the DFT of that complex-valued series, setting all but the lowest-frequency components to zero, then applying the inverse DFT. See page 83 for information on computing the inverse transform.

Recall that each frequency component except the offset, $H(0)$, and the Nyquist frequency, $H(n/2)$, is represented by *two* terms in the Fourier expansion. A frequency of $k$ cycles per revolution has a positive frequency component of $H(k)$ and a negative component of $H(n-k)$. When it is said that harmonics up to $k$ are kept, both the positive and negative components are included.

It should not be surprising that the fundamental frequency alone (positive and negative) gives us nothing more than an ellipse. What may be a bit surprising, though, is that addition of just the second harmonic pulls the shape into a triangle. And with the third harmonic, the T shape is clearly visible. That is a lot of information in a few coefficients! Keeping six harmonics produces a fairly respectable T, and keeping 20 does it perfectly except for a bit of rounding on the corners. It takes another 60 or so harmonics beyond that to square off the corners to a reasonable degree. Whether or not that extra bit of information (perfectly square corners) justifies all that extra data for the network is a problem-dependent decision. When that decision is made, be sure also to consider the amount of noise in those coefficients relative to the amount of information. This author has not yet seen an application that benefits from more than a dozen or so harmonics.

It is particularly important to carefully consider the number of significant harmonics when the sampling rate is high. If the perimeter contains 256 measured points, a typical number, it is extremely unlikely that we would want to present all 256 frequency-domain points to the network. In most cases all we would want is $H(i)$ for $i$ from 1 through 12 and from 244 through 255. This gives us 24 complex inputs, a far cry from 256.

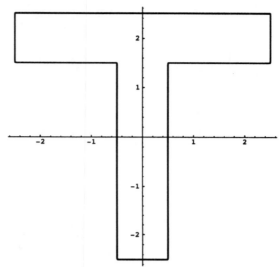

Fig. 4-16: An object for perimeter tracing.

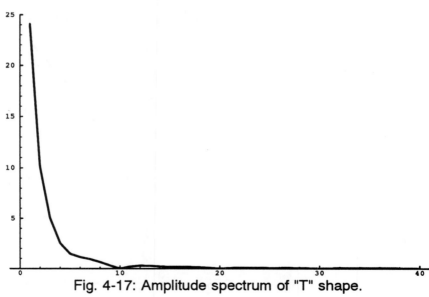

Fig. 4-17: Amplitude spectrum of "T" shape.

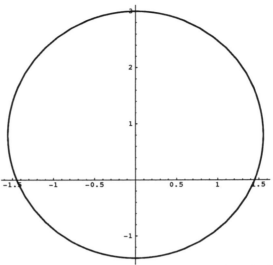

Fig. 4-18: Keeping only fundamentals (CW & CCW).

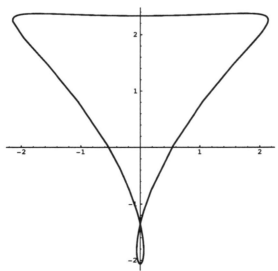

Fig. 4-19: Fundamental plus second harmonic.

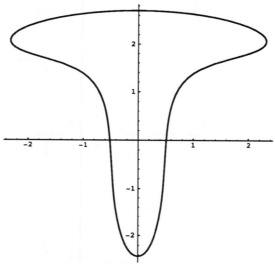

Fig. 4-20: Through third harmonic.

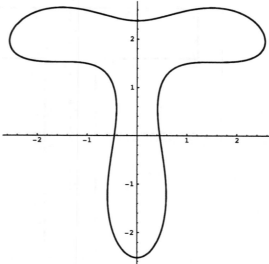

Fig. 4-21: Through sixth harmonic.

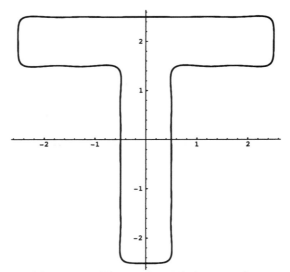

Fig. 4-22: Through 20th harmonic.

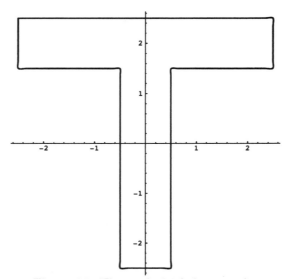

Fig. 4-23: Through 80th harmonic.

## Magnitude and Offset Normalizations

There are many decisions to be made regarding normalization of the Fourier coefficients, and each decision depends strongly on the particular problem at hand.  Therefore, we must closely examine many aspects of normalization.

First, let us review and expand upon the meaning of each of the Fourier coefficients.  $H(0)$ represents the constant offset of the time-domain series.  The real part of $H(0)$ is the sum of the real parts of the time-domain series, and similarly for the imaginary part.  $H(1)$ is the positive frequency component at a rate of one cycle per revolution.  (When we say revolution, we mean one lap around the perimeter, or one period of a periodic function.)  For the definition of the DFT given by Equation (4-3), this corresponds to a clockwise path.  Therefore, when tracing a perimeter, we will always do so in a clockwise direction so as to achieve consistency with the discussion that follows.  $H(n-1)$ is the negative frequency component at a rate of one cycle per revolution.  It corresponds to counterclockwise motion.  We can just as well refer to that component as $H(-1)$ so that the fact that it is a negative frequency is explicit.  Programmers only need to remember to add $n$ to get the subscript in the DFT array.

Now comes a visualization exercise that is tricky, but definitely worth pursuing.  $H(1)$ in the frequency domain corresponds to a clockwise rotation of one cycle per revolution in the time domain.  In other words, the path traced by a point corresponding to that one coefficient would be a perfect circle whose radius is determined by the magnitude of the coefficient.  The phase angle of that coefficient is only related to the starting/stopping point for tracing the circle.  Now suppose that we are about to start tracing that path from a point at the far left of the circle, vertically centered.  Bring $H(-1)$ into the picture.  It wants us to move counterclockwise, while $H(1)$ wants us to move clockwise.  Perhaps the magnitude of $H(-1)$ is such that the vertical motions nearly cancel.  Now, as the tracing begins, both coefficients will be pulling us to the right.  But $H(1)$ will want us to go up (CW), while $H(-1)$ wants us to go down (CCW).  The end result is that we trace out a flattened ellipse.  On the other hand, it may be that the phase of $H(-1)$ is such that the vertical motions reinforce.  In this case, we will get a tall, thin ellipse.  Or they may be only partially out of phase, so we get a slanted ellipse.  Engineers familiar with Lissajous figures should understand this well.

The bottom line is this.  The two fundamental frequency components, one clockwise and the other counterclockwise, work together to provide an elliptical path in the time domain.  $H(1)$ exceeds $H(-1)$ in magnitude if and only if the dominant motion is clockwise.  If either is zero, the ellipse is a perfect circle.  As their magnitudes become closer, the ellipse flattens.  The limiting condition of a straight line is reached when their magnitudes are equal.  The direction of the major axis of the ellipse is determined by the phase angle difference between $H(1)$ and $H(-1)$.

We are now in a position to dispose of several easy scaling decisions.  First, we will nearly always want to discard $H(0)$.  It determines the center of the time-domain path.  The pertinent question is whether or not the position of that center conveys useful information.  In most cases it will not.  But in the rare cases that the location of the center will help the network to find a correct answer, then we should keep it.

We have a related problem when deciding about $H(1)$. Multiplying every Fourier coefficient by a constant has the effect of multiplying the time-domain path by the same constant. It is a simple magnification about the center of the path. In some problems, the scale conveys useful information. The absolute magnitude of the sampled time-domain values may be important to the network's decision. In this case we would keep $H(1)$. Very often, though, we want to remove the effect of a change in scale. This would certainly be the case in shape identification based on perimeter. We do not want the distance separating the camera from the object to affect the decision process. The traditional method for normalizing for scale is to divide every Fourier coefficient by the magnitude of $H(1)$. Since we have agreed to trace the path in a clockwise direction, that coefficient will have the greatest magnitude. At least it will never be zero! (It can fail to have the greatest magnitude if the supposed perimeter crosses itself.)

Note that since this division will always leave the magnitude of $H(1)$ equal to one, we may want to discard that coefficient as well. This question is tied up with the importance of the phase of $H(1)$, which will be discussed in the next section.

Some readers may argue that we may not always have the luxury of following a path in a clockwise direction. We may be following a measured trajectory that determines its own direction. The solution is simple. Compare the magnitudes of $H(1)$ and $H(-1)$. If the former is greater, we are following a clockwise path and all is well. Otherwise we are following a counterclockwise path. In this case we just reverse the rolls of $H(k)$ and $H(n-k)$ for all $k$ of interest.

There is one last scaling decision that should be mentioned, mainly because it contains a dangerous pitfall. It is almost always the case that Fourier coefficients drop off rapidly in magnitude as frequency increases. This may tempt us to multiply the various coefficients by preordained constants, with larger constants being used for larger frequencies. The philosophy behind this action is to approximately equalize the magnitudes of the network's inputs. Superficially, this is admirable. And if the data is very noise-free, with the network being used to detect subtle features in clean data, it may be a good move. However, in most practical situations, the data is contaminated with wide-band random noise. In this case, multiplying higher-frequency Fourier coefficients by larger constants will exaggerate the noise along with the signal, and nothing will be gained. Indeed, something may even be lost if we exaggerate the relative importance of information that does not merit such preferential treatment. This is in direct conflict with our usual goal of equalizing the scale of the network's inputs. But remember that the primary purpose of the equal scaling is to equalize *importance*. What if the inputs do not have equal importance? This question must be carefully answered for each application.

If there is a dangerously large difference in the magnitudes of low-frequency and high-frequency components, we sometimes have a remedy available to us. In some problems we are looking for only subtle changes in shape. One such application will be described on page 130. In this situation, the magnitudes and phases of the low-frequency coefficients will tend to be approximately the same for all cases. In other words, their large magnitude will be due to offset only, not due to much variation. When this is so, it is not only legitimate, but wise, to apply a constant offset to compensate. Collect a representative sample of the possible cases. If any phase normalization is to be done (as described in the next section), do it. Then compute the mean (or median) of the real and imaginary parts of each Fourier coefficient. If any mean is significantly different from zero, subtract the mean from the coefficient to center it. Needless to say, the exact same constant must be used for all training samples, and also during use of the trained network! This simple expedient will equalize magnitudes in a surprisingly large number of applications. Note that this operation is not *theoretically* necessary. The network could in principle learn an appropriate bias that would take care of the offset. However, in practice, this is difficult or impossible for most learning algorithms. Our help is greatly appreciated here.

## Phase Normalizations

The previous section dealt with frequency-domain normalizations that concern the magnitude of the Fourier coefficients. In this section we will discuss normalizations that concern phase. This can be trickier, and sometimes disturbingly arbitrary decisions must be made. It is crucial that the reader thoroughly understands the philosophy behind these normalizations, so that inappropriate actions are not taken blindly. The contents of this section are primarily applicable to problems in which we are tracing the perimeter of an object.

We have already discussed and illustrated the roles of the fundamental frequency and groups of harmonics. For this section we will need to be aware of the contribution of an individual harmonic. Consider $H(k)$ for some $k$ not equal to zero or one. In the time domain, it represents a cyclic variation that repeats itself $k$ times as the perimeter is traversed. (Or it repeats $n-k$ times if $k$ is beyond the Nyquist frequency.) If it represents a positive frequency, it runs in the same direction as the fundamental. Negative frequencies run against the fundamental. Thus, we see that an object whose Fourier transform is dominated by a single positive harmonic of frequency $f$ would have a time-domain shape that is dominated by $f-1$ rotational symmetry. If the Fourier transform is dominated by a single negative harmonic of frequency $-f$, the time-domain shape is dominated by $f+1$ rotational symmetry. (That $f+1$ is not a misprint. The harmonic and the fundamental run against each other. Think about it — it's worth the effort.) For example, Figure 4-24 shows the time-domain shape corresponding to a frequency-domain representation having only $H(1)$ and $H(7)$ nonzero.

The remainder of this discussion will often mention the fundamental frequency and its harmonics. If the sample contains $p=1$ periods of the waveform being sought, as would be the case for traversing a perimeter once, the fundamental frequency will be 1. However, this may not always be the case. If we are sampling a noisy phase-plane path, we may want to collect several periods. On page 97 we discussed the fact that the fundamental is $p$ and the harmonics are $kp$. Components of the Fourier transform not corresponding to integer multiples of $p$ may be safely discarded. Therefore, we will now assume that this has been done, and the transform numbering compressed. In other words, for the remainder of this section, whenever we refer to $H(1)$, we will really mean $H(p)$ in case $p$ exceeds 1. Similarly, $H(k)$ here will mean $H(kp)$ in the DFT array. This

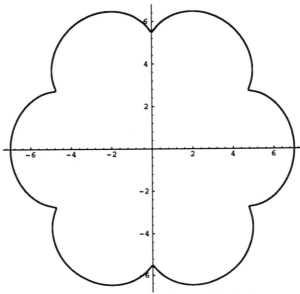

Fig. 4-24: Symmetry when *H*(7) is large.

convention will greatly simplify the discussion and results in absolutely no loss of generality. A simple subroutine for doing this compression is now shown.

```
void keep_harmonics (
   int *n ,                        // Length of input/output vectors
   int p ,                         // Periods in sample
   double *xr ,                    // Real components
   double *xi                      // Imaginary components
   )
{
   int i ;

   if (*n % p) {
      printf ( "\nERROR: n must be a multiple of p" ) ;
      return ;
      }

   *n /= p ;
   for (i=1 ; i<*n ; i++) {
      xr[i] = xr[i*p] ;
      xi[i] = xi[i*p] ;
      }
}
```

After the time-domain data has been collected and transformed to the frequency domain, there are four possible actions that may be taken with regard to phase normalization.

1.  We may do nothing. This is rare, but it happens. For example, each sample may start in response to a trigger event. In this case, all phase information is likely to be important and so should not be disturbed.

2.  We need to normalize against starting-point shift. This is very common. Suppose we are sampling a phase-plane path, looking for a periodic component. Chances are, we have no idea where the point may be when we start collecting data. We must eliminate the effect of starting at unknown times. This was thoroughly discussed on page 98, so it will not be repeated here.

3.  If we are following a perimeter, we could conceivably normalize against rotation of the object (with a fixed starting point). The author has never encountered a situation in which this was called for. The ensuing discussion will show that it can be done by subtracting the phase of $H(1)$ from the phase of all components. But it is highly unlikely that this action will ever be needed, so further discussion will be avoided.

4.  We may want to simultaneously normalize against both starting point shift and rotation of the object. This is the usual situation in shape-recognition problems. When an object comes down the conveyor belt to be photographed for perimeter tracing, we probably neither know nor care what its orientation is. The starting point of our tracing is similarly undefined. This type of normalization will be explored in depth.

Keep in mind that magnitude normalizations (described in the previous section) involving constant multiplication of all of the frequency domain data does not affect phase, so it may be done before or after the phase normalizations described in this section. But normalization done by subtracting means or other constants does affect phase, so such normalizations must be the *last* step performed before sending the data to the neural network.

There is a convention that will be used throughout the remainder of this chapter. *We will not allow implied frequencies beyond the Nyquist limit.* In the previous section, we could afford to be sloppy about this because we were focusing on real-valued data in the time domain. In that case, the Fourier coefficients beyond $n/2$ are superfluous in that they are just the complex conjugates of those below the Nyquist point. But when the time-domain data is complex-valued, the coefficients beyond $n/2$ are meaningful, and we must be careful to remember that they represent negative frequencies. Whenever we write $H(f)$, it will be assumed that $-n/2 < f <= n/2$. For programming purposes, $H(f)$ can be found in the memory location $H(n+f)$ when $f$ is negative. However, all mathematical formulas will refer to the true value of $f$, which may be negative, not the programming subscript. This traditional method of storing negative frequency components was illustrated in Figure 4-2.

At this point, the reader should review starting-point normalization, discussed on page 98. (It is called phase normalization there, because shifting the starting point is equivalent to changing the phase of the signal. Here, where several types of phase shift are significant, we must be more specific.) The effect of shifting the sample's starting point in the time domain by an angle of $\phi$ is to shift the phase of each $H(f)$ by an angle of $f\phi$, as was seen in Equation (4-8).

Let us consider the effect of rotating the time-domain data (e.g., the object whose perimeter is being traced) through an angle of $\rho$. This is trivial if we view a rotation as a multiplication by $e^{j\rho}$. The Fourier transform is a linear operation, so exactly the same thing happens in the frequency domain. The phase of each $H(f)$ changes by an angle of $\rho$.

We can now see the combined effect of rotating the time-domain data by an angle of $\rho$ and simultaneously shifting the starting point by an angle of $\phi$. The phase of each $H(f)$ will be changed by $f\phi + \rho$. In order to compute a normalized Fourier transform that is independent of these two shifts, we must find values of $\phi$ and $\rho$ which, when applied as just shown, leave the frequency-domain data in some "normalized" state.

The question at this point is, "Just what do we mean by a normalized state?" Since there are two numbers to be found ($\phi$ and $\rho$) to take care of two shifts (rotation and starting point), we intuitively know that we must impose two conditions. We might, for example, ask for the phase of $H(3)$ to equal 67 degrees. And we might ask that the phases of $H(7)$ and $H(11)$ be equal. But such conditions are arbitrary nonsense. We would like something meaningful. As will soon be seen, we can deal with the nonsense part, but the arbitrary part is a bit more difficult.

The most important aspect of any normalization scheme is that it must be as unambiguous as possible. Suppose that there are some shapes for which two or more very different normalizations are possible. It would be a disaster if patterns in the training set tended toward one normalized form, while runtime data sometimes followed an alternative normalization path. The neural network would surely fail to perform correctly. Thus, our goal will be to reduce ambiguity as much as possible. It is not an easy goal to attain.

One of the two normality conditions that we will impose has already been discussed. We will insist that the phase of $H(1)$ be 0. As long as we follow the path in a clockwise direction, $H(1)$ will be the dominant force in the frequency domain. (Recall that if $H(-1)$ has the greatest magnitude, we are tracing CCW, so reflect the coefficients about $n/2$.) Since $H(1)$ dominates, its phase will be well-defined and relatively stable. For any fixed rotational position, there is exactly one normalization that zeros the phase of $H(1)$, so this requirement does not detract from uniqueness. The effect of zeroing this phase is to define the phases of all other components of the Fourier transform as being relative to the phase of the fundamental. This is good.

The real problem is finding a second condition to impose. Many authors have proposed a wide variety of suggestions. [Wallace and Wintz, 1980] provide a nice overview. Due to space limitations, we will emphasize only one highly regarded method. That method is to zero the phase of the Fourier coefficient having the maximum magnitude after $H(1)$. The algorithm will be described in such a way that the reader can easily adapt it to zero any other coefficient instead, if that is deemed preferable.

There is an obvious problem with this method. If more than one coefficient is in a near tie for the maximum, there can be ambiguity. Suppose, for example, that the expected path is such that $H(3)$ and $H(4)$ both tend to have nearly the same large magnitude. The training and test normalizations can flip-flop from one to the other, making the network's job extraordinarily difficult. Unfortunately, every normalization scheme has its weaknesses. It has been found in practice that this sort of ambiguity is not often a problem. And there is a definite advantage to this particular choice. The larger a coefficient is in magnitude, the more reliable is its phase. Small coefficients will be overwhelmed by noise, making the phase unstable. Larger coefficients are more likely to reflect the true phase of the data. On the other hand, if the designer knows in advance that some particular frequency component will always be strong, even if not necessarily the strongest, this source of ambiguity can be eliminated by specifying the frequency in advance. *This should be done whenever possible.*

We now have everything we need. We know how to compute the frequency-domain effect of time-domain rotation and starting-point shift: change the phase of each $H(f)$ by $f\phi + \rho$, where $\phi$ and $\rho$ are the shift and rotation, respectively. And we know how we wish to define *normalized*: the phase of $H(1)$ and $H(k)$ are 0 for some specified $k$. All that remains is to compute values for $\phi$ and $\rho$ that will accomplish the normalization. Let $\theta_1$ be the phase of $H(1)$ and $\theta_k$ be the phase of $H(k)$. Some simple manipulation gives the answer.

$$\phi = \frac{\theta_1 - \theta_k}{k - 1}$$

$$\rho = \frac{\theta_k - k\theta_1}{k - 1}$$

(4-11)

As an exercise, let us verify that Equation (4-11) does what we want it to do. First, see what it does to the phase of $H(1)$. For this coefficient, $f = 1$, so we add $\phi + \rho$ to the phase of $H(1)$. That quantity works out to $\theta_1 (1 - k) / (k - 1)$, which is just $-\theta_1$. Since the phase of $H(1)$ is $\theta_1$, that phase is clearly zeroed. The reader should work through the same process for $H(k)$.

Alas, we have one more troublesome ambiguity to deal with. Suppose we have applied the above phase modifications so as to zero the phases of $H(1)$ and $H(k)$. Imagine now that we continuously shift the time-domain starting point throughout a full circle. Figures 4-25 through 4-27 show the effect on the frequency-domain phase of several Fourier coefficients as a function of the starting-point shift. Recall that the phase shift of $H(f)$ is equal to $f$ times the time-domain shift.

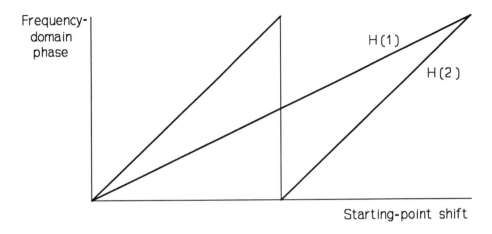

Fig. 4-25: Phase shifts of *H*(1) and *H*(2).

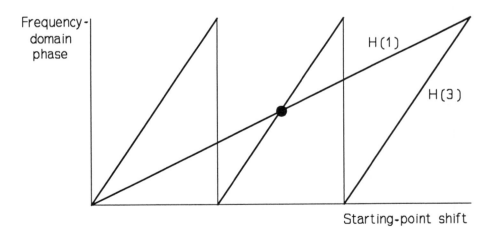

Fig. 4-26: Phase shifts of *H*(1) and *H*(3).

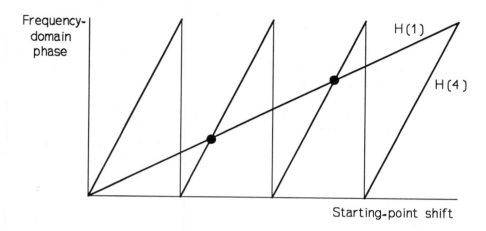

Fig. 4-27: Phase shifts of H(1) and H(4).

Observe that when the time-domain shift is half of a full circle, the phase of $H(3)$ is the same as the phase of $H(1)$. And this happens twice for $H(4)$. Whenever the phases match, we can apply the rotation operator, which subtracts the same angle from all phases, in such a way that the phases of $H(1)$ and $H(k)$ are once again zero. This means that for some values of $k$, the shift/rotation operation that zeroed the two phases is not unique. We can compute the number of possibilities, $m(k)$, using Equation (4-12).

$$m(k) = |k-1| \quad , \quad -n/2 < k \leq n/2 \qquad (4\text{-}12)$$

Whenever we have zeroed the phases of $H(1)$ and $H(k)$, we can shift the starting point by an angle equal to a full circle divided by $m(k)$, then rotate in the opposite direction by the same angle. This pair of operations gives us another normalized transform that still has the phases of $H(1)$ and $H(k)$ equal to zero. (This is easy to prove. Do the cases of $k$ positive and negative separately.) We are forced to make another uncomfortably arbitrary decision as to which of these $m(k)$ normalizations to use. The best case is, of course, when $k = 2$, for $m(2) = 1$. This tells us that if we know in advance that $H(2)$ will always have relatively large magnitude, we would do well to preordain that coefficient as the one whose phase we will zero. *Do this if at all possible*!

Now we need a method for deciding which of the $m(k)$ normalizations to choose. A popular method that seems to work well in most applications is to maximize the *positive real energy*. This is discussed in [Wallace and Wintz, 1980]. For each of the $m(k)$ normalizations, compute the criterion defined in Equation (4-13), and choose whichever maximizes that criterion.

$$E_r = \sum_i Re(H(i)) \, |Re(H(i))| \qquad (4\text{-}13)$$

This section will conclude with an explicit algorithm for normalizing the Fourier transform of time-domain data obtained from a cyclic path in the complex plane. C++ code for accomplishing this will follow.

1.  If the number of periods, $p$, in the time-domain sample is greater than 1, select only the fundamental and integer harmonics in the frequency domain. In other words, from now on work with $H'(f) = H(pf)$. As in the previous section, this assumes that $n$, the number of sample points, is a multiple of $p$. If not, we curse our misfortune and keep the whole transform, abandoning all hope of simple normalization.

2.  If there is any chance that the time-domain data is collected from a counter-clockwise path, check for that by comparing the magnitude of $H(-1)$ with that of $H(1)$. If the latter does not dominate, we must reflect about the Nyquist frequency. From now on, work with $H'(f) = H(-f)$ for nonzero $f$.

3.  If we wish to normalize against changes in scale, divide every Fourier coefficient by the magnitude of $H(1)$.

4.  If all phase information is significant, we are done.

5.  If rotational position is significant, but we wish to normalize against starting-point shift, compute $\theta$, the phase angle of $H(1)$. For each $H(f)$, subtract $f\theta$ from its phase. We are done. Note that if we already normalized for scale in step 3, $H(1)$ is now worthless and may be discarded.

6.  If we get to this step, we want to normalize against both rotation and starting-point shift. Choose the coefficient other than $H(1)$ whose phase we will zero. It must be guaranteed to have a large magnitude for stability. If we know that $H(2)$ will always be reasonably large, choose $k = 2$, as that is the only choice that is entirely free from ambiguity. If we cannot be sure $H(2)$ will be large, but if we can be sure that some other coefficient will always be large, choose that other coefficient. As a last resort, we can, for each case, pick whichever coefficient in that case's transform has the largest magnitude. This is the least desirable alternative, but it is sometimes necessary.

7.  Compute $\phi$ and $\rho$ using Equation (4-11). For each $H(f)$, add $f\phi + \rho$ to its phase. This gives one of the $m(k)$ normalizations that zeros the phase of $H(1)$ and $H(k)$, where $m(k)$ is defined by Equation (4-12). Compute the criterion in Equation (4-13).

8.      Compute each of the other $m(k) - 1$ equivalent normalizations. This is done by repeatedly adding $f\delta - \delta$ to the phase of each $H(f)$, where $\delta = 2\pi/m(k)$. Compute the criterion of Equation (4-13) for each of these normalizations. Choose the one that maximizes the criterion.

There is actually one final normalization step that should always be performed, regardless of whatever else may or may not have been done. We absolutely must present "reasonable" values to the neural network's inputs. These should have a strong tendency to be single-digit numbers. If the inputs are extremely large or small, the network will have difficulty learning effective weights. The usual method for executing this final step is to examine typical data in advance and specify a constant multiplier for all inputs. If, for example, we normalized against scale changes by dividing all coefficients by the magnitude of $H(1)$, it would be appropriate to multiply all normalized coefficients by a constant on the order of ten or so before presentation to the network. We may also want to apply constant offsets as described on page 116.

Code for each of the above operations is now presented. The first step, selecting the fundamental and its harmonics, had its code listed on page 118. The second step, forcing the time-domain data to clockwise motion, is given here.

```
void force_clockwise (
  int n ,                        // Length of vectors
  double *xr ,                   // Real components
  double *xi                     // Imaginary components
  )
{
  int i, nyquist ;
  double mag1, magm1, temp ;

  mag1  = xr[1]   * xr[1]   + xi[1]   * xi[1] ;
  magm1 = xr[n-1] * xr[n-1] + xi[n-1] * xi[n-1] ;
  if (mag1 >= magm1)             // If we are tracing clockwise
    return ;                     // nothing to fix

  nyquist = (n+1) / 2 ;          // N may be odd (unlikely, though!)
  for (i=1 ; i<nyquist ; i++) {  // Reflect about Nyquist
    temp = xr[i] ;
    xr[i] = xr[n-i] ;
    xr[n-i] = temp ;
    temp = xi[i] ;
    xi[i] = xi[n-i] ;
    xi[n-i] = temp ;
    }
}
```

If the overall scale (multiplicative size) of the time-domain path does not convey useful information, we should eliminate its effect. The following subroutine does this by dividing by the magnitude of $H(1)$.

```
void normalize_scale (
  int n ,                          // Length of vectors
  double *xr ,                     // Real components
  double *xi                       // Imaginary components
  )
{
  int i ;
  double factor ;

  factor = xr[1] * xr[1]  +  xi[1] * xi[1] ;
  if (factor == 0.0)
    return ;

  factor = 1.0 / sqrt ( factor ) ;
  for (i=1 ; i<n ; i++) {
    xr[i] *= factor ;
    xi[i] *= factor ;
    }
}
```

When we need to remove the effect of starting the time-domain sampling at unknown times, but the overall position of the path with regard to rotation is significant, we can do that with the following routine. It zeros the phase of $H(1)$.

```
void normalize_phase (
  int n ,                          // Length of vectors
  double *xr ,                     // Real components
  double *xi                       // Imaginary components
  )
{
  int i, nyquist ;
  double mag, phase, theta ;

  mag = sqrt ( xr[1] * xr[1]  +  xi[1] * xi[1] ) ;
  if (mag > 0.0)
    theta = atan2 ( xi[1] , xr[1] ) ;
  else
    theta = 0.0 ;

  nyquist = (n+1) / 2 ;  // N may be odd (unlikely, though!)
```

```
for (i=1 ; i<nyquist ; i++) {

   mag = sqrt ( xr[i] * xr[i]  +  xi[i] * xi[i] ) ;
   if (mag > 0.0)
      phase = atan2 ( xi[i] , xr[i] ) - theta * i ;
   else
      phase = 0.0 ;
   xr[i] = mag * cos(phase) ;
   xi[i] = mag * sin(phase) ;

   mag = sqrt ( xr[n-i] * xr[n-i]  +  xi[n-i] * xi[n-i] ) ;
   if (mag > 0.0)
      phase = atan2 ( xi[n-i] , xr[n-i] ) + theta * i ;
   else
      phase = 0.0 ;
   xr[n-i] = mag * cos(phase) ;
   xi[n-i] = mag * sin(phase) ;
   }
}
```

The most complicated situation is when we need to remove the effects of both starting-point shift and rotation. The following subroutine lets the user specify $k$, the secondary coefficient whose phase we zero. The value of $k$ in this case must be positive, and it reflects the actual array position, not the (possibly negative) frequency. Alternatively, the user may call this routine with $k = 0$, in which case it chooses $k$ as the coefficient having greatest magnitude.

```
void normalize_rotation_shift (
   int n ,                          // Length of vectors
   int k ,                          // Secondary coef to use, 0 to pick max
   double *xr ,                     // Real components
   double *xi                       // Imaginary components
   )
{
   int i, freq, m, mbest ;
   double mag, phase, theta_1, theta_k, phi, rho, crit, best ;
   double real, imag, offset ;

/*
   If the user specified some k>0 we will use it.
   Otherwise, we pick k as the coefficient having max magnitude.
*/
```

```
    if (k <= 0) {  // User wants us to choose k as max
      best = -1.e30 ;
      for (i=2 ; i<n ; i++) {
        mag = xr[i] * xr[i]  +  xi[i] * xi[i] ;
        if (mag > best) {
          best = mag ;
          k = i ;
          }
        }
      } // Choose k

/*
  Compute the phase angles of H(1) and H(k).
*/

    if ((xr[1] != 0.0)  ||  (xi[1] != 0.0))
      theta_1 = atan2 ( xi[1] , xr[1] ) ;
    else
      theta_1 = 0.0 ;

    if ((xr[k] != 0.0)  ||  (xi[k] != 0.0))
      theta_k = atan2 ( xi[k] , xr[k] ) ;
    else
      theta_k = 0.0 ;

/*
  Compute the multiplicity, m, and the normalization constants, phi and rho.
*/

    freq = (k <= n/2) ? k : k-n ;
    m = abs ( freq - 1 ) ;

    phi = (theta_1 - theta_k) / (double) (freq - 1) ;
    rho = (theta_k - freq * theta_1) / (double) (freq - 1) ;

/*
    Apply the normalization, shifting the starting point by phi and rotating by rho.  This gives us
    one of the m normalizations. If m is 1, that is all we need, so save the normalized transform
    directly, then return.  Otherwise, temporarily save the magnitude in xr and the phase in xi.
    Note that for H[1] and H[k] the magnitude is the real part and the phase is zero, so those
    two terms will be OK as is!
*/
```

```
for (i=1 ; i<n ; i++) {
  mag = sqrt ( xr[i] * xr[i]  +  xi[i] * xi[i] ) ;

  if ((i == 1)  ||  (i == k)) {  // Can do these explicitly
    xr[i] = mag ;
    xi[i] = 0.0 ;
    continue ;
    }

  freq = (i <= n/2)  ?  i  :  i-n ;
  if (mag > 0.0)
    phase = atan2 ( xi[i] , xr[i] )  +  freq * phi + rho ;
  else
    phase = 0.0 ;

  if (m == 1) {                     // If only 1 possibility just do it
    xr[i] = mag * cos(phase) ;
    xi[i] = mag * sin(phase) ;
    }
  else {                            // If many, stay in polar for now
    xr[i] = mag ;
    xi[i] = phase ;
    }
  }

if (m == 1)                         // If this was the only possibility
  return ;                          // then we are done

/*
    We have found one of the m normalizations that zero the phase of H(1) and H(k).  Each of
    the others is offset by equal portions of a full circle.  Try them all, keeping track of the one
    that has maximum positive real energy.  Recall that we have stored the magnitude in xr and
    the phase in xi to avoid going back to polar each time.
*/

offset = 2.0 * PI / (double) m ;    // Each is offset this amount
mbest = -1 ;                        // Keep track of best here

while (--m >= 0) {                  // Try each of the m possibilities
  crit = 0.0 ;                      // Will sum criterion here
  for (i=2 ; i<n ; i++) {           // No need to include constant H(1)
    if (i == k)                     // or H(k) in the criterion sum
      continue ;
    freq = (i <= n/2)  ?  i  :  i-n ;
```

```
         real = xr[i] * cos( xi[i]  +  offset * m * (freq-1) ) ;
         crit += real * abs ( real ) ;
         }
      if ((mbest < 0)  ||  (crit > best)) {
        best = crit ;
        mbest = m ;
        }
      }

/*
   We now know which of the m normalizations was best.  Apply it.
   Do not bother with H(1) and H(k), as they are already correct.
*/

   offset *= mbest ;                      // This is the best angle offset

   for (i=2 ; i<n ; i++) {
     if (i == k)
       continue ;
     freq = (i <= n/2)  ?  i  :  i-n ;
     real = xr[i] * cos( xi[i]  +  offset * (freq-1) ) ;
     imag = xr[i] * sin( xi[i]  +  offset * (freq-1) ) ;
     xr[i] = real ;
     xi[i] = imag ;
     }
}
```

## Example of a Phase-Sensitive Problem

Complex-domain neural networks are especially superior to real-domain networks when the information relevant to the solution is primarily embodied in phase relationships. This is often the case in two-dimensional vibration analysis. Suppose that a tiny spot has been marked on a piece of mechanical equipment. The cyclic path of that spot as the machine runs can tell a lot about the state of the machine's health.

An example of this was contrived by parametrically defining a path having a tiny anomaly. The anomaly takes the form of a brief glitch in the otherwise smooth path of the monitored spot. The position of this glitch relative to the overall path is presumed to convey useful information to the operators of the machine. Three examples of this are shown with the glitch exaggerated for visibility. Figure 4-28 depicts one extreme position of the glitch. Look at the upper-right of the path. Its normal position, further to the left, is shown in Figure 4-29, and the other extreme that it may take on is shown in Figure 4-30.

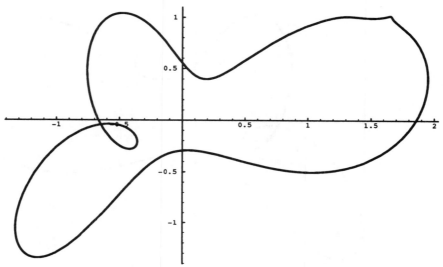

Fig. 4-28: One extreme position of glitch.

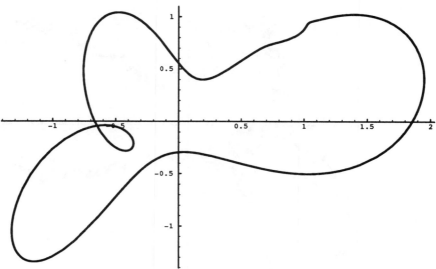

Fig. 4-29: Normal position of glitch.

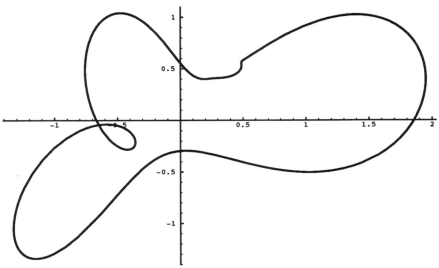

Fig. 4-30: Other extreme position of glitch.

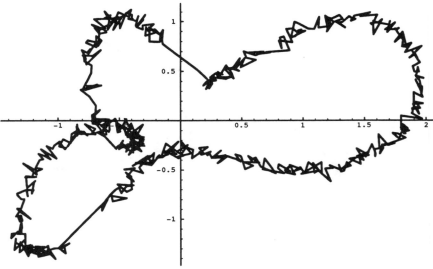

Fig. 4-31: Path as collected with noise.

The actual amplitude of the glitch as used in the neural network tests is half of that shown in those figures. That low level is nearly invisible when plotted, so it was exaggerated in order that the reader can see what the problem involves. To further complicate things, we are forced to work with an abysmal sampling device. It is subject to almost constant random noise contamination. Figure 4-31 shows what a typical training and test sample really looks like. The reader can see that this is an immensely difficult problem to solve. It was deliberately made this hard to tax the networks to the utmost so that their differences would be revealed. With less noise or higher-amplitude glitches, nearly perfect results are obtained too easily for a good comparative test to be made.

The noisiness of the sampling device violates the wraparound assumption of the discrete Fourier transform. To be safe we must use a data window such as the Welch window. To help cancel random noise, we sample four periods of the vibration path. In practice, as many periods as possible should be sampled. A Welch data window is applied to the time-domain data, and the Fourier transform is done. Since $p = 4$, we discard all Fourier coefficients that are not integer multiples of 4 and consecutively renumber those that are kept.

$H(0)$, the constant offset, is meaningless, so it is discarded. It is virtually impossible to consistently agree on a starting point in the time domain. That source of variation is eliminated by zeroing the phase of $H(1)$ (after verifying that the sampled path has a dominant clockwise motion). Because the magnitude of $H(1)$ is not important, it is discarded. However, since the assumed camera in this application is held at a fixed distance from the target, the remaining coefficients are not scale normalized. If the camera were movable, we could compensate for this variation by dividing all coefficients by the magnitude of $H(1)$. It was also decided that only the first 12 harmonics convey useful information. (This is almost always a safe number, but take care nonetheless.) Thus, we will be working with a total of 23 complex numbers: 11 positive frequencies in $H(2)$ through $H(12)$, and 12 negative frequencies in $H(n-12)$ through $H(n-1)$. Remember to choose $n$ large enough that the Nyquist frequency due to the sampling rate exceeds the highest frequency present in the process being sampled.

The final data-preparation step involves bringing the 23 Fourier coefficients to a range suitable for neural network inputs. To do this, several hundred trial samples were made and processed as described earlier, and the mean of each of the 23 complex coefficients found. These means were then carved in stone. They are ordained to be subtracted from every sample case ever presented to that network, both in training and in ultimate use. After doing that, it turned out that the absolute value of the coefficients is quite consistently single-digit numbers. That is perfect, so no multiplication by a constant is needed.

When the sample data was generated, the position of the glitch as it ranged from one of its extreme positions to the other was encoded as a real number ranging continuously from –1 to 1, with the normal, center position at 0. Training and validation sets, each consisting of 500 cases, were generated. The glitch position for each set smoothly covered its entire range.

Both real and complex networks were trained using several numbers of hidden neurons. To equalize the number of weights, we need twice as many hidden neurons in a real-domain network as in a complex-domain network. In other words, a real-domain network having four hidden neurons is equivalent to a complex-domain network having two hidden neurons, since the latter uses complex-valued weight vectors. This should be kept in mind as the experimental results shown in Figures 4-32 through 4-34 are examined.

Figure 4-32 shows the mean squared error for three different size real-domain neural networks. Note the common phenomenon of the training-set error decreasing as more hidden neurons are added, while the validation-set error increases due to the network learning too much specialized information.

Figure 4-33 presents the same thing for equivalent complex-domain neural networks. Observe that this network generalizes much better than its real-domain counterpart.

Finally, Figure 4-34 compares the validation-set error for the real-domain and complex-domain networks. This, of course, is the true test of a network's worth. The winner is obvious.

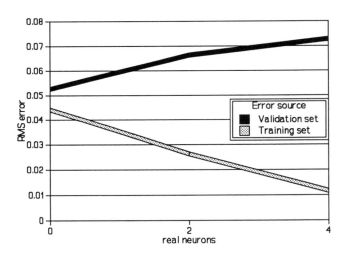

Fig. 4-32: MSE for real networks.

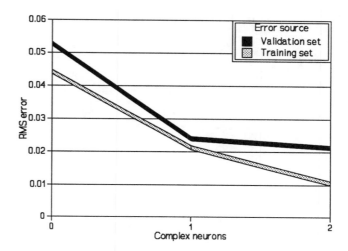

Fig. 4-33: MSE for complex networks.

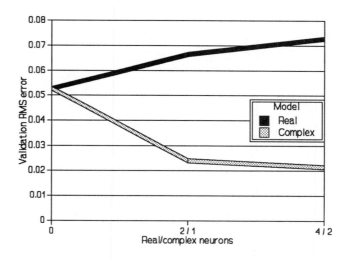

Fig. 4-34: Validation set MSE.

# 5

Time/Frequency Localization

Time series generated by many physical processes are characterized by the appearance of particular frequency components at specific times. If we can find a function that maps a time series to a description of the time/frequency composition of the series, this function may be effective for generating neural network inputs. As will be seen, some of the best candidates map to matrices whose elements are complex numbers. Thus, complex-domain neural networks are appropriate. This chapter will focus on functions that map from real vectors to complex matrices in such a way as to expose the time/frequency structure of the domain vector (the input time series). Generalizations to image processing will be discussed in a later chapter.

## An Intuitive Layout for the Mapping

Whenever one works with a complicated data structure, it helps if the structure can be organized in a manner that is easily grasped by the intuition. There is an established convention for doing this for time/frequency analysis. The function maps the times series vector to a (usually complex-valued) matrix. Each row of this matrix corresponds to a single frequency bin, and each column corresponds to a single time bin. The squared magnitude of an element of this matrix represents the time series' energy in that particular frequency bin at that particular time.

Visual display of this matrix is straightforward. A two-dimensional layout is used. The horizontal axis is usually time, and the vertical axis is frequency. (Some applications are better served by flipping these dimensions. Naturally, this is a convenience issue only.) Three-dimensional plots, with height determined by the magnitude of each matrix element, can be very revealing. Contour plots are also common. If color is available, contour plots that are colored according to magnitude can be both visually striking and useful. Also, three-dimensional plots in which magnitude determines the height and phase determines the color can be wonderful.

It is this time/frequency matrix that provides the inputs to our neural network. Most common mapping functions produce complex-valued matrix elements. For some applications we are able to work with only the magnitude of these elements, discarding the phase. In such cases we can use real-domain neurons. But most of the time, the phase information is too valuable to toss aside. Complex-domain neurons are appropriate.

An example display of a mapping from a time series to a time/frequency representation is shown in Figures 5-1 through 5-4. We have a time series whose frequency steadily increases shown in Figure 5-1. The power spectrum of that series is plotted in Figure 5-2. Note that the spectrum tells us nothing about the fact that the signal is a pure sine wave whose frequency steadily increases over time. There is an infinite number of time series that would have exactly the same power spectrum.

Figure 5-3 much more clearly illustrates the structure of the signal. It shows the relationship between time and frequency. The same data is displayed as a contour plot in Figure 5-4. Methods for computing such transformations will be the topic of this chapter. Neural network applications will appear later.

## The Short-Time Fourier Transform

When we compute and examine the Fourier transform of a time series, we assume that the frequency content of the signal is reasonably constant during the measurement interval. If it is not, our results will be compromised. We will certainly not be able to dissect easily the frequency structure across time. Worse, the frequency shifts may introduce spurious components into the spectrum. If we suspect that the frequency content of the series may change while it is being measured, we may need a different analysis procedure.

One obvious method for handling this problem is to split the time series into sections, analyzing each section separately. This is illustrated in Figure 5-5.

Splitting the series as shown in that figure is a primitive method, and it has a serious problem. Review the material on data windows (page 92). Crude splitting causes intolerable leakage from neighboring frequencies. In other words, localization in the frequency domain is poor. Granted, localization in the time domain is excellent. One hundred percent of the information in each subsection is used, and events outside that precise time interval have no effect whatsoever on the Fourier transform of the data inside the interval. That is excellent. But it comes at a price not worth paying. The value of each Fourier coefficient computed from the data within an interval can be unduly influenced by frequency components significantly distant from the frequency represented by that coefficient. We need something better.

The obvious (and correct) solution is that we need to partition the time series with a smoothly tapering data window, rather than with a brutally abrupt rectangular window. Define a window function $g(t)$ that is large when $t=0$ and that tapers smoothly to 0 as the absolute value of $t$ increases. (For the mathematically inclined, we also need $g(t) \in L^2(\Re)$ and $tg(t) \in L^2(\Re)$. For the rest of us, we need $g(t)$ to go to 0 extremely rapidly as $t$ moves far away from 0.) Then we simply slide this window function along the time axis as is shown in Figure 5-6, letting it carve out areas of interest in the time domain.

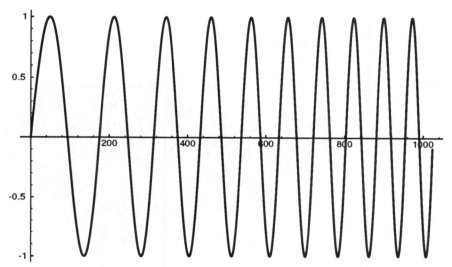

Fig. 5-1: Time series having increasing frequency.

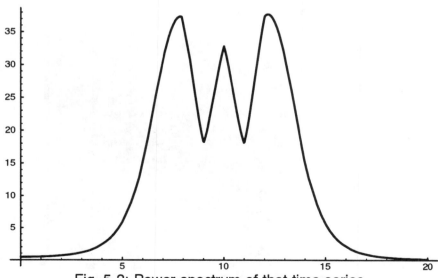

Fig. 5-2: Power spectrum of that time series.

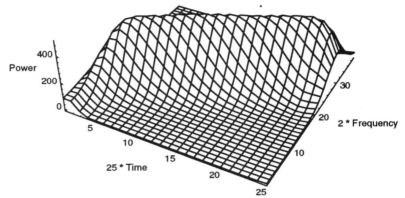

Fig. 5-3: Time/frequency as a 3-D plot.

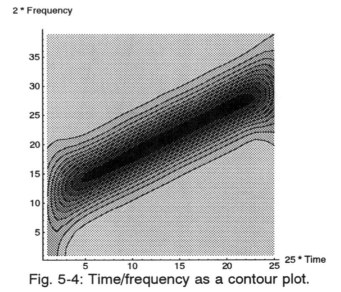

Fig. 5-4: Time/frequency as a contour plot.

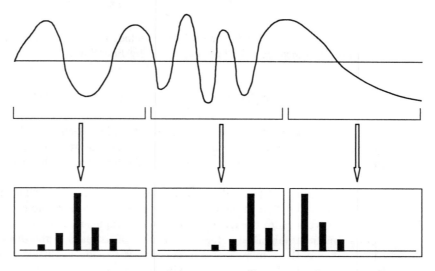

Fig. 5-5: Applying separate Fourier transforms.

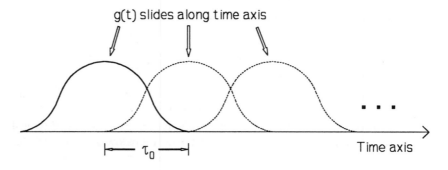

Fig. 5-6: Sliding a smooth window across time.

It is time for some explicit definitions. Suppose that we have a time-domain signal $x(t) \in L^2(\Re)$. The usual physicist's definition of the Fourier transform of that signal is given by Equation (5-1). Electronics engineers often define the Fourier transform using a minus sign in the exponent. Either definition is acceptable as long as consistency is maintained.

$$F_x(\phi) = \int_{-\infty}^{\infty} x(t) e^{2\pi\phi ti} dt \qquad (5-1)$$

That equation is easily modified to include time windowing with $g(t)$. The *short-time Fourier transform (STFT)* is expressed by Equation (5-2).

$$S_x(\tau,\phi) = \int_{-\infty}^{\infty} x(t) g(t-\tau) e^{2\pi\phi ti} dt \qquad (5-2)$$

It can be seen that the presence of $g(t-\tau)$ in Equation (5-2) causes the value of $S_x(\tau,\phi)$ to be most heavily influenced by the values of $x(t)$ in the vicinity of $t=\tau$, with $x(t)$ having essentially no effect when $t$ is far from $\tau$. Therein lies the significance of the STFT.

Equation (5-2) is called the *continuous* version of the STFT because it defines the STFT as a function of continuous time and frequency variables. This does us little good when we want to use the STFT to generate inputs to a neural network. One cannot input a function; numerical values are needed. This leads to the *discrete* version of the STFT. It is nothing more than the continuous version evaluated at discrete time and frequency points. For reasons that will become more clear later, we usually let these points lie on an equispaced *lattice* in the time/frequency plane, although in practice there may be cause for doing otherwise. The spacing in the time dimension is taken to be $\tau_0$, and the spacing in the frequency dimension is $\phi_0$. The discrete STFT can therefore also be seen as a function of two integers. One integer, $p$, specifies the time-dimension coordinate on the lattice. The other, $q$, specifies the frequency-dimension coordinate. This is defined in Equation (5-3).

$$S_x(p,q) = \int_{-\infty}^{\infty} x(t) g(t-p\tau_0) e^{2\pi q\phi_0 ti} dt \qquad (5-3)$$

When we use the STFT to generate neural networks inputs, we will need to define in advance the values of $p$ and $q$ that interest us. This will require either knowledge of the physical processes involved or examination of typical data samples. We will also need to define a shape for the data window $g(t)$. As will be seen, the width of this window directly impacts the lattice spacing, which in turn affects the time and frequency points represented by particular values of $p$ and $q$. These issues will be addressed over the next few sections of this text.

We have been looking at the STFT as a Fourier transform of a windowed time series, conceptually grouping the $x(t)$ and $g(t)$ terms together before multiplying by the complex exponential. For the remainder of this chapter, we will find it more beneficial to change that grouping so that the data window multiplies the complex exponential, producing a damped oscillation. It is the damped oscillation that then multiplies the time series. Of course, the two viewpoints are mathematically equivalent at this point, although later we will sabotage that equivalence by some normalizations. It is purely a convenience for the intuition. We can now view the STFT as convolutions of the time series $x(t)$ with members of a family of functions like those shown in Figures 5-7 and 5-8. The members of this family contain varying amounts of oscillation within the window according to the frequency parameter, and the position of the window is shifted along the time axis according to the time parameter of the STFT. In particular, we can write each member of this family as $g_{pq}(t)$, defined in Equation (5-4).

$$g_{pq}(t) = g(t - p\tau_0) e^{2\pi q\phi_0 t i} \tag{5-4}$$

The discrete STFT can now be written more compactly as Equation (5-5).

$$S_x(p,q) = \int_{-\infty}^{\infty} x(t) g_{pq}(t) \, dt \tag{5-5}$$

If we employ standard scalar product notation, things become even simpler, as shown in Equation (5-6). That notation will be found particularly useful later when we want to consider the possibility of the $g_{pq}(t)$ family comprising a basis by which $x(t)$ can be represented.

$$S_x(p,q) = \langle x, g_{pq} \rangle \tag{5-6}$$

## Frames

*This section contains relatively advanced material that is intended for mathematically inclined readers who desire a more complete presentation. Much of it is not directly relevant to common neural network applications. Some readers may wish to skip to the Summary section on page 149.*

An obvious question at this point arises from discretizing the continuous STFT. It can be proven (though not here; see [Daubechies, 1990]) that under reasonable conditions the continuous STFT $S_x(\tau, \phi)$ captures all of the information in the time series $x(t)$. In other words, if we know $S_x(\tau, \phi)$, we can compute $x(t)$. In fact, it is quite easy to reconstruct $x(t)$ from its continuous STFT. But just how much information about $x(t)$ is lost if we only know the STFT at the discrete lattice points $(p\tau_0, q\phi_0)$? Not surprisingly, it turns out that the answer depends on how close the spacing is (i.e., how small $\tau_0$ and $\phi_0$ are). The closer we space the lattice, the more information we capture. As will be seen later, it is not difficult to capture enough information for practical use by means of quite reasonable lattice spacing.

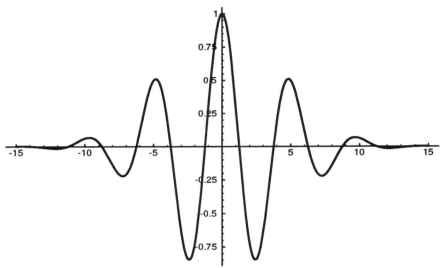

Fig. 5-7: Real part of windowed complex exponential.

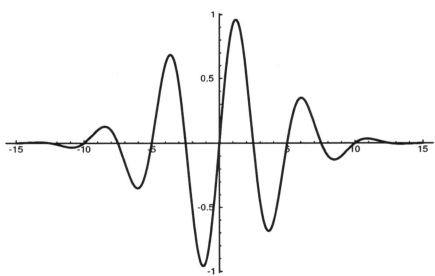

Fig. 5-8: Imaginary part of windowed complex exponential.

When we want to be able to recreate $x(t) \in L^2(\mathfrak{R})$ from $S_x(p, q)$, what we would like is that the set $g_{pq}(t)$ for $p, q \in Z$ constitutes a basis for $L^2(\mathfrak{R})$. In fact, what would really please us is if we were to find some window function $g(t)$ and lattice spacings $\tau_0$ and $\phi_0$ such that we have that holy grail of linear geometry, an orthonormal basis. Unfortunately, no orthonormal basis grounded on the discrete STFT exists if we demand good localization in both time and space. One or the other of these coveted properties must be dropped if we are to find an orthonormal basis using the discrete STFT. Either $tg(t)$ or $fG(f)$, where $G$ is the Fourier transform of $g$, will not be square integrable if we have a basis. On the other hand, under fairly general conditions on $g(t)$, we can find threshold values for $\tau_0$ and $\phi_0$ such that as long as the spacing is at least that close, we can recover the function to arbitrary accuracy with numerically stable procedures. What we will usually experience, though, is that there is a significant amount of redundancy in the information in the discrete STFT. Our family of windowed complex exponentials does not constitute a true basis because there are "too many" of them. Sometimes this is good. For example, we may discover that $x(t)$ can be recovered to four significant digits even though the discrete STFT is computed to only three digits! In other applications we must reduce expensive redundancy as much as possible. That will usually be the situation when the STFT forms the inputs for a neural network. More inputs means more memory and longer training and execution time. Appropriate tradeoffs between localization, thoroughness, and redundancy will be discussed later.

We now understand that not only does $g_{pq}(t)$ for $p, q \in Z$ not constitute an orthonormal basis in any case of interest to us, it does not even generally constitute a basis at all. Any finite subset of these functions will in general be linearly independent, but there will nevertheless be "too many" of them for a basis in that any one of them lies in the closed linear span of the infinitude of others.

In order to deal with this beast that behaves almost like a basis, but is not generally a true basis, we use the concept of *frames* developed in [Duffin and Schaeffer, 1952]. Let $\{\psi_j\}$ be a (possibly infinite) collection of elements of a Hilbert space $\mathcal{H}$. In the current context, the $\{\psi_j\}$ will be the $\{g_{pq}(t)\}$ family already defined. If we remember basic linear geometry, we understand that when $\{\psi_j\}$ is an orthonormal basis for $\mathcal{H}$, the squared length of any element $f$ of $\mathcal{H}$ is equal to the sum of squared projections of $f$ on each $\psi_j$.

$$\|f\|^2 = \sum_j |\langle f, \psi_j \rangle|^2 \tag{5-7}$$

A frame has nearly the same property. It just relaxes the equality a little. It says that if there exist fixed constants $A$ and $B$ that can, for any $f$ in $\mathcal{H}$, bound the sum of squared projections of $f$ on $\{\psi_j\}$, then $\{\psi_j\}$ is said to constitute a frame. $A$ and $B$ are called the frame bounds. The fundamental frame definition is shown in Equation (5-8).

$$A \|f\|^2 \leq \sum_j |\langle f, \psi_j \rangle|^2 \leq B \|f\|^2 \tag{5-8}$$

Frames are important to us because they determine reconstructability of the original time series $x(t)$. Suppose that we have a family of weighted complex exponentials, $\{g_{pq}(t)\}$, on which we base a discrete STFT or some other time/frequency representation. If we know that family is a frame, then we know that we have captured all of the information in $x(t)$. Conversely, if it is not a frame, then we have missed some information. When we are preprocessing our valuable data this way, presenting the discrete STFT to a neural network, we often hope that nothing has been lost.

Obviously, when $\{\psi_j\}$ is an orthonormal basis, it is a frame with $A=B=1$. Similarly, if $\{\psi_j\}$ is a frame with $A=B=1$, and each $\psi_j$ has unit norm, then $\{\psi_j\}$ is an orthonormal basis.

If $\{\psi_j\}$ is a frame with $A=B$, then it is called a *tight frame*. If the elements of a tight frame are unit norm, then $A$ is an indication of the redundancy of the frame. Its minimum value of 1, corresponding to an orthonormal basis, is clearly minimum redundancy (none at all). As has already been mentioned, there is no orthonormal basis $\{g_{pq}(t)\}$ capable of simultaneously localizing in the time and frequency dimensions. Therefore, some redundancy is necessary to provide both localizations.

When $B>A$, the usual case, reconstruction is still possible. It is simply more unstable numerically than it is for a tight frame. The ratio $B/A$ indicates the degree of instability. A more detailed discussion of that issue would take us too far afield. See [Daubechies, 1990] for details. But the reader does need to understand that for critical applications, some effort should be made to minimize $B/A$ even if only characterization, not reconstruction, is the goal. If the ratio of frame bounds is large, the characterization can be unstable. Consult the cited references for means of computing $A$ and $B$ if a major application is at stake and the precise models given later are not followed to the letter.

## When Do We Have a Frame?

Suppose that we have a weight function that we like, and we also have lattice spacings $\tau_0$ and $\phi_0$ that seem reasonable for the task at hand. It may be that we have a priori grounds for these choices, and we do not care whether or not they will lead to the $g_{pq}(t)$ constituting a frame. Perhaps the information that is missed is known to be irrelevant to the task. In that case, our life is simplified. On the other hand, we may not be so fortunate. We may need to be sure that all of the information in $x(t)$ is captured in the discrete STFT that we pass to the neural network. In general, making that determination is not a trivial task. Readers who need full details must be referred to [Daubechies, 1990]. This section will provide some heuristics for rough guidance. Subsequent sections, in which particular weight functions are studied, will provide more explicit guidelines.

The most fundamental requirement for generation of a frame is that the overall density of the lattice must be sufficiently tight to satisfy basic information theoretic demands. It can be shown that if $\tau_0 \cdot \phi_0 > 1$, then there is no chance of having a frame. There is no weight function $g(t)$ that will rescue us. If the product is exactly equal to 1, then it is possible to construct a special weight function that generates a frame.

Unfortunately, this frame is unable to localize in time and frequency simultaneously. In order to localize in both time and frequency, the product of the time-domain spacing times the frequency-domain spacing must be strictly less than one. Understand that this is not a *sufficient* condition. There are many weight functions that will fail to generate a frame even when $\tau_0 \cdot \phi_0 < 1$. It is a necessary condition. If we do not have it, then there is no hope.

What does that condition mean in physical terms? Remember that we will be planting the weight function at discrete positions along the time axis. The space (in time units) between these positions is $\tau_0$. For each planting, we will be computing scalar products involving frequencies of 0, $\phi_0$, $2\phi_0$, $3\phi_0$, ..., cycles per unit time. (If $x(t)$ were complex, we would also have to deal with negative frequencies.) In other words, the smallest positive frequency in the family is $\phi_0$ cycles per unit time. The condition $\tau_0 \cdot \phi_0 < 1$ means that the lowest frequency, $\phi_0$, must complete less than one entire cycle in the time interval $\tau_0$. Equivalently, the period of that frequency must exceed the time spacing of the lattice.

Another condition that must be satisfied in order to have a frame is that there must not be any "holes" in the time domain placements of the window. If the spacing $\tau_0$ is wide, but the window $g(t)$ is narrow, there may be dead spots between window placements that are ignored. If $x(t)$ is free to do whatever it pleases in these areas, without impacting the STFT, then we obviously cannot have a frame. Mathematically, this condition can be stated as shown in Equation (5-9).

$$\sum_p |g(t - p\tau_0)|^2 > 0 \qquad \text{for all } t \qquad (5\text{-}9)$$

One more condition is that the window function $g(t)$ must decay quickly as $t$ goes to infinity. Roughly, it needs to decay at least as fast as $(1 + |t|)^{-3}$.

The three conditions just described are neither complete nor entirely precise. [Daubechies, 1990] is the ultimate reference for readers wanting more rigor. Also, they are not sufficient conditions, only necessary conditions. Methods for generating guaranteed frames will be given later. However, it is hoped that the preceding discussion has given the reader an intuitive feeling for what a frame involves. That is all that a neural network text can hope to offer.

Be aware that the entire preceding discussion has been slightly unrealistic in one sense. It has assumed that $x(t)$ is a continuously measurable series, and that we are able to compute an infinite number of lattice point values for the discrete STFT. Naturally, in practice, we must take samples of $x(t)$ at discrete time intervals. These time intervals, which we may call $t_0$, have no relationship whatsoever to the lattice spacing in the time dimension, $\tau_0$. Rather, they determine the Nyquist limit for high-frequency resolution. In other words, we have thus far implied that the resolved frequencies are $q\phi_0$ for $q = 0$, 1, 2, ... as far as we wish to go. However, the time interval $t_0$ between samples limits how far we *can* go. Recall from page 88 that if we are to avoid aliasing, an extremely serious problem, the series $x(t)$ must not contain any frequencies whose period is less than twice the time interval between samples. When we discretely sample a time series every

$t_0$ time units, the highest frequency that can be properly handled is $1/(2t_0)$ cycles per time unit. Therefore, there is no point in computing the STFT for any values of $q$ such that $q\phi_0$ exceeds the Nyquist limit. Once again, it must be emphasized that this is an entirely different issue from the issue of lattice point spacing in the discrete STFT.

A specific example may help. Suppose that we sample a time series at a rate of one sample per millisecond ($t_0 = 0.001$). We must assume that the underlying series does not contain any frequencies beyond the Nyquist limit of $1./0.002 = 500$ cycles per second or we are in big trouble due to aliasing. We decide to use a bell-shaped window, with the discrete STFT spaced every 15 milliseconds in the time domain. We make the window wide enough to produce significant overlap. Three contiguous windows are shown as dotted lines in Figure 5-9. (Ignore for now the solid line oscillations.)

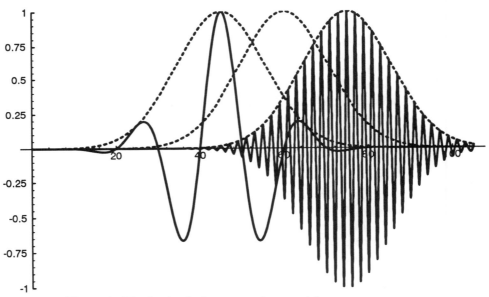

Fig. 5-9: Typical window spacing and frequency range.

Since the time-domain spacing is 15 milliseconds, we know that the frequency-domain spacing must be less than $1/0.015 = 66.67$ cycles per second if we are to satisfy the fundamental frame requirement. To be fairly safe, we decide to go with a period of 20 milliseconds, giving $\phi_0 = 50$ cycles per second. The frame element $g_{3,1}$, having a time offset of $3 \cdot 15 = 45$ milliseconds and having the lowest frequency of 50 cps, is shown in conjunction with the leftmost bell curve in Figure 5-9. For visual clarity, the center bell curve, corresponding to $p = 4$, is shown alone. The rightmost bell curve accompanies $g_{5,10}$, which is the highest frequency that we can handle. In the theoretical case of $x(t)$ being continuous, there is no upper limit to the frequency. But because we are sampling the signal once per millisecond, we hit the Nyquist frequency at $q = 500/50$.

It is interesting to see what happens if we push everything to the critical sampling limit in a discrete-sampled, band-limited, nonoverlapping case. Suppose that we have $N$ measured samples of $x(t)$, with one sample taken per second. We use a rectangular window whose width is $n$ samples. This gives us $N/n$ window placements (values of $p$). The frequency resolution at the critical frame limit is $1/n$ cycles per second, and the Nyquist limit is 0.5 cycles per second. Thus, $q$ will take on the values 0, 1, ..., $0.5n$. The first and last of these will be real (review page 86), while the remainder will be complex. Therefore, for each window placement, the frequency components will be represented by exactly $n$ real numbers. Since there are $N/n$ window placements, the entire time/frequency representation of the $N$-component time series will contain $N$ components! There is no redundancy; we have a basis. However, frequency-domain localization is terrible (page 92). If we keep the same time and frequency spacing but smooth the corners of the rectangular data window, frequency localization improves as we lose reconstructability. To get back to a frame, we must decrease either the frequency spacing or the time spacing. But now we have redundancy. Ponder these relationships to arrive at a clearer perception of critical sampling.

## Summary

This section has provided a broad introduction to the concept of frames. We project the time series $x(t)$ onto members of the family of functions $g_{pq}(t)$ in order to obtain the series' time/frequency components. These are the values that comprise the inputs to our neural network (after suitable scaling, of course).

In all cases of practical interest, the family $g_{pq}(t)$ for $p$, $q \in Z$ is redundant. It contains too many members to be a basis in the mathematical sense. Therefore, the concept of basis is generalized to the concept of frames. The set of projections of $x(t)$ on all of these family members captures all of the information in $x(t)$ (i.e., allows $x(t)$ to be accurately reconstructed from the projections) if and only if the family $g_{pq}(t)$ constitutes a frame.

In some applications we do not care whether or not we have a frame. We may know, for example, that only the information in certain frequency bands, and/or in certain relative time areas, is important to us. In this case, we simply compute what we need and blissfully ignore the rest. Often, though, we must care. We may not be blessed with a priori information. When that is the case, we must take care that our family of functions is a frame.

This section has not presented any *sufficient* conditions on $g(t)$, $\tau_0$, and $\phi_0$ in order to guarantee a frame. Those conditions are quite complicated and can be found in the cited references. Later, we will provide particular examples of practical frames. However, this section has provided several *necessary* conditions for having a frame. The most important is that $\tau_0 \cdot \phi_0 < 1$. The other two conditions are that the windows must overlap in the time domain, and they must quickly vanish outside their area of influence. The reader wishing to experiment without the benefit of explicitly verifying frame properties can be assured that if these three necessary conditions are *very conservatively* followed, especially keeping $\tau_0 \cdot \phi_0 < 0.5$, a frame will likely result. Will the Mathematics Police kindly refrain from hanging me for that statement.

There is a quite unrelated but important issue that must be addressed. It is the issue of the Nyquist limit due to discrete sampling of the time series. Theoretically, there is no upper limit imposed on $q$, the frequency identifier for the family $g_{pq}(t)$. But since in practice $x(t)$ cannot be continuously measured, we must take care that it does not contain any frequency components whose period is less than twice the sample interval. Therefore, there is no point in letting $q$ exceed the value that would correspond to that frequency.

Finally, it should be pointed out that much of the intuitive foundation for this section was based on the short-time Fourier transform. However, careful reading will show that it is far more general than that. Later, we will come back to frames when the wavelet transform is discussed.

# The Gabor Transform

Almost nothing has been said of the role of the data window, $g(t)$, in the STFT. We have stated that in order to promote good frequency localization, it should taper smoothly to 0, in contrast to the sharp corners of a rectangular window. We have also stated that as the absolute value of $t$ becomes large, $g(t)$ should approach 0 extremely rapidly. But what about other aspects of its shape? Are some weight functions better than others? It turns out that the answer is an emphatic *yes*. There is one weight function, the Gaussian function, which can be shown to be optimal in a very important way. A short-time Fourier transform employing the Gaussian weight function is called a Gabor transform, and that is the subject of this section.

## Resolution Rectangles

The position of the data window along the time axis defines a range of influence in the time domain. If the window height drops to exactly 0 at some distance from its center, then the range of influence has a definite boundary. Such would be the case for a rectangular data window. However, as we have seen, most functions of interest taper off smoothly. In fact, although for all practical purposes they rapidly go to 0, the most important window functions never actually reach 0. Therefore, we need to be able to rigorously define a range in the time domain over which a given data window $g(t)$ allows the time series $x(t)$ to exert influence on the STFT at that time slot. We need an indicator of the "center" of that area of influence, and we need a measure of the distance from that center over which the influence extends. For example, we may say that our window is centered at a time of 35 seconds from start, and that it encompasses signal points plus or minus 8 seconds from that center. Understand that this is not an all-or-nothing situation. We do not have series points seven seconds from the center strongly influencing results, while points nine seconds away have no influence at all. Since the window tapers smoothly, influence likewise tapers off smoothly. Rather, this window extent measure is a convenient way of describing the size of the region that carries the bulk of the responsibility for determining STFT results for any particular time point.

It is easy to define a center for the area of influence created by a window. This measure should be analogous to the center of gravity of an object. It need not necessarily locate the position of *maximum* influence (like the mode in statistics), although in most cases of practical interest the shape of the window will be such that its center corresponds to its point of maximum influence. If we square the weight function so that we work with energy, the most common measure of central influence should look very familiar to most physicists and engineers. It is given in Equation (5-10). That equation uses the absolute value of the function. Since in our application the function is real and nonnegative, this is not needed. It is included in the spirit of full generality.

$$\mu_t = \frac{\int_{-\infty}^{\infty} t \, |g(t)|^2 \, dt}{\int_{-\infty}^{\infty} |g(t)|^2 \, dt} \qquad (5\text{-}10)$$

There is, of course, an infinite number of ways of defining a measure of window size. The one most commonly used in the context of time/frequency localization is the root-mean-square (RMS) duration of the window, a simple extension of the above centroid obtained by squaring $t$. This quantity, sometimes called the window's *radius*, is used by physicists to measure the duration of wave packets, and by electrical engineers to measure the bandwidth of a signal (using a frequency-domain equivalent version). See, for example, [Bracewell, 1986]. Its traditional symbol is the Greek letter *delta*, with a subscript indicating the domain variable, as shown in Equation (5-11).

$$\Delta_t = \sqrt{\frac{\int_{-\infty}^{\infty} (t - \mu_t)^2 \, |g(t)|^2 \, dt}{\int_{-\infty}^{\infty} |g(t)|^2 \, dt}} \qquad (5\text{-}11)$$

For the weight functions that are the subject of this chapter, and for many others as well, the above integrals can be explicitly computed. If the reader experiments with custom weight functions, standard numerical techniques for evaluating definite integrals invariably work well due to the fact that the window function goes to 0 rapidly. (It had better, or you will not have a frame, as discussed earlier.) But remember that while the computed center and radius have some interpretive value, nothing beats an actual plot of the window function when it comes to intuitive understanding.

So far, we have considered only areas of influence in the time domain. We have a time series $x(t)$, and we have computed a component of its discrete STFT, $S_x(p, q)$. (We could just as well be concerned with a term in the continuous STFT, $S_x(\tau, \phi)$, but we will largely ignore that more theoretical case.) We have been answering two questions. The area of influence of $x(t)$ on the computed STFT term is centered at what value of $t$? How far (in time) from that center does the influence of $x(t)$ extend? These are obviously important questions, but they are only half of what we need answered. That STFT term also is influenced by the frequency content of $x(t)$ within the data window. We have made it so by including a complex exponential of frequency $\phi = q\phi_0$ in the associated frame function, $g_{pq}(t)$. Our hope is that components of $x(t)$ whose frequency is around $\phi$ will most strongly influence this STFT term, while components with more distant frequencies will exert diminishing influence. We need to compute the center and radius of this implicit window in the frequency domain.

Those readers who are familiar with convolution already know how to find the frequency-domain window associated with a given time-domain window. It is simply the Fourier transform of the time-domain window. The previous formulas for the center and radius of the window then apply just as well. In particular, let $G(f)$ be the Fourier transform of $g(t)$. Then we have

$$\mu_f = \frac{\int_{-\infty}^{\infty} f\,|G(f)|^2\,df}{\int_{-\infty}^{\infty} |G(f)|^2\,df} \qquad (5\text{-}12)$$

$$\Delta_f = \sqrt{\frac{\int_{-\infty}^{\infty} (f-\mu_f)^2\,|G(f)|^2\,df}{\int_{-\infty}^{\infty} |G(f)|^2\,df}} \qquad (5\text{-}13)$$

For the windows that we will be studying in this chapter, and for many others, both the Fourier transform and the center and radius can be explicitly computed. In case the reader experiments with other windows, great care must be exercised in numerically evaluating those integrals in the frequency domain. Things may not always be as easy as they were in the time domain. It is not unlikely to find that the frequency-domain window of a carelessly chosen time-domain window goes to zero so slowly that numerical techniques encounter difficulty. And by all means, make the time-domain window symmetric about 0 so that its Fourier transform is real!

Now that we can talk about a central location and a size for windows in both the time and frequency domains, we can visualize them in two dimensions. This is done in Figure 5-10.

Recall that the time-window size measure given by Equation (5-10) (and similarly for frequency) is a radius about its center. Therefore, when we refer to the *window width*, we traditionally double the radius. That is indicated in Figure 5-10.

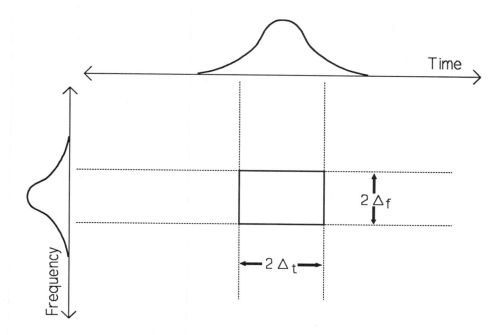

Fig. 5-10: A resolution rectangle.

## The Heisenberg Uncertainty Principle

We can control the shape of the resolution rectangles by the shape of the window function $g(t)$. It can be shown that if we make the time-domain window narrower, the width of the frequency-domain window increases, and vice-versa. If we are careless about the shape of $g(t)$, the size of the rectangles may be needlessly large. Unfortunately, we cannot shrink the rectangle to an arbitrarily small size. It would be nice to be able to resolve the time/frequency components of a time series to arbitrary accuracy, but a basic physical principle prevents us from doing so. The *Heisenberg Uncertainty Principle* places a firm lower limit on the area of the resolution rectangle. In particular, it is stated in Equation (5-14).

$$\Delta_t \Delta_f \geq \frac{1}{4\pi} \qquad (5\text{-}14)$$

We are free to control the shape of the rectangle, trading off resolution in one domain for resolution in the other domain. But we cannot have high resolution in both domains simultaneously. There is no weight function $g(t)$, and no lattice spacings ($\tau_0$, $\phi_0$) that allow us to circumvent this limit. The laws of physics are quite firm here.

There is now a question begging to be answered. A limit has been placed on our resolution capabilities. Can we attain this limit? We certainly can. And it easy to do so. Simply use a Gaussian function, of the form shown in Equation (5-15), for the weight function.

$$g(t) = e^{-t^2} \qquad (5\text{-}15)$$

It turns out that as long as we use a function having that shape, no matter how we scale its height or width, our resolution rectangles will always meet the Heisenberg limit. That is a powerful argument for limiting ourselves to the Gaussian shape.

There is another interesting aspect to using a Gaussian time window. The Fourier transform of a Gaussian function is just another Gaussian function! This means that the frequency-domain window will have the same nice bell shape as the time-domain window. In particular, examine Equation (5-16), which shows the Fourier transform of a scaled Gaussian function.

$$\int_{-\infty}^{\infty} e^{-\pi (at)^2} e^{2\pi i f t} dt = \frac{1}{|a|} e^{-\pi (\frac{f}{a})^2} \qquad (5\text{-}16)$$

That equation shows how the scaling of the two windows is related. Note the scale factor $a$. It is multiplicative in the time domain, and it divides in the frequency domain. This tells us that if we make the time-domain window twice as wide, the frequency-domain window will become half as wide. (Of course, knowing that exact equality is had, Equation (5-14) tells us the same thing.) Later, though, we will see that in order to preserve good frame properties, we must exercise some restraint in our choice of this tradeoff.

## Resolution of the Gabor Transform

There are two small problems with the simple expression of the weight function given by Equation (5-15). One problem is that it doesn't include provisions for scaling it to arbitrary widths. We need that ability in order to customize the tradeoff between time and frequency resolution. The other problem is that some sort of normalization would be nice in order to achieve consistent results regardless of the scaling. The first problem is easily solved by including a scale factor as a time divisor. The second problem is traditionally solved by asking that the weight function have unit energy, as expressed in Equation (5-17). This is achieved by including an appropriate multiplicative factor.

$$\int_{-\infty}^{\infty} [g(t)]^2 dt = 1 \qquad (5\text{-}17)$$

When both of these accommodations are made, the final form of the (now scaled) weight function $g_o(t)$ is given by Equation (5-18). The Fourier transform of that weight function, similarly scaled to have unit energy, is shown in Equation (5-19).

$$g_\sigma(t) = \frac{1}{\pi^{1/4}\sqrt{\sigma}} e^{\left(-\frac{t^2}{2\sigma^2}\right)} \qquad (5\text{-}18)$$

$$G_\sigma(f) = \sqrt{2\sigma}\,\pi^{1/4} e^{-2\pi^2\sigma^2 f^2} \qquad (5\text{-}19)$$

Just as we would hope, the center of each of these windows is zero, meaning that the center of the resolution rectangle for the discrete STFT component $S_x(p, q)$ is $(p\tau_0, q\phi_0)$. This should not be surprising, since the windows are symmetric. We can also explicitly write the radius of the time and frequency windows. This is shown in Equation (5-20). It is also interesting to multiply these quantities to verify that the Heisenberg limit is reached.

$$\begin{aligned} \Delta_t &= \frac{\sigma}{\sqrt{2}} \\[2ex] \Delta_f &= \frac{1}{2\sqrt{2}\,\pi\,\sigma} \end{aligned} \qquad (5\text{-}20)$$

$$\Delta_t\Delta_f = \left(\frac{\sigma}{\sqrt{2}}\right)\left(\frac{1}{2\sqrt{2}\,\pi\,\sigma}\right) = \frac{1}{4\pi} \qquad (5\text{-}21)$$

## Practical Limits on the Resolution Tradeoff

We have seen that it is legal to freely vary the scale parameter, $\sigma$, in the data window. This trades time-domain resolution for frequency-domain resolution, always keeping the radius product constant at the Heisenberg limit. But how much freedom do we really have in practical problems? Certainly we cannot make the window narrow relative to $\tau_0$, the lattice spacing in the time dimension. If we did, there would be gaps along the time line during which the window function is practically 0, causing those areas of the time series to be essentially ignored. The converse is just as much of a problem, although it is more difficult to visualize. If we make the data window too wide, the frequency-domain resolution $\Delta_f$ may become too fine relative to the lattice spacing $\phi_0$. There will be gaps such that frequency components of $x(t)$ that fall between the windows centered at the $q\phi_0$ points are practically ignored. This is just as bad as gaps in the time domain. (In many practical problems, it is actually worse.)

It is worthwhile noting that as long as we satisfy the fundamental lattice spacing limit, $\tau_0 \cdot \phi_0 < 1$, the Gaussian weight function still provides us with a frame no matter how much we skew the resolution rectangle. It is just that it may be an extraordinarily poor frame, utterly useless in practice. There may be regions of the time/frequency domain that for all practical purposes are 0, even though the weight function is not *exactly* 0 anywhere. There is still a tiny amount of information being captured. But the ratio of the frame bounds, $B/A$, can be so gigantic that the transform is worthless for reconstruction purposes due to overwhelming numerical instability. And even if our goal is not reconstruction, it is probably still worthless to us due to the missing information in one of the domains.

Let us further pursue the matter of the shape of the resolution rectangle. There are three interrelated parameters here, but only one degree of freedom. The quality of the frame is determined by the time-domain lattice spacing, $\tau_0$, the frequency-domain spacing, $\phi_0$, and the data window scale, $\sigma$. But since we can always rescale them all along the time dimension, it is only the relationship between these quantities that matters. For any given value of $\sigma$, we need the resulting $\Delta_t$ as large as possible relative to $\tau_0$ for thorough overlap. But we simultaneously need the resulting $\Delta_f$ as large as possible relative to $\phi_0$. If we fail in either domain, we will miss information that may be important. Intuition leads us to equalize the two ratios $\Delta_t / \tau_0$ and $\Delta_f / \phi_0$. By doing so, we will maximize their minimum value. Any other shape for the resolution rectangle would cause the ratio in one of the dimensions to become worse. It becomes further apparent that this is a reasonable goal when one visualizes the effect in the time/frequency plane. It means that the shape of the resolution rectangles is the same as the layout of the lattice points defining the centers of influence. Look at Figures 5-11 and 5-12. The upper figure shows the layout of the resolution rectangles in the time/frequency plane when $\sigma$ is chosen so that the rectangle shape is the same as the lattice spacing ratio. Coverage is uniform and sensible. The lower figure shows what happens when the shape of the resolution rectangles is a poor match to the lattice-point spacing. There is much unnecessary overlap in one dimension, and wide gaps in the other dimension. Not good.

It is not difficult to compute the value of $\sigma$ that generates resolution rectangles having this optimal shape. Let $K = \tau_0 / \phi_0$ be the lattice-point spacing ratio. If we set this equal to the resolution-rectangle ratio, using the values of $\Delta_t$ and $\Delta_f$ given by Equation (5-20), and solve for $\sigma$, we get the formula shown in Equation (5-22).

$$\sigma = \sqrt{\frac{K}{2\pi}} \qquad (5\text{-}22)$$

How important is it that we use this optimal shape? Perhaps we have a practical reason for wanting a different tradeoff. There are two aspects to the answer to this question. First, the amount of departure from optimal shape that can be tolerated depends very much on how close we are to critical spacing of the time/frequency lattice points. If the resolution product $\tau_0 \cdot \phi_0$ is close to its upper limit of 1, we are bound by tight constraints. If, on the other hand, the resolution product is small, giving us great redundancy, we have tremendous latitude in choosing the shape. The second aspect of the answer to the shape question is that it may not matter anyway. We sometimes have other recourse. These two aspects will now be discussed.

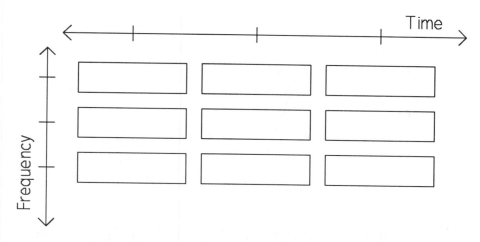

Fig. 5-11: Sensible shape for rectangles.

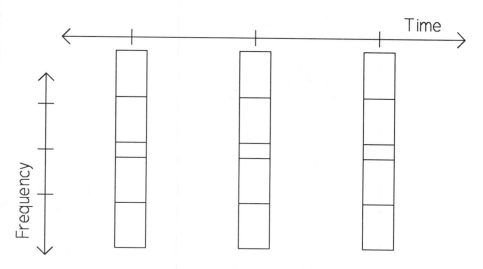

Fig. 5-12: Poor shape for rectangles.

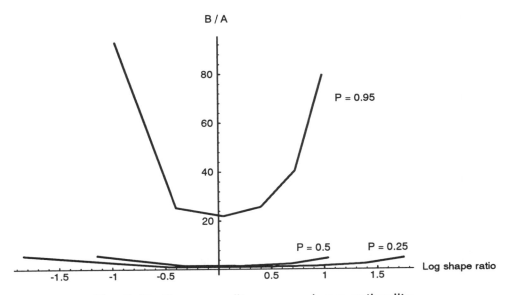

Fig. 5-13: Frame quality versus shape optimality.

Recall why the shape of the resolution rectangle affects the quality of the frame. The quality deteriorates when the width of the window in either the time ($\Delta_t$) or frequency ($\Delta_f$) dimension is narrow compared to the lattice point spacing ($\tau_0$ or $\phi_0$) in that dimension. When that happens, there are gaps in which information is ignored. But if we shrink the lattice point spacing, decreasing the resolution product $\tau_0 \cdot \phi_0$, those gaps will not become a problem until the rectangle becomes severely misproportioned. That effect can be seen in Figure 5-13.

That figure is based on data tabled in [Daubechies, 1990]. She computes the values of $A$ and $B$ for various values of $\tau_0$ and $\phi_0$, using a fixed window width. A good measure of the compatibility between the window shape and the lattice spacing ratio is $\log(\tau_0) - \log(\phi_0) - \log(\Delta_t) + \log(\Delta_f)$. This quantity will equal 0 when the shapes are identical. Using that shape similarity as the horizontal axis variable, the frame quality expressed as $B/A$ is plotted for three values of the resolution product, $\tau_0 \cdot \phi_0$. The uppermost graph is produced when the resolution product is near its limit of 1. Note how poor the frame is even when the shapes are identical, and how quickly it deteriorates as the shapes become mismatched. The two lower graphs demonstrate the liberty provided by having a small resolution product.

In many applications, we need to get the resolution in one of the dimensions down to a level smaller than that obtained by using the optimal $\sigma$ of Equation (5-22). If the resolution product is close to the critical limit of 1, our options are severely limited. But if it is small, we have a lot of room for trading resolution in the two dimensions.

There are two apparent problems with using tight lattice spacing. The first is that additional computation is involved. Each STFT point $S_x(p, q)$ on the lattice is found by computing the dot product of $x(t)$ with the corresponding frame element $g_{pq}(t)$. Even though we limit the extent of the dot product to the nonzero (for all practical purposes) portion of $g_{pq}(t)$, that can still be an expensive operation. If we must operate in real time, the density of the lattice cannot be any finer than absolutely necessary. The second problem is that we always want to minimize the number of inputs to our neural network. Luckily, that problem sometimes has a simple solution. The solution also ties in with the idea that it may not pay to worry too much about the window shape. Take a look at Figure 5-14.

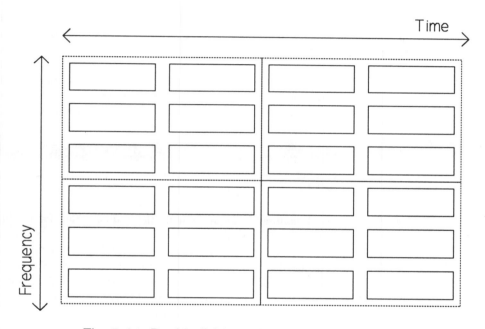

Fig. 5-14: Pool individuals for arbitrary windows.

That figure illustrates a common technique for simultaneously capturing all of the information in $x(t)$, achieving a desired window size and shape, and minimizing the number of inputs to the neural network. This solution is only available if we are interested in just the power in each frequency band, and we are able to ignore the phase information. Sometimes we have that luxury, and sometimes we don't. And most often, we won't know unless we try it. But if we can discard the phase, there is a simple solution.

What we do is choose $\tau_0$ and $\phi_0$ as small as possible subject to available computation time. With any luck at all, the resolution product $\tau_0 \cdot \phi_0$ can be kept very comfortably under 1. We then choose $\sigma$ using Equation (5-22) so as to get the best frame possible. Finally, group the discrete STFT points in such a way as to provide whatever final resolution rectangles we want. Each new point is simply the sum of the power (squared modulus) of its component points. If the resolution product is small enough, we should have significant leeway to adjust $\sigma$ to fine tune the shape of the grand rectangle if we want to do so.

Be warned that we *cannot* simply add the complex values of the components, obtaining a complex-valued sum. The result would be rubbish. If we want to supply our neural network with phase information, this technique cannot be used. It is only appropriate if we compute the squared modulus of each component, adding those. In many applications it is safe to work with power only. But sometimes the phase contains information that is too valuable to discard. When this is the case, we are forced to juggle $\tau_0$, $\phi_0$, and $\sigma$ to achieve a suitable compromise between computation time, number of input neurons, window shape, and frame quality. The procedure often employed by the author is the following:

1.  Choose $\tau_0$ and $\phi_0$ as appropriate for the physical process being studied. If in doubt about a good value for one of them, first choose the one that is more easily chosen. Then compute the other as part of the second step shown below.

2.  Verify that their product is well under 1. The author likes to make the product at most 0.5, and even 0.25 if possible. If necessary, reduce whichever one can be most easily reduced, or perhaps reduce both.

3.  Use Equation (5-22) to compute the optimal value for $\sigma$. In the unlikely event that the resulting window is disagreeable, adjust $\sigma$ as needed. But then compute $\Delta_t$ and $\Delta_f$ using Equation (5-20) and compare them to $\tau_0$ and $\phi_0$ respectively. It may be necessary to further reduce $\tau_0$ or $\phi_0$. That is the price paid for being picky about the window shape. *As a rough rule of thumb, you will be in trouble if the lattice spacing in a dimension exceeds about three times the $\Delta$ for that dimension.*

If, in your application, you find that better time resolution is needed, then you must decrease $\sigma$ so that a satisfactory value of $\Delta_t$ is obtained from Equation (5-20). Do not let it be under about one-third of $\tau_0$, and preferably greater. Decrease $\tau_0$ if necessary. And realize that decreasing $\sigma$ will increase $\Delta_f$. Verify that it is still acceptable.

If, on the other hand, you need better resolution in the frequency domain, then you must increase $\sigma$ to obtain from Equation (5-20) whatever $\Delta_f$ you need. Again, it must not be less than about one-third of $\phi_0$. Decrease that spacing if necessary. And check on $\Delta_t$, which just increased.

Finally, if you need better resolution in both the time and frequency domains, you're outta luck. You have just come head-to-head with the Heisenberg Uncertainty Principle, and it won. There is absolutely nothing that can be done about it. Shrinking $\tau_0$ and $\phi_0$ insures that you collect sufficient information about the signal. By making either of these quantities arbitrarily small, you will be able to choose $\sigma$ so as to achieve

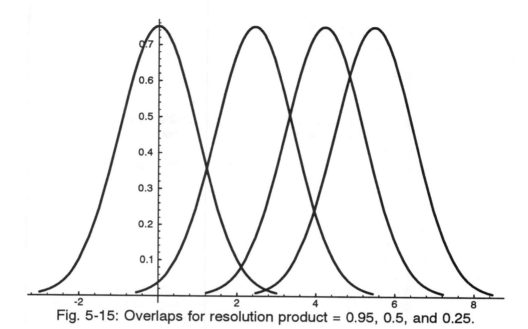

Fig. 5-15: Overlaps for resolution product = 0.95, 0.5, and 0.25.

arbitrarily fine resolution in that dimension. But the resolution that is *simultaneously* available is firmly fixed, even if you make both $\tau_0$ and $\phi_0$ arbitrarily small.

It is instructive to see typical overlaps graphically portrayed. Figure 5-15 shows the overlaps corresponding to optimal resolution rectangles for the same three resolution products used in Figure 5-13.

There are four windows shown in Figure 5-15. Since we are assuming optimal rectangle shape ($\sigma$ computed using Equation (5-22)), these window relationships apply to both the time and frequency domains (except, of course, for constant scaling that does not affect relative proportions). The leftmost pair of windows corresponds to the resolution product $\tau_0 \cdot \phi_0$ being 0.95, which is very close to the critical point. The center pair of windows is for the product being 0.5, and the rightmost pair has a resolution product of 0.25, a typical conservative value. It must be emphasized that, except for constant scaling, this illustration of relative window overlap is independent of the actual values for any of the parameters. It depends only on the resolution product.

## Code for the Gabor Transform

This section presents complete source code for the Gabor transform. There are three routines. The constructor allocates memory for and computes the vectors comprising the family of frame functions, $g_{pq}(t)$. The destructor frees that memory. The transformation routine performs the actual Gabor transform and may be applied to as many data sets as desired. Let us start with several static constants and the class header.

```
static double two_pi = 2. * 3.141592653589793 ;
static double root2 = sqrt ( 2.0 ) ;

class Gabor {

public:
   Gabor ( int tau0 , double phi0 , double *sigma , double *delta_t ,
        double *delta_f , int *border , int *nfreqs , int *ok ) ;
   ~Gabor () ;
   void transform ( int nx , double *x , int *nt , double *rt , double *it ) ;

private:
   int t0 ;                        // Tau0
   double f0 ;                     // Phi0
   double sig ;                    // Sigma
   double dt ;                     // Delta_t
   double df ;                     // Delta_f
   int hl ;                        // Filter half length (=border)
   int nf ;                        // Number of frequencies resolved (=nfreqs)
   double gconst ;                 // Normalizing constant for Gaussian
   double *coefs ;                 // FIR coefficients for all frequencies
} ;
```

The constructor computes and returns sigma ($\sigma$), delta_t ($\Delta_t$), and delta_f ($\Delta_f$) as discussed in the following. It also allocates memory for holding the FIR filter coefficients that define the frame functions. The sample interval defines the time unit throughout this routine. The following parameters are passed to the constructor:

tau0    This is the number of samples separating each lattice point in the time dimension.

phi0    This is the frequency increment separating the lattice points in the frequency dimension. Its theoretical upper limit to have a frame is 1/tau0, but it really should be half that or less for good coverage.

sigma, delta_t, delta_f These are the scale factor and window radii. If all three are input negative, sigma is computed as its optimal value (which produces a resolution window whose shape matches the lattice spacing). If sigma is input positive (regardless of the deltas), that value is used. In either case, the deltas are computed according to sigma. If sigma is input negative and delta_t is input positive, that value of delta_t is used, and sigma and delta_f are returned accordingly. If sigma and delta_t are input negative, and delta_f is input positive, that value will be used to compute sigma and delta_t.

border  This is returned as the number of points that will be skipped at the beginning and end of the input series. The g(t) vector that is dotted with the input series has length = 2 * border + 1. So that the entire vector is used, the transform will start at x[border] and will go no further than x[nx-border-1], centering the filter at every tau0'th point.

nfreqs  This is returned as the number of frequency bands that will be resolved. It is 1 + 0.5 / phi0, truncated down to an integer. This takes us as high as we can go without passing the Nyquist limit.

ok  This is returned 1 if all went well, and 0 if there was insufficient memory for storing the vectors.

The constructor will need to allocate a work area to hold the filter coefficients. This will be 2 * nfreqs * border doubles. (The filter is complex, symmetric, and the center coefficient is the same for all frequencies, so it is not stored here.)

```
Gabor::Gabor (
   int tau0 ,                    // Lattice spacing in time dimension
   double phi0 ,                 // Lattice spacing in frequency dimension
   double *sigma ,               // Data window scale factor
   double *delta_t ,             // Radius of time-domain window
   double *delta_f ,             // Radius of frequency-domain window
   int *border ,                 // Filter half-length (skip this at ends)
   int *nfreqs ,                 // Number of frequencies resolved
   int *ok                       // Memory allocation ok?
   )
{
   int ifreq, ipos ;
   double freq, weight, fac, *cptr ;

   t0 = tau0 ;                   // Keep a private copy of these parameters
   f0 = phi0 ;

/*
   Compute (as needed) sigma, delta_t, and delta_f
*/

   if (*sigma <= 0.0) {
     if (*delta_t <= 0.0) {
```

```
      if (*delta_f <= 0.0) {
        *sigma = sig = sqrt ( ( (double) tau0 / (phi0 * two_pi) ) ) ;
        *delta_t = dt = sig / root2 ;
        *delta_f = df = 1.0 / (two_pi * root2 * sig) ;
        }
      else {
        df = *delta_f ;
        *sigma = sig = 1.0 / (two_pi * root2 * df) ;
        *delta_t = dt = sig / root2 ;
        }
      }
    else {
      dt = *delta_t ;
      *sigma = sig = root2 * dt ;
      *delta_f = df = 1.0 / (two_pi * root2 * sig) ;
      }
    }
  else {
    sig = *sigma ;
    *delta_t = dt = sig / root2 ;
    *delta_f = df = 1.0 / (two_pi * root2 * sig) ;
    }

/*
    Compute the filter half-length such that the weight at the end of the filter goes to about
    1.e-12 times its max.  Compute the number of resolved frequencies such that we go as high
    as possible, without exceeding the Nyquist frequency of 0.5.
*/

  *border = hl = 1 + 7.4 * sig ;  // Goes to about 1.e-12
  *nfreqs = nf = 1 + 0.5 / f0 ;

  coefs = (double *) malloc ( 2 * nf * hl * sizeof(double) ) ;
  if (coefs == NULL) {
    *ok = 0 ;
    return ;
    }
  *ok = 1 ;

/*
    Compute the filter coefficients.  The center coefficient for all real parts is gconst, the
    gaussian multiplier, regardless of frequency.  The center of the imaginary part is always 0
    (sin 0).
*/
```

```
gconst = exp ( 0.25 * log ( 3.141592653589793 )) ;// Pi to the 1/4 power
gconst = 1.0 / (gconst * sqrt ( sig )) ;              // Actual normalizer for Gaussian

fac = -0.5 / (sig * sig) ;                           // Common factor in exponent

for (ifreq=0 ; ifreq<nf ; ifreq++) {      // Covers 0 to Nyquist
   freq = ifreq * f0 * two_pi ;                     // Freq term in complex exp
   cptr = coefs + ifreq * 2 * hl ;                  // Point to this freq's coef area
   for (ipos=1 ; ipos<=hl ; ipos++) {   // Right half of filter
      weight = gconst * exp ( fac * ipos * ipos ) ;// Data window
      *cptr++ = cos ( ipos * freq ) * weight ;      // Real part of filter
      *cptr++ = sin ( ipos * freq ) * weight ;      // And imaginary part
      }
   }
}
```

The destructor is trivial. All that it does is delete the memory used to store the filter coefficients.

```
Gabor::~Gabor ()
{
   if (coefs != NULL)
      free ( coefs ) ;
}
```

The last routine is the one that does the actual Gabor transform. Its operation is quite simple. For each lattice point, it starts by initializing the sum to the center coefficient times the current input center point. Since the cosine of 0 is 1 and the sine of 0 is 0, the center coefficient of the real part is the normalizing constant in Equation (5-18), and the center coefficient of the imaginary part is 0. The constructor did not store those coefficients, which are the same for all frequencies, in the coefficient matrix. Next it cumulates the right half of the filter. Finally, it cumulates the left half. Because the cosine function is even, its coefficients are exactly symmetric about 0. The sine is odd, so its coefficients are symmetric except for a sign reversal. The following parameters are in the transform parameter list:

nx      This is the length of the time series input vector x.

x       Input vector

nt      This is returned as the number of time slots that will be resolved. The
        transformable extent of x runs from x[border] through x[nx-border-1]. The
        transformation filter will be applied first to x[border], then to x[border+tau0],

continuing through x[border+(nt-1)*tau0]. Thus, nt = 1 + (nx-2*border-1) / tau0.

rt        Real part of output transform.  It contains ntimes * nfreqs elements. The first nfreqs elements correspond to x[border] for frequencies from 0 through (nfreqs-1)*phi0. The next nfreqs are the same frequencies for x[border+tau0]. The last nfreqs are the same frequencies for x[border+(nt-1)*tau0].

it        Imaginary part of transform as above.

```
void Gabor::transform (
  int nx ,                              // Length of input vector
  double *x ,                           // Input vector
  int *nt ,                             // Number of time points resolved
  double *rt ,                          // Real output
  double *it                            // Imaginary output
  )
{
  int ifreq, itime, ipos ;
  double *cptr, *xptr, rsum, isum ;

  *nt = 1 + (nx - 2 * hl - 1) / t0 ;                    // This many times resolved

  for (itime=0 ; itime<*nt ; itime++) {   // All time slots
    xptr = x + hl + itime * t0 ;                        // Center of this filter position

    for (ifreq=0 ; ifreq<nf ; ifreq++) {               // Covers 0 to Nyquist
      rsum = gconst * *xptr ;                           // Center real coef is gconst
      isum = 0.0 ;                                      // Center imaginary coef is 0

      cptr = coefs + ifreq * 2 * hl ;                  // Point to this freq's coef area
      for (ipos=1 ; ipos<=hl ; ipos++) {               // Right half of filter
        rsum += *cptr++ * xptr[ipos] ; // Sum real part of right half
        isum += *cptr++ * xptr[ipos] ; // And imaginary part
        }

      cptr = coefs + ifreq * 2 * hl ;                  // Point to this freq's coef area
      for (ipos=1 ; ipos<=hl ; ipos++) {               // Left half of filter
        rsum += *cptr++ * xptr[-ipos] ;                // Sum real part of left half
        isum -= *cptr++ * xptr[-ipos] ; // And imaginary part
        }
```

```
        *rt++ = rsum ;              // Output real part of Gabor transform
        *it++ = isum ;              // And imaginary part
      }
    }
}
```

# Fixed versus Variable Resolution

The entire preceding discussion of the short-time Fourier transform and its particular form, the Gabor transform, has focused on the use of identical resolution windows throughout the time/frequency plane. Look back at Figure 5-11 on page 157. Every rectangle is exactly the same shape. In many cases this is perfectly reasonable, perhaps even desirable. But in a large number of practical problems, a very different shape scheme is better. This section will present an intuitive justification for a popular alternative to constant-shape resolution rectangles and will simultaneously lay a foundation for the wavelet transform that will appear in the next section.

Suppose that we need to locate (in the time dimension) a well-defined event. Perhaps it is a single pulse, or it may be the onset of a sine wave. Look ahead at Figures 5-19 through 5-22 starting on page 171. Those figures show the real part of four different members of the Gabor family of frame functions, ranging from a low frequency to a high frequency. Which of those would most accurately locate (in time) the event of interest? Certainly, the lowest-frequency member shown in Figure 5-19 would be the best, assuming that the scale of that member is consistent with the scale of the event. Higher-frequency members, like that shown in Figure 5-22, would have more than one position in the time dimension that produce a high response. The exact peak would be less evident. Of course, use of a member shaped like that in Figure 5-22 would give great specificity in the frequency domain. The price of the time accuracy attained with Figure 5-19 is very poor frequency resolution, so there is a definite tradeoff involved. But it is often a trade worth making.

The above example is one specific manifestation of a general principle. *In a large number of practical problems, we would like to be able to locate higher-frequency events more accurately in time than we need for lower-frequency events.* In other words, the resolution rectangles should not be laid out as shown in Figure 5-11 on page 157. Rather, they should look more like Figure 5-16.

In that figure, the width of each resolution rectangle in the frequency dimension is proportional to its center frequency, while its width in the time dimension is inversely proportional to its center frequency. One implication of this policy is that we use closer lattice-point spacing in the time domain for points corresponding to higher frequencies. This insures that there are no information gaps, as would be the case if the lattice points were laid out in a uniform grid.

Frequency

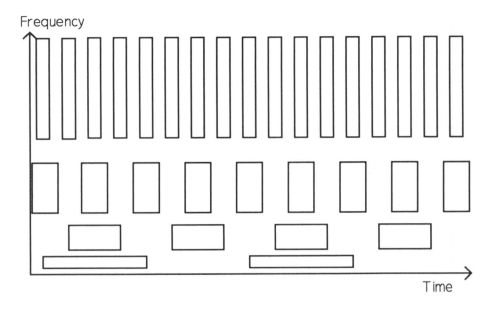

Time

Fig. 5-16: Variable-shape resolution rectangles.

The concept of wanting finer time resolution for higher frequencies may initially seem curious, but it is actually a common phenomenon throughout life. For example, suppose that we are interested in a coastline. Viewed from a satellite, the gross outline would be represented by very low frequencies. We would probably be satisfied if we could locate to within a few miles the major curves that will be used to generate a map in an atlas. On the other hand, what if we were building a lighthouse to mark the presence of a series of huge boulders stretching outward into the sea? This protrusion from the coast would be represented by much higher-frequency information than the general coastal outline. We would want to locate these boulders to within a few yards. Finally, suppose that we are strolling along the coast. Our every twist and turn is governed by still higher-frequency components in the coastline path, and we need to place our feet quite accurately, lest we fall into the sea. The higher the frequency of coastline variation being considered, the finer resolution we desire.

The concept underlying this principle is that of *scale*. When we take in a wide view of a phenomenon, we are usually looking at low-frequency components and examining features that cover a lot of time or space. But when we peer at a tiny portion of that same phenomenon through a magnifying glass, the scale is smaller. Details are governed by higher frequencies, and the distances that concern us are smaller. It helps us if our tools follow the same pattern of scaling. The Gabor transform does not do so. Examination of Figures 5-17 and 5-18 will help us to understand this concept as it relates to the Gabor transform and to the alternative that will soon be proposed.

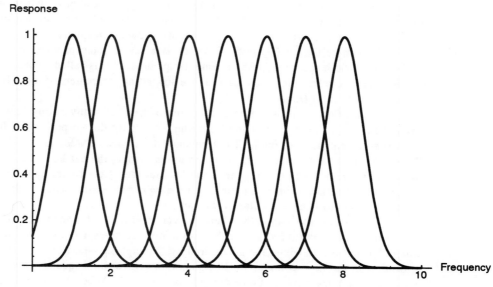

Fig. 5-17: Constant bandwidth of Gabor transform.

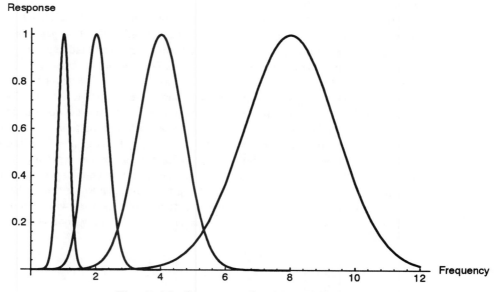

Fig. 5-18: Constant Q of Morlet wavelet.

Figure 5-17 shows a typical collection of adjacent Gabor transform windows in the frequency domain. Notice that they are equally spaced, and they are all the same width. Electronics engineers call this constant-bandwidth response.

Compare this to Figure 5-18, which is a set of adjacent windows obtained by scaling their width according to their center frequency. This is called constant-Q response in engineering circles. Note that the distance separating the centers of these filters is also proportional to their frequency.

The effect seen in Figure 5-18 is reminiscent of many natural and man-made phenomena. One example is human hearing. Not only do we perceive loudness on a logarithmic scale, but also pitch. The frequency difference between middle C and middle D is about 31 cycles per second. If we go up an octave, the ear hears the same relative pitch difference between that higher C and D. But their frequencies and their difference are double what they were at middle C. If we were processing acoustic data, we might want our filters to behave similar to the ear.

It must be understood that there is a very real tradeoff made when one adopts the procedure of changing the shape of the resolution rectangles according to the frequency. Our old nemesis, the Heisenberg Uncertainty Principle, guarantees that when we gain resolution in the time domain, we lose just as much in the frequency domain. On the other hand, this tradeoff is often exactly what we need. When we gain time resolution at higher frequencies, we are willing to give up frequency resolution in accordance with the acoustical principles just discussed. Similarly, when we demand sharp frequency resolution at low frequencies, some looseness in the time domain is acceptable.

So how do we go about modifying the Gabor transform to produce windows whose width in the frequency domain is proportional to their center frequency? Recall that the foundation for everything we have done and much of what we will do is the short-time Fourier transform (STFT) defined in Equation (5-2) on page 142. In order to produce the Gabor transform, that general formulation was made specific by letting the window function $g(t)$ be the normalized Gaussian function defined in Equation (5-18). Since we already know that using a Gaussian window blesses us with an optimal resolution product (right to the Heisenberg limit), we will stick with it. We will simply vary its scale factor $\sigma$ according to the frequency. The Gabor transform keeps $\sigma$ fixed. We then converted the impractical continuous Gabor transform to the more useful discrete form by discretizing the parameters of its frame functions $g_{pq}(t)$ according to Equation (5-4). We will need to do a similar thing, with our aim being to produce lattice spacing like that in Figure 5-16 on page 167, rather than the equal spacing produced with Equation (5-4). The exact formula for doing this will have to wait until the next section, but one interesting property of the resulting family of functions will be loosely explored on an intuitive level. Rigor lurks a few pages ahead. For now, look at the graphs of four functions.

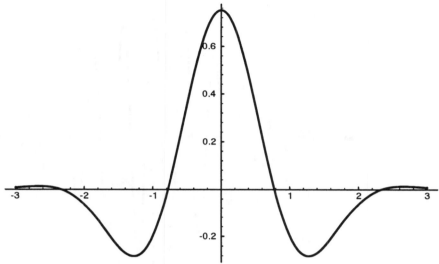

Fig. 5-19: Width = 0.7 times frequency.

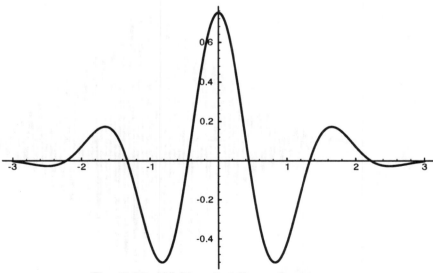

Fig. 5-20: Width = 0.4 times frequency.

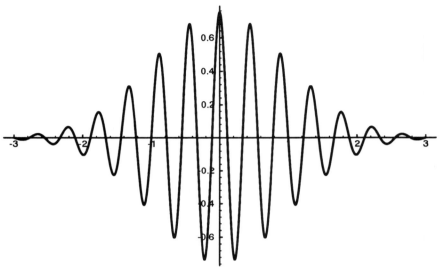

Fig. 5-21: Width = 0.1 times frequency.

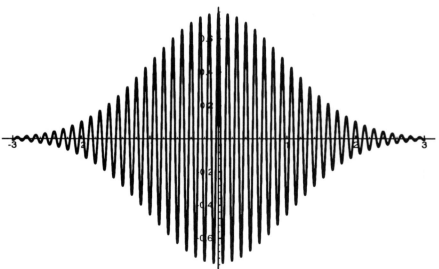

Fig. 5-22: Width = 0.03 times frequency.

Suppose that we are in possession of some member of our desired family of variable-width frame functions. Perhaps its real part resembles one of the functions shown in Figures 5-19 through 5-22. Suppose now that we want to produce another member of this family of frame functions. In particular, let the center frequency of the function we have be $f$, and let the center frequency of our desired new function be $cf$, where $c$ is some constant multiplier. By design, we want the frequency-domain window width of the new function to be $c$ times that of the old function. Look back at Equation (5-20) on page 155 to see what we must do to $\sigma$ to effect this change in window width. We must divide $\sigma$ by $c$. This will cause the time-domain window width to be divided by $c$, no great surprise if we remember that the resolution product remains constant at the Heisenberg limit. For both the old and the new functions, when we venture away from zero (the center) by a distance equal to the time-domain radius, the weight function will decrease to a fixed fraction of its maximum height. (Look at Equations (5-18) and (5-20) if that is not clear.) During that time, the trigonometric component will complete some number (perhaps fractional) of periods. The new function will complete $c$ times as many periods in any time distance as the old function, but the time-domain window radius of the new function is $1/c$ times that of the old function. We move $1/c$ times as far in the new function to get out to the time-domain window radius, and we oscillate $c$ times faster during that interval. The net result is that for both the old and the new frame functions, the trigonometric component completes exactly the same number of periods getting out to the time-domain radius. In other words, *the old and the new frame functions are identical in appearance except for linear scaling.* It is exactly as if someone took the old function and spread it out by tugging at its ends (for lower frequencies) or pressed it in like an accordion (for higher frequencies). The real parts of three such family members are shown in Figure 5-23.

Remarkable. We wanted a family of frame functions that would allow us to explore detail at different scales of magnification, and here we have a family that is nothing but scaled versions of one "mother" shape! The wonders never cease. This valuable family of variable-width frame functions has a name. They are called *Morlet wavelets.* This brings us to the next topic.

## Wavelets

This text will not attempt to provide an exact definition of what a wavelet is, for to do so would take us to mathematical depths best left to texts devoted to that subject. Interested readers can find brutal rigor in [Chui, 1992]. On the other hand, the reader of this text will be introduced to the most important aspects of wavelets and guided into some of their most useful neural network applications. The preceding section provided an intuitive foundation for wavelets by deriving the Morlet wavelet from the Gabor transform. We will now make that derivation more mathematically precise by explicitly presenting a general formulation for wavelets, then showing how the Morlet wavelet fits into the mold.

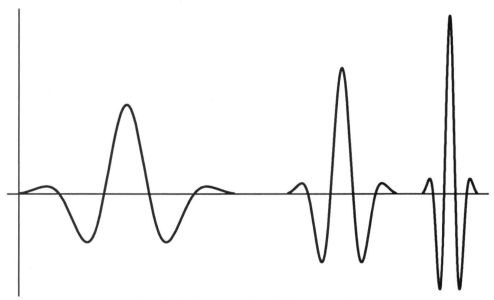

Fig. 5-23: Three Morlet wavelets.

If one needed to characterize wavelets in one word, a good candidate for that single word would be *scale*. Everything in the world of wavelets is based on scale. When we take in the big picture of something, we are observing it at a large scale. When we examine its local features, we are at a finer scale. Except for magnification, our tools and methods of examination are identical, no matter what the scale. The short-time Fourier transform and its special form, the Gabor transform, made use of a frequency dimension. Although we may sometimes do the same for wavelets, we will in a great many cases find it more useful to consider a scale dimension as an alternative to frequency. Throughout the remainder of this chapter, we will use whichever of these two options is more convenient at the time.

Back on page 143 we used Equations (5-5) and (5-6) to characterize the discrete STFT of a signal $x(t)$ by means of a scalar product of that signal with a family of functions, $g_{pq}(t)$. The integer parameters $p$ and $q$ determined discrete lattice points within a continuous domain having a time dimension ($\tau$) and a frequency dimension ($\phi$). We will do nearly the same thing for wavelets, with only two differences. First, in place of frequency, $\phi$, we will use scale, $\xi$. Second, to help the reader keep the notation straight, we will use $h$ rather than $g$ for the frame functions. We shall start with the continuous-parameter version, analogous to Equation (5-2) on page 142. For clarity, we will use superscripts for the continuous-parameter version, and subscripts for the discrete-parameter version to be defined later. The *continuous wavelet transform* is expressed in Equation (5-23).

$$CWT_x(\tau,\xi) = \int_{-\infty}^{\infty} x(t)\, h^{\tau,\xi}(t)\, dt \qquad (5\text{-}23)$$

Each function $h^{\tau,\xi}$ is understood to be primarily sensitive to events that occur around time $\tau$ and at a scale of about $\xi$, where those time and scale parameters can take on any continuous value. Soon we will show how to convert that impractical continuous version into the more practical discrete version, $h_{pq}$, like we did for the STFT in Equation (5-3).

In the conclusion of the intuitive discussion of the Morlet wavelet (on page 173), we saw that the shape of every member of that family is identical to every other member except for a scale change. They can all be derived from one mother function by nothing more than pushing and pulling on the ends to squash or to expand the mother function. And just like the STFT, we generate different members along the time dimension by simply shifting in time.

We can make this mathematically explicit. We have one canonical function, the mother function, from which all members of its wavelet family are derived. Let us call this $h(t)$. Each family member $h^{\tau,\xi}$ will be obtained from $h$ by shifting in time by $\tau$ and scaling by $\xi$, as shown in Equation (5-24).

$$h^{\tau,\xi}(t) = \frac{1}{\sqrt{\xi}} h\left(\frac{t-\tau}{\xi}\right) \tag{5-24}$$

The factor of the square root of $\xi$ appears in that equation so that the energy of the wavelet remains constant. It is traditional to scale wavelets so that they are normalized according to Equation (5-25). Note that normalizing the continuous wavelet to unit energy as shown is the continuous equivalent of normalizing a vector to unit length. And even if the mother wavelet is not so normalized, Equation (5-24) will preserve the energy integral at a constant value across all members of the family.

$$\int_{-\infty}^{\infty} |h(t)|^2 dt = 1 \tag{5-25}$$

While on the subject of normalization, there is one more thing to worry about that was not a serious issue for the STFT. In the case of wavelets, we need the mother (and all family members) to be centered about zero. This is expressed in Equation (5-26).

$$\int_{-\infty}^{\infty} h(t) dt = 0 \tag{5-26}$$

The reason for this requirement is that we need small-scale wavelets to be insensitive to large-scale events and constant offsets. When a wavelet is itself offset from zero, computing its scalar product with a series that is also offset from zero will introduce a bias. We did not worry about this with the STFT (and Gabor transform in particular) because the higher-frequency family members, where offsets would be a problem, are for all practical purposes automatically centered at zero. But it can be a problem for wavelets if we are not careful. We will return to this issue later when we dig more deeply into Morlet wavelets.

## Discrete Wavelet Parameters

Given a suitable mother wavelet, we now know how to find wavelet family members at any time offset and scaling. But, as was the case for the STFT, having an infinitude of continuous-parameter wavelets available does us little good in practice. Our neural network has only a finite (and probably small) number of input neurons. We need to find a subset of that continuous domain, a discrete lattice of points corresponding to particular values of $\tau$ and $\xi$. We would like that lattice to be as sparse as possible, to reduce the amount of data that must be handled. At the same time, we often want it to be complete in the sense of providing us with a frame so that all of the information in the time series is captured. Conditions for having a frame will be discussed later. Right now we will intuitively deduce the most common way of laying out the time/scale lattice points.

Look back at Figure 5-16 and see what can be ascertained. First, note that when the frequency increases by a factor of $c$ (or, equivalently, the scale decreases by a factor of $c$), the size of the resolution rectangle in the frequency dimension also increases by that same factor. This tells us that the lattice points should be equally spaced on a logarithmic scale in the frequency (inverse scale) dimension. Next, for any fixed frequency, the shape of the resolution rectangles remains constant across time. Therefore, we are led to space the lattice points equally in the time dimension, with the spacing determined by the rectangle's width in that dimension.

As we did for the STFT, we will let $p$ be an integer that indexes lattice points in the time dimension, and we let $q$ do likewise in the scale dimension. We have a basic time spacing $\tau_0$, and a basic scale step size $\xi_0 > 1$. The equal-logarithmic spacing in the scale dimension is shown in Equation (5-27).

$$\xi = \xi_0^q \, , \qquad\qquad q \in \mathbf{Z} \qquad\qquad (5\text{-}27)$$

For any fixed scale, we space the lattice points in the time dimension equally, separated by a distance proportional to the scale. If we use the scale just defined in Equation (5-27), the time-dimension spacing is governed by Equation (5-28).

$$\tau = p\tau_0\xi_0^q \, , \qquad\qquad p \in \mathbf{Z} \qquad\qquad (5\text{-}28)$$

Finally, we join those two lattice-spacing equations with the fundamental wavelet relationship, Equation (5-24). This gives us what we have been working toward, the formula for any wavelet function corresponding to a point in the time/scale lattice. This is shown in Equation (5-29). That equation is analogous to Equation (5-4) on page 143 for the STFT.

$$h_{pq}(t) = \frac{1}{\xi_0^{q/2}} h\left(\frac{t}{\xi_0^q} - p\tau_0\right) \qquad (5\text{-}29)$$

On first glance at that equation, it may appear that the time-domain lattice spacing is the same at all scales, which we know is not true. But observe that the raw time variable is scaled. Ponder this equation carefully. Make sketches of it for different scales and time offsets if needed. That will help a lot. The entire essence of wavelets is embodied in this one equation. Do not go on until it is clear.

In nearly all practical problems, it is most convenient to choose $\xi_0 = 2$. This is physically meaningful, as then the frequencies corresponding to different discrete scale parameters are in octave relationships. That is most relevant to acoustical phenomena, but it is often meaningful in other situations as well. Probably more important in practice is that the resulting translation steps, $\tau_0 \cdot 2^q$, always "line up" with the discretely sampled data. In other words, the smallest-scale wavelet might be evaluated every sample point. The next in the scale hierarchy would be evaluated every second sample point. The next would be placed at every fourth point, and so on. Values of $\xi_0$ other than 2 could make our life very difficult with regard to sampling!

## Wavelet Frames

*This section contains relatively advanced material that can be safely skipped if desired. Before tackling this section, the reader should be thoroughly familiar with the material starting on page 143.*

We have already seen the most important conditions that are necessary for a family of STFT functions to constitute a frame. Things are a little different for the discrete wavelet transform, as the time/scale lattice is not equally spaced. But the main principles are the same. The magnitude of the mother wavelet function must drop to zero very quickly as the distance from the center increases beyond a moderate distance. The shape of the function and the lattice-point layout must be such that there are no holes in the time/scale plane. And last but certainly not least, the lattice points must be sufficiently close to satisfy basic information-theoretic needs. In practice, it is this last condition that is most subject to experimental variability and most easily violated.

It turns out that the notion of closeness is far more subtle in the wavelet case than it was for the STFT. There is no clear-cut necessary condition analogous to requiring $\tau_0 \cdot \phi_0 < 1$ in the STFT. That is because of the infinite number of scaling possibilities. In particular, suppose that the wavelet family $h_{pq}$ for some $h(t)$, $\tau_0$, and $\xi_0$ constitutes a frame. Then it can be shown that the wavelet family based on the mother wavelet and basic parameters shown in Equation (5-30) is also a frame. In fact, [Daubechies, 1990] shows how to construct a frame for *any* $\tau_0$, and $\xi_0$.

$$h_c(t) = \sqrt{c}\, h(ct)$$
$$\xi_{c;0} = \xi_0 \qquad\qquad (5\text{-}30)$$
$$\tau_{c;0} = c^{-1}\tau_0$$

Let us now ponder the implications of Equation (5-30). First, we see that there is a trivial tradeoff between the time-domain lattice spacing and the scaling of the mother wavelet, with the frequency-domain spacing not involved in that tradeoff. So we must obviously consider $\tau_0$ in conjunction with a fixed $h(t)$. It is apparent that for a fixed $h(t)$, decreasing $\tau_0$ will capture more information and hence improve our situation as regards frame quality. Second, we see that the implications of $\xi_0$ as regards frame quality depend only on the shape of $h(t)$ (regardless of scaling) as long as $\tau_0$ has not been set. Since we have already agreed that in practice we should use $\xi_0 = 2$, we may have trouble sometimes. In the STFT case, we can to good degree make up for large spacing in one dimension by shrinking the spacing in the other dimension, since it is the resolution *product* that is critical. For wavelets, things are not so straightforward. Of course, shrinking $\tau_0$ will always improve the frame quality. But for some mother wavelet functions, that is just not enough. The time-domain spacing in many cases must be made impracticably small to get even mediocre frames when $\xi_0 = 2$. A solution to that dilemma will be discussed in the next section. The bottom line is that knowing if we have a frame, and knowing the quality of that frame, is actually quite difficult in general. The best approach is to stick with mother wavelets whose frame quality is known. [Daubechies, 1990] provides very useful tables that show, for many common mother wavelets, the frame quality associated with various values of $\tau_0$. All wavelets that are discussed in this text will also come with guidance on choice of suitable lattice spacing.

## Wavelet Voices

It has been pointed out that for some mother wavelets, $\xi_0 = 2$ is too large to generate good frames. The time-domain spacing, $\tau_0$, would have to be far smaller than is practical in order to produce even a moderately good frame. If we were working with a function $x(t)$ that is continuous and can be sampled cheaply whenever desired, the obvious solution would be to use a scale factor smaller than 2. But we are rarely so fortunate. Economics, physics, or both, nearly always dictate that we sample at equally spaced intervals. This makes fractional scale factors impractical. What to do?

The answer lies in using more than one mother wavelet. These are typically dilated versions of one mother wavelet, though in principle they can be different shapes. The idea is to have each one's center frequency at a different position, to minimize the gaps in the frequency domain. The dilation is nearly always geometric, thus preserving the logarithmic spacing that is traditional for wavelet lattices.

The general form of the voices $h^0$, ..., $h^{N-1}$ is shown in Equation (5-31).

$$h^j(t) = 2^{-j/N} h(2^{-j/N} t), \qquad j = 0, ..., N-1 \qquad (5\text{-}31)$$

When we use this expanded family, $\{h^i_{pq}\}$, what we get is N lattices that line up in the time dimension but are shifted in the frequency dimension. Note that this is different from what we would have if we were to use one family based on a scaling of $\xi_0 = 2^{1/N}$. Because our N lattices all line up in time, we are able to use time-domain lattice points that lie on the same sample points. At the same time, we have closed in the gaps in the frequency dimension. Elegant.

Astute readers will notice that even if the mother wavelet $h(t)$ is normalized to unit energy as shown in Equation (5-25), the voices other than $h^0$ will not be. This is correct and compensates for the unconventional layout of the lattice points. In particular, if the mother wavelet is normalized to unit energy, the energy (squared length for vectors) of $h^i$ is equal to $2^{-j/N}$. The physical meaning of this can be seen by examining Figure 5-24, which shows typical lattice-point spacing. As usual, $\xi_0 = 2$, and there are N = 4 voices. Each voice is shown with a different type of filled circle. The top row is the basic wavelet ($j = 0$) at a relatively high frequency (small scale). The next row down is that same scale wavelet, but voice $j = 1$. The next two rows are for the other two voices. The fifth row down is voice $j = 0$ for the next larger scale, twice the previous scale. Notice that the time-domain spacing is now doubled. Compare that to the fourth row down. The scales of these two rows are very close, differing by a factor of the fourth root of two. But the time spacing of the fourth row is twice as dense as that of the fifth row. If the energy components of the signal are to be equitably represented by these two wavelets, the contributions of the largest voice of the lower scale (row 4) must be reduced to compensate for the fact that there are twice as many of them.

## The Mexican Hat Wavelet

The Morlet wavelet is an immensely useful complex-valued wavelet and is the featured wavelet of this chapter. However, so as to avoid misleading the reader, it should be pointed out that many or most wavelets in use today are real-valued functions. Therefore, in due homage to that fact, a short section will be devoted to what is probably the most commonly used real wavelet. It is (except for sign reversal) the second derivative of the Gaussian function. But a quick look at its shape, shown in Figure 5-25, shows why it is generally called the *Mexican hat wavelet*. The positive part of its Fourier transform is shown in Figure 5-26. The equation of the Mexican hat wavelet and its Fourier transform are shown in Equations (5-32) and (5-33) respectively. These equations satisfy the normalization criteria of Equations (5-25) and (5-26).

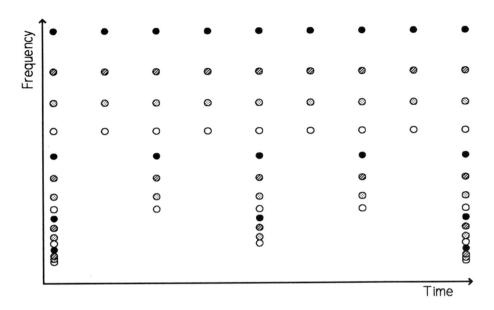

Fig. 5-24: Lattice points with four voices.

$$h(t) = \frac{2}{\sqrt{3}} \pi^{-\frac{1}{4}} (1 - t^2) e^{-\frac{t^2}{2}} \tag{5-32}$$

$$H(f) = \frac{2}{\sqrt{3}} \pi^{-\frac{1}{4}} f^2 e^{-\frac{f^2}{2}} \tag{5-33}$$

The Mexican hat wavelet contains very little in the way of cyclic variation in its effective time window. This leads us to believe (correctly) that its window width in the frequency dimension is relatively wide. That in turn implies that when we use the traditional $\xi_0 = 2$, multiple voices are not vital to effective coverage of the information in that dimension. As long as the time-domain spacing allows significant overlap, this wavelet will provide a decent frame. In particular, we should have $\tau_0$ less than 1.25 or so. Values of 1.0 or less are even better. Even with several voices, frame quality deteriorates rapidly as $\tau_0$ moves past 1.75. This is because serious gaps appear in the time dimension that all the voices in the universe cannot repair.

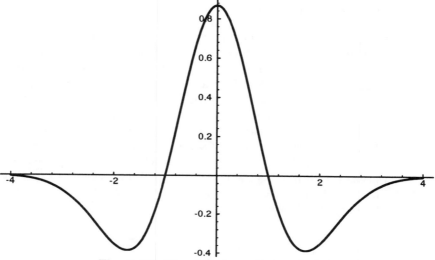

Fig. 5-25: The Mexican hat wavelet.

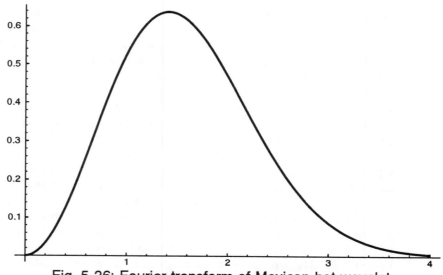

Fig. 5-26: Fourier transform of Mexican hat wavelet.

## The Morlet Wavelet

In the intuitive introduction to wavelets (starting on page 167), the Morlet wavelet was introduced as a member of the Gabor transform family. Actually, that definition contains a tiny little white lie. It glosses over one small but important detail. We have already emphasized that it is vital for wavelets to have a mean of zero, as stated in Equation (5-26). If the wavelet does not integrate to zero, then constant offsets and low-frequency signal components will leak into bins reserved for high-frequency components. The low-frequency Gabor functions, typically used as wavelet mother functions, are almost, but not quite, centered at zero. This offset, though small, cannot be tolerated. Therefore, we must very slightly modify the Gabor function to exactly center it. The true Morlet wavelet is shown in Equation (5-34). Note that the complex exponent is positive, while Morlet used a negative exponent. This is to agree with more modern usage and is of no practical consequence.

$$h(t) = \pi^{-\frac{1}{4}} \left( e^{2\pi i k t} - e^{-k^2/2} \right) e^{-\frac{t^2}{2}} \qquad (5\text{-}34)$$

Observe that this is practically the same as a Gabor function. The only difference is the subtraction of the exponential term involving $k$. This is the magic term that centers the function. Also observe that except for small values of $k$, that term is nearly insignificant.

We have deliberately chosen to use the letter $k$ rather than the letter $f$ in that equation in order to clarify semantics. The Gabor function uses $f$ in the complex exponential because that value determines the center frequency of the function's response. Here, we are not so concerned with frequency. We are primarily dealing in terms of scale. All members of the particular wavelet family defined by a mother wavelet of the form in Equation (5-34) will be derived from that one function by nothing more than rescaling. Therefore, the role of $k$ is to determine the shape of the mother wavelet. It is better to downplay the frequency aspects of $k$.

That said, let us now briefly touch on one frequency aspect of $k$ that is of some slight interest. Since the Morlet wavelet is almost a Gabor function, we can use our previous formulas to approximate some of its properties. In particular, let us consider the width of this function's window in the frequency dimension. Looking back at Equation (5-18) on page 155, we see that the Morlet wavelet is nearly a Gabor function with $\sigma = 1$. Equation (5-20) then tells us that its radius is approximately equal to 0.113. We have a rule of thumb that the distance separating adjacent frequencies should not exceed about three times the window radius. Of course, that rule applies to equally spaced lattice points like we had with the STFT. Wavelets use geometric spacing. Nevertheless, the scaling applies to both the center frequency and to the width, so this rule of thumb should very roughly hold true. (Think about that until it becomes clear. Realize that the center frequency spacing here applies to the neighbor in one direction only.) To be safe, we can reduce the limit from 3 to 2. It is thus apparent that if we use $k > 0.226$ or so, we are putting our frame quality in jeopardy unless we compensate by using multiple voices. We will soon see that we nearly always want to use $k$ considerably larger than that, so multiple voices will be the rule. It can be interesting to look at Morlet wavelets parameterized by the window width (twice the radius) expressed as a fraction of $k$. The real part of several of these functions are graphed in Figures 5-19 through 5-22 starting on page 171.

We are assuming (and will continue to assume) that the traditional scale factor $\xi_0 = 2$ is used, as other values severely complicate our lives in regard to sampling. Any lack of frame quality will be made up for by using multiple voices, if indeed a frame is even required for our application. Thus, the members of our family of wavelet functions will be based on scale factors of $2^0$, $2^1$, $2^{-1}$, $2^2$, $2^{-2}$, and so on. How far can we go in these directions? It should be apparent that when the scale becomes so large that the effective time-domain width of the wavelet is an appreciable fraction of the total length of the series being studied, we have reached a practical limit. Nearly always, we will stop well before that limit is reached due to the fact that we simply are not interested in very-low-frequency (large-scale) phenomena. In the other direction, we have an extremely firm limit: the Nyquist limit. Review the section starting on page 88. The sampling rate (number of samples per unit time) must be at least twice the maximum frequency to be resolved. For the Morlet wavelet, that frequency is $k/\xi$. Thus, we cannot make the scale so small that the Nyquist limit is violated.

When a wavelet transform is applied to discretely sampled data, as is virtually always the case, there is a traditional scaling convention. The mother wavelet function is defined in such a way that it is the smallest-scale member of that wavelet family. It is never scaled downward for finer time resolution. If the scale factors that generate the family are $2^q$, then $q$ is never negative. It will be zero for the smallest scale (the mother wavelet's scale is $2^0 = 1$), and it will be positive integers for all other, larger, scales. If our data were continuous, sampled at infinite resolution, then there would be no physical limit on how small the scale could be. But when discrete sampling imposes this limit, we can simplify things by defining the mother wavelet in terms of the limit. Naturally, it is not mandatory that we resolve down to the Nyquist limit. In many cases we are not interested in events at that small a scale. In such cases, we are best off defining the mother wavelet in terms of the smallest resolution that we want to attain, so that the integer scale parameter $q$ still runs from zero through the largest positive integer that interests us.

We must exercise great caution when attempting to resolve detail near the Nyquist sampling limit. When we use only a few points along the effective time window of the mother wavelet, but are supposedly representing the full detail of the continuous mother wavelet by means of those few points, the shape implied by the discrete points may not resemble the true shape. Worse, the discretely sampled wavelet may not remain centered. Just because the full, continuous wavelet integrates to zero, that does not mean that any equispaced sampling of that function will automatically sum to zero! And the lack of centering will appear where it does the most harm: in the smallest scale components. As higher scales employ more points along the wavelet, the appropriately scaled mean will approach the integral, zero. But the approximation will be least accurate when we are near the Nyquist limit, where relatively few points lie inside the effective time window. This is best illustrated graphically, which we will do very shortly. But first let us point out that the situation may not be quite as bad as it seems.

Every now and then, a mathematical result appears that is both strongly counter-intuitive and wonderfully valuable. Mathematicians live for these moments. The Morlet wavelet gives us such a result. It turns out that as long as $k$ exceeds approximately 0.7, discrete sampling does not significantly distort the shape of the mother wavelet. Moreover, the normalizations of zero center and unit energy are preserved well *at all scales*, right down to almost the Nyquist limit. That is actually quite a remarkable situation. Here we have a continuous function whose integral is zero and whose energy is one. If we sample this function at equally spaced points, the sum of the function at those points is nearly zero, and the (domain-weighted) sum of squared absolute values is nearly one, *no matter what the spacing is*. The implication of this dramatic result is that we are free to choose any $k > 0.7$ and sample rate at or above the Nyquist limit, without worrying about troublesome interactions between the oscillations of the wavelet and the spacing of the samples — impressive, indeed.

Distortion due to discrete sampling is best illustrated with an example. J. Morlet favored the mother wavelet having $k$ approximately equal to 0.849. The real and imaginary parts of this function are shown in Figures 5-27 and 5-28, respectively.

The ratio of the center frequency, 0.849, to the frequency-domain window radius, 0.113, certainly exceeds our conservative rule-of-thumb value of 2. Therefore, we need multiple voices if we are to overlap frequency bands sufficiently to have a frame. Using four voices reduces the geometric spacing to one-quarter of its single-voice value, and so four voices are clearly sufficient. That is what we shall use for this example.

With $k = 0.849$, the Nyquist sampling limit is twice that, or 1.698 points per unit time. So that limiting behavior is most clearly portrayed, we will use that minimum sampling rate. Figure 5-29 shows the real part of this Morlet wavelet sampled at the Nyquist rate. The discrete points are spaced 1.0 / 1.698 = 0.589 apart, and these points are connected with lines to guide the eye. Note that the shape of the continuous wavelet (shown in Figure 5-27) is well preserved. The imaginary part of that same wavelet is graphed in Figure 5-30. No, that is not a misprint. Remember that the imaginary part of the Fourier transform of real data is identically zero at the Nyquist frequency. The same principle applies here. All of the points are zero. Look carefully at the continuous form shown in Figure 5-28. Our sample points all lie at the points where the function crosses the time axis. This is not a distortion or error. This is a simple reflection of the fact that we cannot resolve these events at the Nyquist limit.

According to Equation (5-31), the scale of the first voice beyond the basic one is approximately 1.189 (the fourth root of two). Thus, the spacing of the sample points is 0.589 / 1.189, or approximately 0.495. The real and imaginary parts of that voice are shown in Figures 5-31 and 5-32, respectively. Note that the discrete sampling does introduce some distortion for the imaginary part, but not anything terribly serious. And the real part is very clean.

The other two voices are shown in Figures 5-33 through 5-36. Observe how the quality of representation rapidly increases with relatively small increases in the sampling rate.

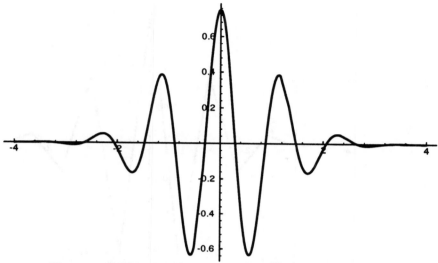

Fig. 5-27: Real part of Morlet wavelet with k=0.849.

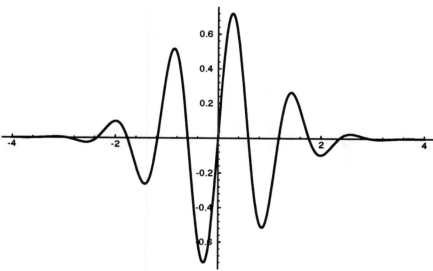

Fig. 5-28: Imaginary part.

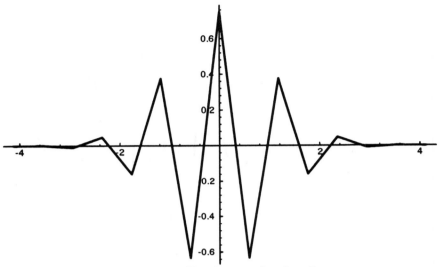

Fig. 5-29: Real part of voice 0.

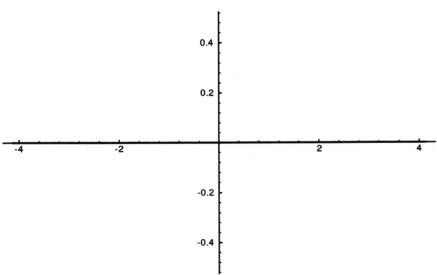

Fig. 5-30: Imaginary part of voice 0.

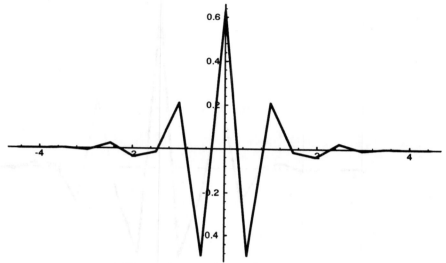

Fig. 5-31: Real part of voice 1.

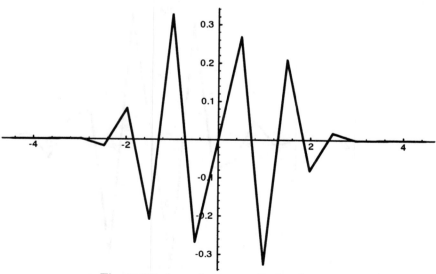

Fig. 5-32: Imaginary part of voice 1.

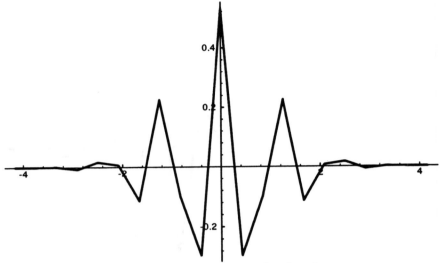

Fig. 5-33: Real part of voice 2.

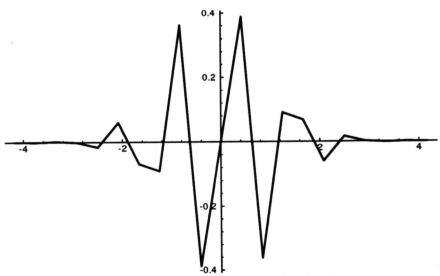

Fig. 5-34: Imaginary part of voice 2.

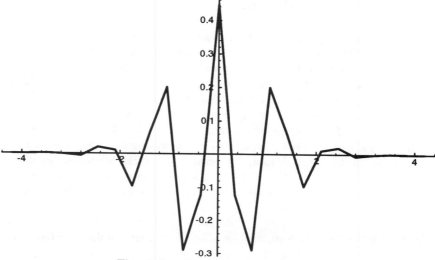

Fig. 5-35: Real part of voice 3.

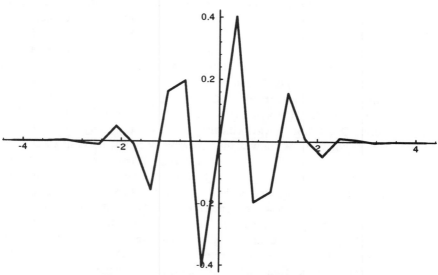

Fig. 5-36: Imaginary part of voice 3.

## Code for the Morlet Wavelet

This section presents complete source code for computation of a wavelet transform based on the Morlet wavelet. There are three listings: the constructor, the destructor, and the transformation routine. The constructor computes and saves all wavelet coefficients. It also returns several parameters that the user may find useful. The destructor frees the coefficient memory. The function transform actually performs the transform. It may be called as many times as desired after the constructor has been called.

The most important decision to be made by the user of a Morlet wavelet is the shape of the mother wavelet. This is determined by the parameter $k$ in Equation (5-34) on page 182. Larger values of $k$ cause the wavelet to oscillate more frequently during the effective time window. This makes frequency-domain resolution sharper, necessitating more voices. See Figures 5-19 through 5-22 starting on page 171 for illustrations of this varying oscillation.

An intuitively meaningful way of expressing this shape is as the ratio of the center frequency to the radius of the frequency window. We will use this ratio as the shape parameter in the program that follows. As we saw on page 182, that radius is fixed at 0.5 / (sqrt(2) * pi), or about 0.1125. When we express the wavelet's shape this way, we have an immediate feeling for the sharpness of the frequency response of each member of the wavelet family. Remember that the center frequency and the radius change scale *together* for the members of the family, so this ratio is constant for all members. Also remember that since the frequency doubles for each scale halving, the center frequency for each wavelet is the distance out to the next frequency point in the lattice (except for any voices filling in the gaps). Thus, if the center frequency is many times the window radius, we know that many voices will be needed to fill in the gaps, and that each voice will resolve sharply in the frequency domain. Conversely, if the center frequency is a small multiple of the window radius, the spacing of the lattice points in the frequency dimension will not be terribly wide relative to the extent of each filter's response. Frame quality can be maintained with very few voices. If in doubt as to the choice of a value for the shape parameter, Morlet's choice of 7.54 is generally recommended. That value is best served with about four voices, although Morlet used more. One can get away with three voices if necessary. For the user's protection, the constructor will not allow a shape parameter less than 6.2 if resolution near the Nyquist limit is requested. Experienced and daring programmers may want to remove those few lines of code near the beginning of the constructor, but that is dangerous. Normalization, especially centering, deteriorates rapidly for smaller shape values. This can lead to serious bleeding of low-frequency components into the high-frequency bins. Do not take this warning lightly. Such bleeding is an extremely serious event.

The highest frequency (smallest scale event) that can be resolved depends on how often the physical process is sampled. If we try to force the scale of the wavelet smaller than the Nyquist sampling limit, nothing will be gained. Normalization will be lost, the sampled shape of the wavelet will no longer represent the continuous mother wavelet, and information at larger scales will find its way into the excessively small-scale wavelet bins. In fact, if possible, we should not even try to resolve right down to the limit. Sometimes we must deal with physical or economic constraints on the sampling method, while still needing every bit of information that we can squeeze out of the data. But whenever possible, it is better to sample faster than the Nyquist limit, then start the wavelet scale family at a level larger than the limit. The constructor facilitates this with the rate parameter. This is the multiple of the Nyquist limit for the smallest-scale wavelet. Naturally, the minimum value for this parameter is 1. We can gain a significant safety factor with surprisingly small increases of the rate. Even specifying 1.5 helps tremendously. Essentially perfect representation and normalization is obtained when rate >= 2. *Do not confuse the* rate *parameter with the sample rate.* This parameter is the *multiple* by which the actual sample rate exceeds the Nyquist limit needed to accommodate the smallest-scale wavelet. The effect of a rate parameter exceeding 1 is that the smallest-scale wavelet generated by the constructor has a lower center frequency (larger scale) than the sample rate could theoretically support.

The implication is that the period (sample points per cycle) of the center frequency corresponding to the smallest-scale wavelet is equal to twice the rate parameter. That fact is useful when designing wavelet families customized to a particular application when we are interested in events having a frequency that is known in advance. For example, suppose that our application primarily involves events having a fundamental frequency of 120 cycles per second. The minimum sample rate needed to resolve this frequency is double that, or 240 samples per second. (Review page 88 if that is not clear.) But suppose that we also anticipate several harmonics. If we sampled at the minimum rate, those harmonics would cause aliasing problems. Therefore, to be safe, we collect our data at 1000 samples per second. If we called the constructor with rate = 1000 / (2 * 120) = 4.1667, the smallest-scale wavelet would be optimally tuned to respond to the fundamental frequency. Using a rate of half of that would let us detect the first harmonic with the smallest-scale wavelet and the fundamental with the next scale (double). If we are not interested in components having frequencies between these amounts, or if they are not even present, we would not need to bother with intermediate voices.

We must tell the constructor how many voices to generate. This is done with the nvoices parameter. Specifying a value of one, the minimum legal number, generates just the basic wavelet family with scales at powers of two. In general, more voices are needed for larger values of the shape parameter. Since the relative width of the frequency window is fixed at any scale, it may be overkill to use more voices than are needed to cover the space between lattice points. For typical shape values of seven to ten, using about four or five voices is generally the best compromise between economy of data and thorough coverage of the frequency domain.

The nscales parameter tells the constructor the maximum number of power-of-two scales that will be possible to compute from data supplied later. This count does not include voices within each scale. It does include the minimum-scale mother wavelet, so nscales must be at least equal to one. If the data to which the wavelet transformation is applied does not contain enough sample points, all of these scales may not be computed. Each unit increase in the number of scales specified here will approximately double the amount of memory required to store the filter coefficients computed by the constructor. It will also double the time required to compute the stored wavelet coefficients. The total number of scales that can be resolved at any time is nscales times nvoices.

For each scale, the filter required to transform the data for the wavelet at that scale theoretically extends to infinity. But since the negative exponential term in the mother wavelet defined by Equation (5-34) on page 182 vanishes rapidly away from the time center, we are able to treat this as a *finite impulse response* (*FIR*) filter. We extend the filter until the magnitude of each term is approximately 1.e–12 times the magnitude of the center term. (This is extremely conservative. Some users may wish to use a shorter filter.) The computed half-length of the filter for the largest voice of the smallest-scale wavelet is returned as the border parameter. When computing the minimum-scale wavelet, this many sample points will be skipped at the start (and end) of the input data. Since the half-length doubles for each succeeding scale, more data points will be skipped for larger scales. The transformation routine will return full information. See its description starting on page 196 for more details.

We already know that in order to avoid frame-damaging gaps in the time domain, we must space the lattice points closely enough. But we also do not want to generate more numbers than are needed by spacing them too closely. We can be helped in choosing the optimal spacing by means of the samprate parameter. This is returned equal to the number of points per unit time in the basic wavelet equation. [Daubechies, 1990] shows that the spacing in the time domain must not be much more than that value. A factor of about 1.75 times samprate is an absolute upper limit if even moderate frame quality is desired. It would not be unreasonable to make the spacing approximately equal to samprate. This quantity is equal to the Nyquist sample rate limit, $2k$, times the user's multiple, rate.

We can easily compute the factor by which the period corresponding to the center frequency of each successive voice increases. It is the nvoices'th root of two. But it may be easier to let the constructor do it for us. That quantity is returned as the voicefac parameter.

Finally, the constructor needs to inform us as to whether or not it was able to construct a set of wavelet transform filters. Perhaps the user specified illegal parameters. Or perhaps there was too little memory available to compute the number of scales, nscales, requested. If there were problems, ok is returned equal to zero. Otherwise, it is returned equal to one.

The constructor will need to allocate a work area to hold the FIR filter coefficients. Each coefficient has a real and an imaginary part. The filter is symmetric, so only the right half is stored. The center coefficient is the same for all scales, so it will not be stored in the array. There will be a separate filter for each voice. Only the largest-scale filter will be computed and stored. Smaller-scale filters will be derived by decimation. Thus, the total memory requirement is 2 * nvoices * border * 2**(nscales-1) doubles. The class declaration and code for the constructor and destructor now appear.

```
static double two_pi = 2. * 3.141592653589793 ;
static double root2 = sqrt ( 2.0 ) ;

class Morlet {

public:
   Morlet ( double shape , double rate , int nvoices , int nscales ,
         int *border , double *samprate , double *voicefac , int *ok ) ;
   ~Morlet () ;
   void transform ( int nx , double *x , int spacing , int *starts ,
               int *counts , int *nout , double *rt , double *it ) ;

private:
   double kparam ;              // "Frequency" shape parameter k
   double srate ;               // Sample rate (samples per unit time)
   int nv ;                     // Number of voices
   int ns ;                     // Number of scales
   int hl ;                     // Shortest filter half length (=border)
   int npv ;                    // Number of coefs per voice (for longest filter)
   double gconst ;              // Normalizing constant for Gaussian
   double *coefs ;              // FIR coefficients for all frequencies
} ;

/*
   Constructor
*/

Morlet::Morlet (
   double shape ,               // f/delta, Morlet used 7.54
   double rate ,                // Multiple of Nyquist, at least 1.0
   int nvoices ,                // Number of voices, at least 1
   int nscales ,                // Number of scales to prepare, at least 1
   int *border ,                // Half-length of shortest filter
   double *samprate ,           // Samples per unit time
   double *voicefac ,           // Factor period is multiplied by for voices
   int *ok                      // Parameters and memory allocation ok?
   )
{
   int i, iv ;
   double window_radius, weight, *cptr, fac, x, con, vfac, rfac ;

   coefs = NULL ;  // So destructor doesn't do bad free if failure here
```

```
/*
        Verify that parameters are legal.  If the shape-determining frequency is very low, normaliza-
        tion is poor unless the sample rate is well above the Nyquist limit.  (This is a crude check,
        but it inspires more care than no check at all.)  The specified rate factor must be at least 1.0
        times that limit.  The number of voices and scales must be at least 1.
*/

   *ok = 0 ;
   if ((shape < 6.2)  &&  (rate < 2.0))
     return ;
   if (rate < 1.0)
     return ;
   if (nvoices < 1)
     return ;
   if (nscales < 1)
     return ;

/*
        The user specified the shape as a multiple of the frequency-dimension window radius.  Use
        that to find k.  Multiply k by 2 pi right now to avoid having to do it later for trig functions.  The
        sample rate is the Nyquist frequency (twice the shape frequency) times the user's multiple
        (which must be at least 1.0).
*/

   window_radius = 1.0 / (root2 * two_pi) ;           // Equation (5-20)
   kparam = window_radius * shape * two_pi ;          // Our definition
   *samprate = srate = window_radius * shape * 2.0 * rate ; // Nyquist limit
   *voicefac = pow ( 2.0 , 1.0 / (double) nvoices ) ; // Equation (5-31)

/*
        Save other user parameters in private storage area.  Compute the filter half-length such that
        the weight goes to about 1.e-12 times its max by the end of the filter.  If multiple voices are
        used, remember that we must accomodate the largest voice.  Feel free to change the factor
        of 7.4 if a different tradeoff of filter length versus accuracy is desired.
*/

   nv = nvoices ;
   ns = nscales ;

   *border = hl = 1 + 7.4 * srate * pow ( 2.0 , (double) (nv-1) / (double) nv );

/*
        Allocate memory for the FIR filter coefficients that make up the wavelet family.  We store
        each voice separately.  For each voice, store the coefficients for the largest scale member of
```

that family. Smaller scale members will be derived by decimation. The number of coefficients needed for that longest filter is the half-length of the shortest filter times the largest scale (2 ** (ns-1)). Then we have a set for each voice, and real and imaginary parts.
```
*/

   npv = hl ;                              // Length of shortest filter
   i = ns ;                                // Number of scales
   while (--i)                             // Compute 2 ** (ns-1)
     npv *= 2 ;                            // to get length of longest filter

   coefs = (double *) malloc ( 2 * nv * npv * sizeof(double) ) ;
   if (coefs == NULL)
     return ;                              // We already initialized ok to 0
   *ok = 1 ;                               // Allocs fine, so OK now

/*
   Compute the filter coefficients.  The center coefficient for all real parts is the voice factor
   times gconst, the Gaussian multiplier.  The center of the imaginary part is always 0 (sin 0).
   The sample rate for a voice is the smallest-scale sample rate, srate, times the scale of the
   largest-scale member (since we compute and save only its coefs) times the rate factor for
   that voice.  The unit time change per point is the reciprocal of that sample rate.
*/

   fac = 1.0 / (srate * pow ( 2.0 , (double) (ns-1))) ;// Largest-scale rate
   rfac = sqrt ( fac ) ;
   gconst = pow ( 3.141592653589793 , -0.25 ) ;        // Pi to the -1/4 power
   gconst *= rfac ;                                      // Equation (5-24)
   con = exp ( -0.5 * kparam * kparam ) ;               // Equation (5-34) centering

   for (iv=0 ; iv<nv ; iv++) {             // For all voices
     vfac = pow ( 2.0 , -(double) iv / (double) nv ) ; // Equation (5-31)
     cptr = coefs + iv * 2 * npv ;         // Point to this voice's coef area
     for (i=1 ; i<=npv ; i++) {            // Right half of filter only
       x = vfac * fac * i ;                // Scale for family (fac) and voice (vfac)
       weight = vfac * gconst * exp ( -0.5 * x * x ) ; // Equation (5-34) right factor
       *cptr++ = (cos ( kparam * x ) - con) * weight ; // Same equation, real part
       *cptr++ = sin ( kparam * x ) * weight ;          // And imaginary part
       }
     }
   }
```

```
/*
  Destructor
*/

Morlet::~Morlet ()
{
  if (coefs != NULL)
    free ( coefs ) ;
}
```

      The actual Morlet wavelet transform is performed by the transform member function. This function may be called as often as desired on different data sets. The data to be transformed is pointed to by the x parameter. The length of that array is nx. Figure 5-37 shows how the filters used to compute each transform output are placed on the data array.

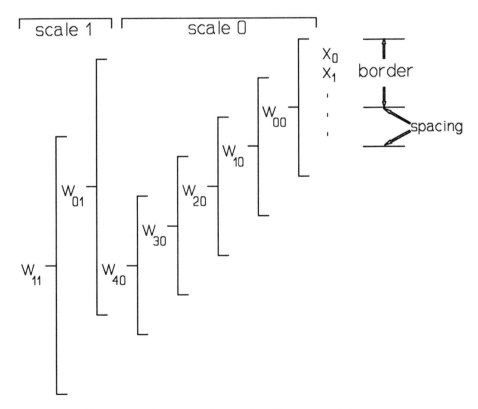

Fig. 5-37: Layout of wavelet filters on data array.

The time-domain lattice-point spacing is specified by spacing. The smallest-scale wavelet is applied to the data every spacing samples. Twice that spacing is used for the next larger scale, with successive doubling for all scales. All voices within a given scale are positioned with identical centers. (See Figure 5-24 on page 180.) Setting the spacing parameter equal to the samprate returned by the constructor, rounded, of course, is a good idea. If the spacing exceeds about 1.75 times samprate, a very poor frame will result from excessive gaps in the time domain. Values much smaller than samprate generate an overabundance of data. It is more difficult to deal with large quantities of data, so spacing should be made as large as possible.

The smallest-scale wavelet will be centered as early in the data as possible. This is exactly border samples from the beginning, the filter's half-length. Successive-scale wavelets will have longer half-lengths (they double with each scale doubling), so those wavelets will start later and later in the series. Their starting point may exceed their filter half-lengths, as wavelets at any scale will always line up with the centers of previous, smaller-scale wavelets. Figure 5-37 shows how the first wavelet at the second scale, $w_{01}$, starts slightly further into the data array than is needed for its half-length in order to make its center line up with $w_{20}$. All voices of a given scale family will have the same centers in the data.

The starting point (filter center in the data) for each scale wavelet is returned in the parameter vector starts. This is a vector nscales long. For each scale (0 being the smallest), this specifies the subscript in the x array where the first wavelet of that scale is centered. Starts[0] will equal border, that filter's half-length, because the first wavelet starts as soon as possible. The filter half-length doubles for each successive scale, so the starting point must move in at least that much. It may be even more, as each longer scale starts at a starting point of the previous (smaller) scale to assure uniformity. Thus, starts[1] will be at least 2 * border, starts[2] will be at least 4 * border, and so on.

The number of time-domain filter centers for each scale depends on the length of the data. This will be returned in the nscales-long vector counts. For each scale i, (0 being the smallest), counts[i] specifies the number of wavelet voice sets computed at that scale. It is possible that one or more of the last elements of this vector are zero. This will happen if the data array is too short to compute the largest-scale wavelets. This routine also returns nout, the total of all of the elements of counts. Thus, the total number of complex outputs is nout times nvoices.

The real and imaginary parts of the output transform are returned as separate arrays, each nout * nvoices long. The real part is rt, and the imaginary part is it. All counts[0] elements of the smallest-scale wavelet (all voices) appear first in the output. The counts[1] elements of the next (doubled) scale appear next. This continues, with the largest-scale wavelets last. Within each scale, elements computed at successive centers along the time dimension (moving through the input data) appear. Finally, for each time center, the voices for that center appear, from shortest to longest scale. In other words, the voices change fastest, the time centers change next, and the scales change slowest. For example, look at Figure 5-37. The first output, rt[0] and it[0], will be voice 0 of $w_{00}$. The next output will be voice 1. This continues until all nvoices voices are output. Then voice 0 of $w_{10}$ will appear, followed by its other voices. This pattern repeats until all wavelets at that smallest scale are output, for a total of counts[0] times nvoices outputs. Then the same pattern is used for the next larger scale, starting with $w_{01}$. The subscript in the data array of the center of $w_{01}$ is returned in starts[1]. Outputs are generated in this manner until either the scale is so large that at any aligned center the filter extends past the end of the data array, or until nscales is reached.

```
void Morlet::transform (
   int nx ,                          // Length of input data vector
   double *x ,                       // Input data vector
   int spacing ,                     // Space between initial time lattice points
   int *starts ,                     // Starts[i]=subscript of center of i-scale output
   int *counts ,                     // Counts[i]=number of outs/voice at scale i
   int *nout ,                       // Total number of outs/voice (sum of counts)
   double *rt ,                      // Real outputs
   double *it                        // Imaginary outputs
   )
{
   int i, n, iscale, iv, itime, decim, flen ;
   double *cptr, *xptr, rsum, isum, vfac, scafac ;

   for (iscale=0 ; iscale<ns ; iscale++)   // Init to all 0 in case x is
      starts[iscale] = counts[iscale] = 0 ; // too short to do all scales
   *nout = 0 ;

   decim = npv / hl ;                // Decimate longest filter to get current filter
   flen = hl ;                       // Half-length of current filter
   scafac = pow ( 2.0 , 0.5 * (double) (ns-1) ) ; // Equation (5-29)

   for (iscale=0 ; iscale<ns ; iscale++) { // Start with shortest filter

      if (iscale) {  // This start must line up with previous and be >= flen
         for (n=starts[iscale-1] ; n<flen ; n+=spacing/2) ; // Previous centers
         starts[iscale] = n ;
         }
      else         // First scale's start is as early as possible
         starts[0] = flen ;

      n = nx - starts[iscale] - flen ;  // This many x's filterable
      if (n > 0) {
         n = counts[iscale] = 1 + (n - 1) / spacing ; // This many resolved
         *nout += n ;
         }
```

```
      else
         break ;  // Input is too short given filter length


      for (itime=0 ; itime<n ; itime++) {              // All time slots
         xptr = x + starts[iscale] + itime * spacing ;  // Center of this filter

         for (iv=0 ; iv<nv ; iv++) {                    // All voices
            vfac = pow ( 2.0 , -(double) iv / (double) nv ) ; // Equation (5-31)
            rsum = vfac * gconst * *xptr ;              // Center real coef is constant
            isum = 0.0 ;                                // Center imaginary coef is 0

            cptr = coefs + iv * 2 * npv - 2 ;           // Point to this voice coef area
            for (i=1 ; i<=flen ; i++) {                 // Right half of filter
               cptr += 2 * decim ;                      // Coef 0 not saved in coefs!
               rsum += *cptr * xptr[i] ;                // Sum real part of right half
               isum += *(cptr+1) * xptr[i] ;            // And imaginary part
               }

            cptr = coefs + iv * 2 * npv - 2 ;           // Point to this voice coef area
            for (i=1 ; i<=flen ; i++) {                 // Left half of filter
               cptr += 2 * decim ;                      // Coef 0 not saved in coefs!
               rsum += *cptr * xptr[-i] ;               // Sum real part of left half
               isum -= *(cptr+1) * xptr[-i] ;           // And imaginary part
               }

            *rt++ = scafac * rsum ;                     // Output real part of wavelet transform
            *it++ = scafac * isum ;                     // And imaginary part
            }
         }

      decim /= 2 ;                                      // Decimation for stored filter coefs
      flen *= 2 ;                                       // Half-length of current filter
      spacing *= 2 ;                                    // Time-domain spacing of filters
      scafac *= sqrt ( 0.5 ) ;                          // Scale factor for h(t)
      }
   }
```

# Fast Filtering in the Frequency Domain

By now there is probably a small but significant group of readers who are writhing in agony over the way the Gabor and wavelet filter algorithms have been presented. We have done all of the filtering in the time domain, explicitly computing the filter coefficients and convolving that vector with the sampled data. That approach is sometimes frowned upon in the general signal processing community because in many cases it can be terribly slow. However, there are several good reasons for taking that approach. First, it is the most straightforward method, and hence it is appropriate for a tutorial presentation. Second, there are times when the logistics of the application demand a time-domain solution. Most importantly, in a large number of practical implementations, this method actually *is* the fastest approach. If the filters are fairly short and if only occasional filtered samples are needed, the low overhead of time-domain filtering makes it the method of choice. On the other hand, we would be remiss if we failed to include a discussion of an algorithm that can be orders of magnitude faster in many applications.

Both the Gabor transform and the Morlet wavelet discussed earlier are nothing more than filter pairs (in-phase and in-quadrature) that have an exceedingly simple frequency response. We saw in Equation (5-19) on page 155 that their frequency response is an impulse function convolved with a Gaussian weight function. Thus, we are led to wonder if we can filter our series by computing its Fourier transform, multiplying the transform by the Fourier transform of the filter, then transforming back to the time domain. The answer is that as long as we take a few simple precautions, we certainly can. And if we need a lot of filters and a lot of responses, this method will usually be much faster than directly convolving each filter with the signal in the time domain. (If this concept is unfamiliar to any readers, look up the *convolution theorem* in the Fourier transform section of any good signal processing text. This subject is so widely discussed that its mathematical details will be skirted in favor of its practical implementation details.)

Let us begin by examining the Fourier transform of the filters in more detail than we needed previously. We are actually applying two filters. One is an in-phase filter based on a cosine wave. A typical example in the time domain is shown in Figure 5-7 on page 144. The filter is real and symmetric, so its Fourier transform will also be real and symmetric. In particular, it would resemble Figure 5-38 on page 203. That graph shows the entire extent of the Fourier transform. Note the symmetry about the Nyquist point, 0.5.

The second filter is the imaginary *i* times a windowed sine function, an in-quadrature filter. An example is shown in Figure 5-8 on page 144. Its Fourier transform is also real, but antisymmetric. That transform is similar to Figure 5-39 on page 203. The actual Gabor or Morlet filter is the sum of the two time domain filters, so the transform of the actual filter is the sum of the two transforms. When we compute that sum, the coefficients to the left of the Nyquist point will cancel, while those to the right will reinforce each other. The net result looks like Figure 5-40. In other words, to implement a Gabor or Morlet wavelet filter in the frequency domain, we multiply the part of the series' transform that is above the Nyquist frequency by a Gaussian function. The remainder of the transformed series is zeroed. (This result assumes that the filter widths are reasonably narrow and not too close to the DC (0 frequency) and Nyquist points, as is usual in practice. Readers interested in more rigor should consult signal-processing texts, looking under the topic of *quadrature-pair filters*.)

When we implement a filter in the frequency domain, we may be curious as to what the time-domain filter function looks like. That is the function that is effectively convolved with the time-domain series as a consequence of our frequency-domain filtering. In particular, we will see later that it is important to find out just how far the filter extends in the time domain. This question is easily answered. Simply compute the (inverse) Fourier transform of the frequency-domain response function. For example, consider the frequency-domain filter shown in Figure 5-40. It is real but not symmetric. Therefore, its transform is complex and (conjugate) symmetric. The real part of that transform is shown in Figure 5-41 and the imaginary part in Figure 5-42. Note that it exhibits the usual wraparound of the discrete Fourier transform. The center coefficient of the time-domain filter is at the far left. Coefficients to the left of the center are at the far right of the graph. It can be seen that coefficients beyond about 40 samples from the filter center are effectively zero. This dwindling can be seen best in Figure 5-43, which is the magnitude of the coefficients.

Filtering in the frequency domain is very straightforward: transform, multiply by the frequency window, then transform back. However, there is one small complication. Recall that an implicit assumption of the discrete Fourier transform is that the transformed series is periodic, with the end of the sample wrapping around to the beginning. This is of no consequence when we are well into the interior of the series. But it can be a problem near the ends. Consider computation of the leftmost filtered output point. It is the sum of many products. We (implicitly) multiply the first original point by the leftmost filter point in Figure 5-41 (this is the filter center) for the real output, and the same point in Figure 5-42 for the imaginary output. Then we multiply the second point in the input by the second point in the filters in those figures, and so on. The leftmost output point contains the sum of all of those products. For the right half of the filter this is exactly what we want. And we also observe that the implicit filter coefficients drop to essentially zero after we get about 40 points into the series, again just what we want and expect. But what about the left half of the filter? Oops. Ideally, we would like it to pick up points prior to the leftmost point. But that is obviously impossible, as we do not possess those points. Generally, the next-best choice is for those points to be all zero so that they contribute nothing to the sum. Unfortunately, that is not what happens. Instead, this method of filtering uses the rightmost points in the series as if they were wrapped around to the left of the first point. Except in the rarest of cases, that is not good. What to do?

Actually, the solution is trivial. We pad the end of the input series with zeros. That way, when it implicitly wraps around, only zeros will wrap. How many zeros do we need? That depends on the window width. Recall from Equation (5-21) on page 155 that the time-domain width and the frequency-domain width are inversely related. We can always invert the frequency-domain Gaussian window as we did when computing Figures 5-41 and 5-42, and then inspect the coefficients to see how far from the center we need to be before their magnitude becomes negligible. That is how many zeros we need to append. But later we will see that there is a simple rule-of-thumb that is completely reliable.

There is one more consideration related to appending zeros to the series. The commonly used mixed-radix FFT performs fastest if the length of the series is the product of small prime factors. The FFT given in this text demands a power of two. Thus, we often have to lengthen the series anyway, just to get it up to a good size. So we usually treat the number of zeros appended to avoid wraparound problems as a minimum number, and we almost always append even more in deference to the FFT algorithm.

This concept may become more clear with the aid of an example. Figure 5-44 shows a sine wave that repeats six times over 512 sample points. Another 512 points, all zero, are appended to this series. We compute its Fourier transform, multiply that transform by a function similar to that shown in Figure 5-40, then transform back to the time domain. The center frequency of this filter is equal to the frequency of the input signal. The real part of the result is shown in Figure 5-45 and the imaginary part in Figure 5-46. The amplitude appears in Figure 5-47.

Observe that the center region of the filtered signal is exactly as expected. However, the amplitude of the left end is diminished. This is due to the inclusion of the rightmost appended zeros. The rightmost part of the original signal, which is in the center of these figures, is similarly diminished due to the leftmost appended zeros. Most interesting, a signal magically appears at the far right end of the output where the input is entirely zero. This is due to the right side of the implicit time-domain filter wrapping around to the left side of the input series. But this is not a problem at all. We will only use the left side of the filtered output signal, the part that corresponds to the original input before zeros were appended. The right part, where the zeros were placed, is discarded. Certainly the use of zeros to pad the input series introduced error in the form of diminished amplitude. Nevertheless, that is virtually always superior to avoiding padding and implicitly using the other end of the series!

## Implementing the Filter

We will now define a frequency-domain filter that can be used for both the Gabor and Morlet wavelet transforms, and we will provide code for that filter. We will also take this as an opportunity to introduce an alternative means of scaling the magnitude of the transforms. We will scale this family of filters in such a way that the amplitude of signal components whose frequency is exactly equal to the center frequency of the filter will remain unchanged, while all other frequencies will be diminished. This is a very intuitive alternative to the unit energy scaling previously used.

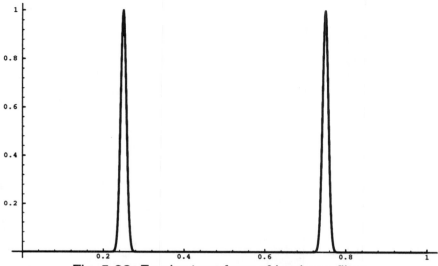

Fig. 5-38: Fourier transform of in-phase filter.

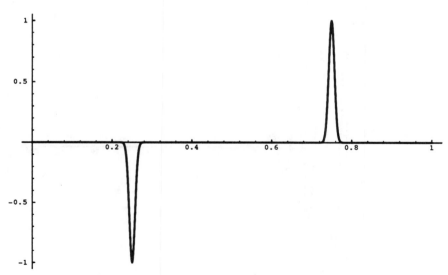

Fig. 5-39: Fourier transform of in-quadrature filter.

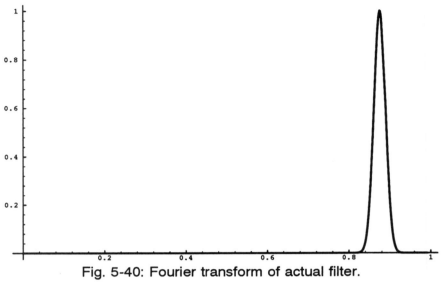

Fig. 5-40: Fourier transform of actual filter.

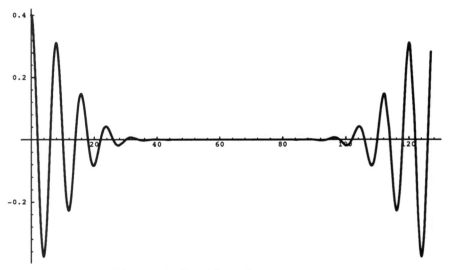

Fig. 5-41: Real impulse response.

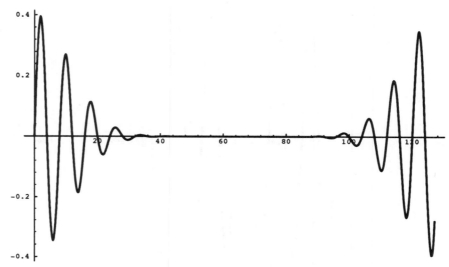

Fig. 5-42: Imaginary impulse response.

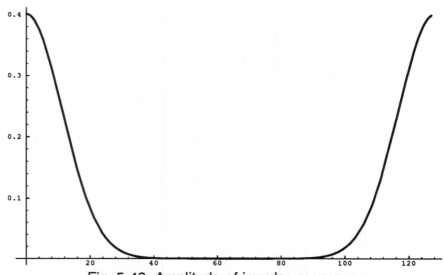

Fig. 5-43: Amplitude of impulse response.

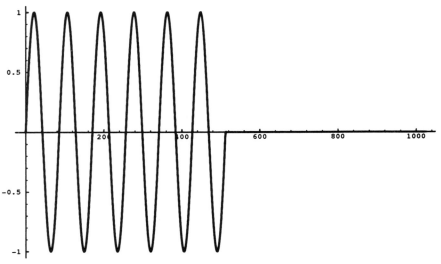

Fig. 5-44: Original signal, zero extended.

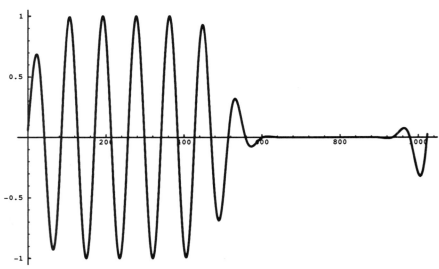

Fig. 5-45: Real (in-phase) filtered signal.

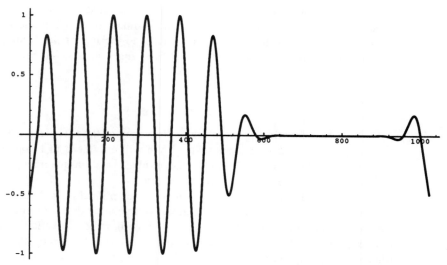

Fig. 5-46: Imaginary (in-quadrature) filtered signal.

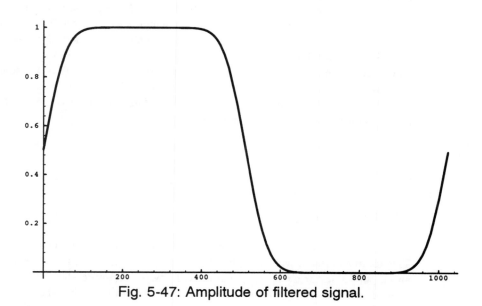

Fig. 5-47: Amplitude of filtered signal.

Scaling the magnitude so as to preserve the input amplitude is trivial. We simply make the maximum of each of the frequency-domain filters (in-phase and in-quadrature) equal to 1. Recalling that these filters cancel to the left of the Nyquist frequency and reinforce to the right, we see that the net filter function is as shown in Equation (5-35).

$$h(f) = 2\, e^{-\left(\frac{f-f_0}{s}\right)^2} \tag{5-35}$$

In that equation, $f_0$ is the center frequency of the filter, and it is greater than 0.5 (the Nyquist frequency) and less than 1. We will only need to compute this value for $f$ from 0.5 to 1.0, as cancellation of the in-phase and in-quadrature filters causes all multipliers below the Nyquist point to be zero.

The width in the frequency domain is determined by $s$. When we use this width measure, the practical implication of any given value is easier to understand than when we use the more traditional and theoretical width based on Equation (5-13) on page 152. For example, we know that an input component at a frequency that differs from the filter's center frequency by exactly $s$ will be attenuated by a factor of $e^{-1} = .36788$, while those differing by 3 times $s$ will be reduced by $e^{-9} = 0.0001234$. The 50 percent attenuation point is at sqrt (-log (0.5)) = 0.83 times $s$.

Earlier, we discussed the need to append zeros to the input signal so as to avoid wraparound problems. We pointed out that the minimum number of zeros needed could be estimated by evaluating the frequency-domain filter, Equation (5-35), at a number of points and then transforming that vector to the time domain. That lets us inspect the time-domain filter that is implicitly convolved with the input series, and we can easily determine the distance from its center after which all of the coefficients are essentially zero. Luckily, there is a much easier way. Remember that as long as the filter is based on a Gaussian window, the product of the time-domain width times the frequency-domain width is constant. Considering this fact, a universal rule of thumb is possible. The author computes the minimum number of zeros using Equation (5-36). It is quite conservative, so there is little practical need to use a constant larger than 0.8 or so. In fact, a slightly smaller constant is acceptable in many cases.

$$m = \frac{0.8}{s} \tag{5-36}$$

## Code for a Fast Filter

We can now present a simple subroutine for implementing the fast filtering algorithm. This routine is fine if we are applying a single filter to the series, and we want to compute an output value for every input point. Otherwise, the more sophisticated routine given later should be used. This routine is supplied mainly so that the reader can see the algorithm in an uncluttered environment.

```
void fastfilt (
  int n ,                          // Length of input/output series
  double *xr ,                     // Raw input, in-phase filtered output
  double *xi ,                     // In-quadrature filtered output
  double freq ,                    // Center frequency, 0-.5
  double s                         // Scale (width) factor in frequency domain
  )

{
  int i, halfn ;
  double f, dist, wt ;

/*
  Pack the real series for the short-cut FFT method
*/

  halfn = n / 2 ;                  // We only need to transform half n
  for (i=0 ; i<halfn ; i++) {      // Because we take advantage of real input
    xr[i] = xr[2*i] ;              // Pack even terms in real part
    xi[i] = xr[2*i+1] ;            // And odd in imaginary part
    }
  real_fft ( halfn , xr , xi ) ;

/*
    Unpack the transformed real array.  It is conjugate symmetric.  Recall that the (real) Nyquist
    point was placed in the imaginary DC point, which is actually zero.  While we are at it, apply
    the frequency-domain filter to the points above the Nyquist frequency.  Simultaneously zero
    the points to the left, as well as the DC and Nyquist points.  Also, there is conjugate symme-
    try, so we should flip the sign of the imaginary parts to the right of the Nyquist point.  But
    when we do the inverse transform later we would just flip it again.  So avoid that double flip.
    Similarly, include in wt the factor of 1/n that we would impose later for the inverse transform.
*/

  for (i=1 ; i<halfn ; i++) {
    f = (double) i / (double) n ;  // Frequency of this point
    dist = (f - freq) / s ;        // Weighted distance from center freq
    wt = exp ( - dist * dist ) ;   // Gaussian weighting function
    wt *= 2.0 ;                    // Real and imaginary parts reinforce
    wt /= (double) n ;             // Include this factor for the inverse transform
    xr[n-i] = wt * xr[i] ;         // Since input was real, transform is symmetric
    xi[n-i] = wt * xi[i] ;         // Conjugate symmetry, but avoid double flip
    xr[i] = 0.0 ;                  // Zero out left part so that we get
    xi[i] = 0.0 ;                  // Both in-phase and in-quadrature outputs
    }
```

```
xr[0] = xi[0] = xr[halfn] = xi[halfn] = 0.0 ; // Also zero these

/*

    Do the inverse Fourier transform.  Note that the first step, flipping the sign of the imaginary
    parts, was implicitly done above when we did not flip them for conjugate symmetry.  Also,
    the division by n was done above in conjunction with the Gaussian weighting.  So we can
    avoid the first two steps!  Just go ahead with the FFT.  Then do the final step of flipping
    imaginary signs.
*/

    fft ( n , xr , xi ) ;
    for (i=0 ; i<n ; i++)
      xi[i] = -xi[i] ;
}
```

The above code is a straightforward implementation of the filter already described. We start by packing the real input signal into a complex array of half that length. That lets us use the more efficient real_fft routine given on page 86. We apply the frequency-domain filter when the transformed array is unpacked. Recall that performing an inverse transform with a forward-transform routine is a three-step process. We flip the signs of the imaginary parts, transform, then flip those signs again. A division by $n$ is also needed somewhere along the way. (It doesn't matter where, as all of those operations are linear.) We choose to do it along with the unpacking, since we are already multiplying by the Gaussian function at that time. Note that when we unpack, we must flip the sign of the imaginary parts above the Nyquist point. But that is also the first step of the inverse transform, so we simply do not flip at all. Then we transform and do the final complex conjugation.

## A Better Implementation

The subroutine given above is perfectly good if we want to apply only one filter to the data and if we want the filtered output computed for every sample point. However, that is not usually the case. We almost always want to apply several filters, and for the lower frequencies, we may even want to decimate the output to reduce the amount of information that must be processed. In this section we will present code that uses the same algorithm as fastfilt but that allows us more versatility and efficiency.

There are two problems with fastfilt. The first is that if we apply more than one filter by repeatedly calling fastfilt, the input series is transformed to the frequency domain every time. It only needs to be done once! The second problem is that the Nyquist sampling theorem tells us that we can get away with a lower sample rate at lower frequencies. In order to avoid aliasing, we only need a sample rate equal to twice the highest frequency present. (This was discussed on page 88.) Evaluating the filter output at every input point when we have removed the high-frequency components is wasted

effort. Not only does it increase the computation time for the filter, but it loads us down
with much excess baggage in the form of redundant information. Neither we nor our
neural network needs that. The code given here will address both of these issues.

The issue of unnecessary repeat transforms is resolved by defining a FastFilter
object. The constructor for this class takes an input series and transforms it, saving the
transform in a private area. The factor of $2/n$ is included at this time to avoid having to
include it every time a different filter is applied. Note that since the time-domain data
is real, the frequency-domain data is conjugate symmetric. We pack it as before, and only
save the packed form. The class declarations, constructor, and destructor now follow.

```
class FastFilter {

    public:
      FastFilter ( int npts , double *signal ) ;
      ~FastFilter () ;
      void filter ( double freq , double s , int decim ,
                double *real , double *imag ) ;

    private:
      int n ;
      double *xr, *xi ;
  } ;

/*
  Constructor, destructor
*/

FastFilter::FastFilter (
    int npts ,                    // Length of input signal (power of 2)
    double *signal                // Input signal
    )
{
  int i, halfn ;
  double wt ;

  n = 0 ;                         // Cheap insurance
  xr = NULL ;                     // So destructor does not err

  if (npts <= 0)                  // Protect from careless user
    return ;

  xr = (double *) malloc ( npts * sizeof(double) ) ;
  if (xr == NULL)
    return ;
```

```
    n = npts ;
    halfn = n / 2 ;
    xi = xr + halfn ;

/*
    Pack the real series for the short-cut FFT method.  Also include the factor of two that results
    from the in-phase and in-quadrature filters reinforcing, and 1/n for the inverse transform.
    Doing it once now saves doing it every time 'filter' is called later.
*/

    wt = 2.0 / (double) n ;
    for (i=0 ; i<halfn ; i++) {
       xr[i] = wt * signal[2*i] ;
       xi[i] = wt * signal[2*i+1] ;
       }
    real_fft ( halfn , xr , xi ) ;
}

FastFilter::~FastFilter ()
{
   if (xr != NULL)
       free ( xr ) ;
}
```

The filtering algorithm that we will use in the upcoming code is nearly identical to that used in *fastfilt*. The only difference involves decimation. If there is no information present above some frequency, there is no need to compute output points at a sample rate more than twice that frequency. The way we accomplish this decimation is to physically remove the frequency-domain points that correspond to unneeded high frequencies. These are (presumably) zero, so no information is lost. Recall how the discrete Fourier transform wraps around, with the highest frequencies in the middle. Thus, we must remove the central points, retaining the ends that represent lower frequencies. Our application of this general technique is even simpler because we do not have to deal with the lower end, as it is forced by the filter to be all zeros. We simply copy the upper end of the transformed data, representing small negative frequencies, to the upper end of a shorter array. When we transform that short array back to the time domain, we get a time series whose sample rate is smaller than the original by the same factor that the shortened frequency-domain array is shorter than the original. Figure 5-48 illustrates this technique. The original series had 16 points. Its Fourier transform is shown at the top of that figure. To decimate by a factor of two, we create a new complex array 8 points long. We discard the inner 8 points of the top array. The upper set of the remaining points, excluding the new Nyquist frequency in deference to the filter, becomes the upper half of the new array. The rest of the new array is set to zero in accordance with our filtering method. This is the first step in the code for *FastFilter::filter*.

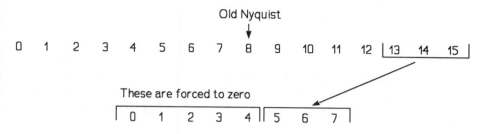

Fig. 5-48: Decimation as part of filtering.

```
void FastFilter::filter (
   double freq ,                    // Center frequency, 0-.5
   double s ,                       // Scale (width) factor in frequency domain
   int decim ,                      // Decimation ratio (power of 2)
   double *real ,                   // Real output (in-phase)
   double *imag                     // Imaginary output (in-quadrature)
   )
{
   int i, newn, nn ;
   double f, dist, wt ;

/*
   The constructor provided us with the packed transform of the input (xr, xi).  Copy it to the
   output area, unpacking and removing the center if there is any decimation.  Recall that the
   DC and Nyquist terms, as well as the entire left half (below the Nyquist) of this data is zero.
   So we only fill in the right half.  Also, normal unpacking demands that we flip the signs of the
   imaginary parts in the right half.  But the first step in the inverse transform is to flip those
   same signs, so avoid the double flip by not doing it at all.  We might as well apply the
   Gaussian function too.
*/

   nn = newn = n / decim ;

   for (i=1 ; i<newn/2 ; i++) {
      f = (double) i / (double) n ;        // Frequency of this point
      dist = (f - freq) / s ;              // Weighted distance from center freq
      wt = exp ( - dist * dist ) ;         // Gaussian weighting function
      real[--nn] = wt * xr[i] ;
      imag[nn] = wt * xi[i] ;
      }
```

```
    while (nn--)
      real[nn] = imag[nn] = 0.0 ;

/*
      Do the inverse Fourier transform.  Note that the first step, flipping the sign of the imaginary
      parts, was implicitly done above when we did not flip them for conjugate symmetry.  Also,
      the division by n was done in the constructor.  So we can avoid the first two steps.  Just go
      ahead with the FFT.  Then do the final step of flipping imaginary signs.
*/

    fft ( newn , real , imag ) ;
    for (i=0 ; i<newn ; i++)
      imag[i] = -imag[i] ;
}
```

## Fast Filter Examples

We will now give several specific examples of how fast filters can be implemented in the frequency domain.  If only a few bands are of interest, and if the data is to be processed in occasional chunks, then the time-domain methods presented earlier are appropriate.  On the other hand, if many filters are to be applied and if numerous output values are needed, DFT filtering in the frequency domain is almost always faster — sometimes by several orders of magnitude!

Let us start out with a simple problem.  We are studying the vocalizations of a marine animal.  We want to be able to associate particular sound patterns with other behavior patterns.  We need to know when the creature is emitting its characteristic sounds.  The start time, intensity, and duration of each possible sound will be used as neural network inputs.  Suppose we know that three sounds are in its vocabulary.  One is ultrasonic, occurring in the range of 25,000 to 26,000 Hertz (cycles per second).  The second is a little lower, running from 18,000 to 20,000.  The final is at a very low frequency, from 100 to 200 Hertz.

The first step is to determine the required sample rate.  The absolute lower limit is twice the highest expected frequency, which would mean sampling 52,000 points per second.  However, there are at least two reasons for sampling at a higher rate.  The obvious reason is that we want as large a safety buffer as possible, since aliasing can cause serious problems.  We certainly don't want unexpected sounds producing artifacts in a region of interest.  A more subtle reason concerns our frequency-domain filter that is shaped like a Gaussian function.  We want it to taper smoothly to zero by the time it gets to the Nyquist frequency.  If it suddenly drops from a significant level to zero at that point, the result will be ringing in the implicit time-domain filter.  Wraparound effects will trouble us unless we pad with more zeros than those indicated by Equation (5-36).  There are ways around this problem, but they are beyond the scope of this text.  If at all possible, keep things simple by sampling fast enough to allow a smooth taper at the high end. We will tentatively agree to sample at a rate of 60,000 times per second, and we will confirm this decision soon.

It is easiest if we work in standard frequency units, with the Nyquist frequency at 0.5. Thus, the highest frequency that we are interested in is 26,000 / 60,000 = 0.4333. The lower end of this band-of-interest is 25,000 / 60000 = 0.4167. This tells us to center our filter at 0.425. Let us assume that we are willing to tolerate an attenuation to 80 percent of the input level at the extreme frequencies. Using the method shown on page 208, we see that the filter will drop to this level at 0.472 $s$. This corresponds to a deviation of 0.4333 − 0.425 = 0.00833. Dividing that difference by 0.472 gives $s$ = 0.0176. Finally, Equation (5-36) tells us that we will need to pad the input with at least 45 zeros to avoid wraparound problems with this filter.

Since this is the highest frequency filter, it behooves us to check its behavior at the Nyquist limit, verifying that it has tapered down nicely. The normalized distance there is (0.5 − 0.425) / 0.0176 = 4.26. When we square that and find the exponential of its negative, in accordance with the filter shown in Equation (5-35), we see that it has indeed dropped to an insignificant value.

We leave it as an exercise for the reader to verify that the filter for the somewhat lower frequency band has a center frequency of 0.3167, and $s$ = 0.0353, requiring at least 23 zeros.

The lowest frequency band is potentially the most difficult. We need it to be wide enough to capture the full range of frequency variation. At the same time, we want it to drop smoothly to zero by the time it gets to the DC point (0 frequency). We must force the filter response to zero at that point in order to avoid leakage from a constant or slowly varying offset in the data. But we must also avoid sharply chopping the response, or we will violate the Gaussian assumption of Equation (5-36), making many more zeros necessary. Let's see what we can do.

The center frequency is 150 / 60,000 = 0.0025, with $s$ = 0.00177. The normalized distance at a frequency of 0 is 0.0025 / 0.00177 = 1.412. The height of the Gaussian there is 0.136. We force that to zero, so there is a small but significant precipice in the response function. That is not good, but it is not a disaster. We have several options. The most simple-minded approach is to use Equation (5-36) to compute the minimum number of zeros, 452, then use many more, perhaps 1000. That's a lot of zeros to append, and we aren't even sure that it's enough, as we just guessed. Let's keep looking. We computed $s$ based on the amplitude dropping to 80 percent at the frequency extremes. Perhaps we can tolerate more loss. We have a 50 percent loss at 0.83 $s$ (see page 208). If we can live with that, we can use $s$ = 0.001. Now, the height of the Gaussian at 0 will be 0.002. That should present no problem whatsoever. We may even be able to back up a little. We need to append 800 zeros, which is a lot. But at least we can be sure of that number; there was no guesswork involved. Finally, we could custom design a filter for this band. Rather than use a Gaussian response, we could use any of the standard filters discussed in signal-processing texts. That way we could get a flatter response in the region of interest while still avoiding DC leakage and padding guesses. Such filters are beyond the scope of this text. Luckily, one should not need them often.

Since this frequency band is so low, we may want to decimate to save processing time and stored output data. The ratio of the original Nyquist frequency to the highest frequency in this band is 0.5 / 0.00333 = 150, so we could conceivably decimate by that factor. However, there are two other considerations. We must make sure that the high end of the filter has dropped to zero by the time it gets to the new Nyquist frequency. And we must accommodate length limitations of our FFT algorithm. The programs given in this text demand a power of two. Thus, we might choose to decimate by 128. That gives a new Nyquist frequency of 0.5 / 128 = 0.00391. The normalized distance of that point from the filter center is (0.00391 − 0.0025) / 0.001 = 1.41. The height of the Gaussian there is about 0.14, which is a little too high for comfort. We had better decimate by 64, which gives us a very comfortable margin.

The last example in this section is a wavelet application. Rather than knowing in advance that there are a few bands of interest, we want to capture a very broad spectrum of information in logarithmic scales. Perhaps we are analyzing audible artifacts in a musical sound. The lowest audible frequency is generally agreed to be 20 Hertz, so we will use that as our lowest frequency. We assume that the instrument will not produce any frequencies beyond 30,000 Hertz, so we sample at 60,000 points per second.

Morlet wavelet applications generally use four voices, although fewer can be tolerated if very wide frequency-domain windows are used and high frame quality is not demanded. Following the discussion on page 178, the voice frequencies will be separated by factors of the fourth root of two. Thus, the other three voices have their lowest frequencies at 23.78, 28.28, and 33.64, respectively. Again following the standard rule-of-thumb that the windows in the frequency domain should overlap to a considerable degree, we decide that the next scale of the lowest voice, 40 Hertz, should be down by a factor of 0.7 relative to the lowest scale of the highest voice, 33.64 Hertz. (That is the worst case. All other adjacent pairs will overlap even more.) That factor of 0.7 is attained at about $0.6\ s$. Divide the frequencies by the sample rate to standardize the units. Thus, we let $s = (0.000667 − 0.000561) / 0.6 = 0.000177$. Oh no! We have just encountered what looks like an unusual difficulty. Equation (5-36) tells us that the implicit time-domain filter used to effect this frequency-domain filter extends about 4500 points in each direction from the center! That seems like a lot of zeros to append. Worse, it tells us that we need at least several times that many points of data. That's a lot of data. This needs closer study.

Things settle down a bit when we look at this quantity in the perspective of the sample rate. We are sampling 60,000 points per second. Thus, 4500 points is less than one-tenth of a second of actual time. That's probably not long at all in our grand scheme. And consider what we are asking of this filter. We are trying to resolve between frequencies of 33.64 and 40 Hertz. That is going to require a significant time interval. At such a high sample rate, that is a lot of points. So we have several choices. The best choice is to understand that once the initial shock wears off, 4500 is really not such a bad figure anyhow. We will almost certainly be working with a lot more points than 4500; 60,000 if we collect even one second. Another choice is to decide that we do not really need data at such a low frequency. The higher we start, the fewer points we will need to resolve the filters. Finally, we could choose to use fewer voices with wider filters.

Depending on the application, resolution with two voices, or even just one (dangerous in regard to frame quality) may be adequate.

Now that we know how to design the lowest frequency filters, we automatically have all of the others. Remember that (in cases of practical interest) the filters scale by powers of two. The lowest voice will have center frequencies of 20, 40, 80, 160, and so forth. The other voices will do the same, and the scales, $s$, will also double. We do need to decide how high to go. This depends on the application, the sample rate, and the width of the filter. As always, we need to take care that the filter for the highest frequency tapers to zero by the time it gets to the Nyquist frequency. The Nyquist limit here is 30,000 Hertz. The highest voice starts at 33.64 Hertz. The highest power of two multiple of that under the Nyquist frequency is 17,224 Hertz. The scale is also multiplied by 512, so that filter will have $s = 0.0906$. The normalized distance of the center of that filter from the Nyquist frequency is $(0.5 - 0.287) / 0.0906 = 2.35$. The height of the Gaussian there is about 0.004, so we are very safe. If the Gaussian were too high, we would either have to increase the sample rate or stop at the next frequency down, 8612 Hertz.

The final issue is decimation. That is trivial with wavelets. It just doubles each time the frequency is halved. The conservative approach is to keep every point at the highest frequency of all four voices. When we cut the frequency and scale factor in half, we decimate by two. When we cut in half again, we decimate by four, and so on. It is left as an exercise for the reader to verify that as long as the smooth decline of the highest frequency filter is achieved as demonstrated in the previous paragraph, all other filters will behave similarly as their decimated Nyquist frequencies are approached.

# 6

## Time/Frequency Applications

This chapter will present a few examples of how time/frequency localization might be used in conjunction with neural networks. These examples are not meant to represent actual applications in detail, although they are very similar to problems that may be encountered in professional practice. Their main purpose is to illustrate various aspects of the procedures that must be followed in order to apply this technique to real problems.

### Speech Understanding and the Gabor Transform

There is tremendous interest in developing algorithms that are capable of understanding spoken words. Great progress is being made, and applications currently exist that have near perfect accuracy for quite a large vocabulary. The best algorithms use extremely sophisticated methods for delineating the beginning and end of phonemes, words, and phrases, that being one of the most difficult and important aspects of the problem. They also extract meaningful features in very clever ways. The simplistic approach presented here is not meant to be a substitute for those truly excellent algorithms. However, it does illustrate the power of neural networks to perform well even when casually applied to sparse data. And more importantly, it demonstrates some of the decisions that must be made in regard to the lattice spacing, resolution, and other parameters of the Gabor transform.

Let us begin by studying Figures 6-1 through 6-8. These are the Gabor transforms of the spoken words *one*, *two*, *three*, and *four*. The value plotted is the log of the power (squared modulus) of each element of the Gabor time/frequency matrix, with a noise floor in place to prevent large negative values. The upper plot is a 3D perspective view, with time running toward the upper-right corner. Zero frequency is at the far upper left. That unusual view is designed to reveal as much detail as possible by placing the low-frequency mountains at the rear of the plot. The lower figure is a contour plot of the same data. In this case, the layout follows the more conventional approach. Time runs from left to right, and frequency increases from front to back.

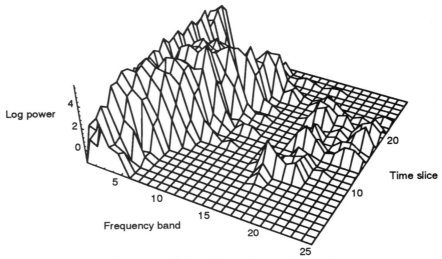

Fig. 6-1: Gabor transform of "one".

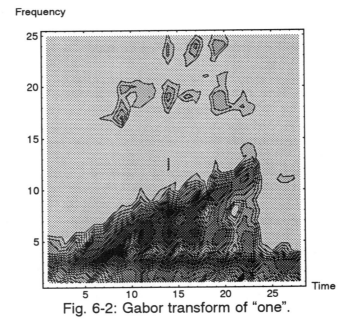

Fig. 6-2: Gabor transform of "one".

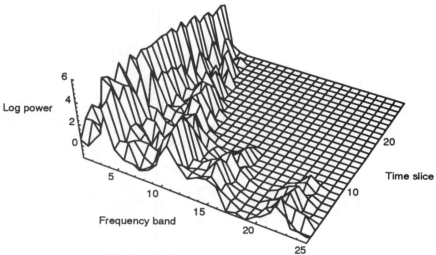

Fig. 6-3: Gabor transform of "two".

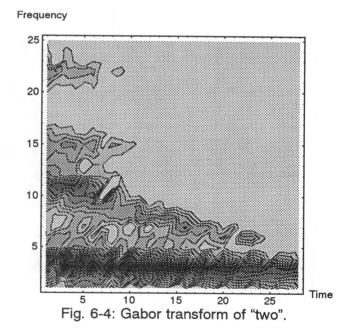

Fig. 6-4: Gabor transform of "two".

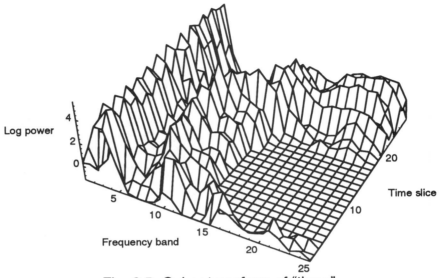

Fig. 6-5: Gabor transform of "three".

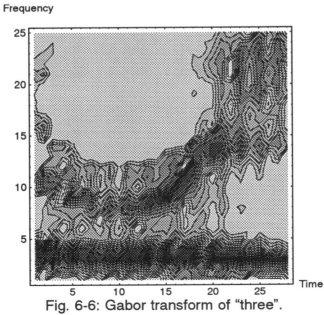

Fig. 6-6: Gabor transform of "three".

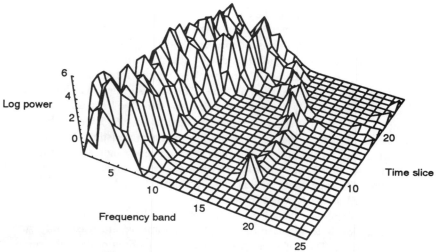

Fig. 6-7: Gabor transform of "four".

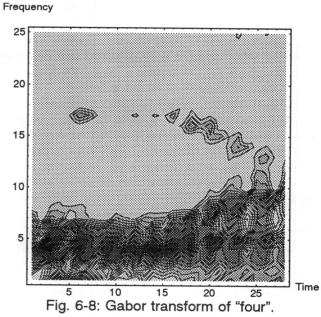

Fig. 6-8: Gabor transform of "four".

It is interesting to study the characteristics of each plot. This tends to be easier with the contour plot. Look first at the structure of *one*. The pronunciation of this word starts with a short "oo" sound, visible in the first three time slots. This is rapidly followed by "wuh" as the cheeks pull back and the mouth opens. The resulting increasing harmonic content can be seen starting at about time 5. The tongue contacts the hard palate at time 18, then suddenly drops at time 24, producing the terminal "nuh" sound. The now cavernous mouth is deeply resonant, producing essentially no harmonics.

The word *two* is simpler. The rush of air at the beginning produces many high frequency components until time 7 or so. As the lips close into a round shape, the harmonics rapidly disappear, leaving the pure "oo" sound.

*Three* is very complex. Once again we have a short rush of air from the beginning "th". The "err" sound has the characteristic pattern seen from time 5 through time 15. As the cheeks pull back, the frequency increases, until the "ee" sound appears at time 20. That sound is rich in high frequencies. Serious researchers would worry about the appearance of this plot. We always want the level of frequency components to drop off as the Nyquist frequency (slot 25 here) is approached. That does not happen in this example, so there is a significant possibility of aliasing. The correct response to this condition is to increase the sample rate until we are reasonably sure that no aliasing is occurring. Then, if necessary to reduce the amount of data, the series would be low-pass filtered and decimated. We will not concern ourselves with that here.

*Four* is relatively simple. The only notable feature is the cheeks relaxing at time 20, giving rise to the "r" sound. The very short rush of air at the beginning, resulting from the "f" sound, is not visible due to the relatively low resolution used.

The first step toward solving the problem of recognizing these sounds is to choose the parameters of the Gabor transform. This is often very arbitrary, and the solution presented here is no exception. There is much room for differing opinions. Nevertheless, there are a few basic guidelines that are followed.

We do not need to demand a frame. Producing a good frame imposes a heavy burden in terms of the amount of data generated. We are not attempting to reconstruct the series, so throwing away some information is not unforgivable. Furthermore, it is safe to assume that the composition of the series changes fairly slowly over time. We do not need to meticulously examine every millisecond, looking for some shortlived event. The general evolution of the spoken word is all that we really need to capture. Therefore, we can safely space the lattice points widely in the time domain. The typical length of a word is about 3000 sample points with the device used by the author. Setting $\tau_0 = 100$ would then provide about 30 time intervals — more than enough.

A similarly arbitrary decision is needed for the lattice-point spacing in the frequency domain. Dividing the range from 0 to the Nyquist frequency (0.5) into 25 intervals seems reasonable. Thus, the frequency spacing, $\phi_0$, is $0.5 / 25 = 0.02$.

At this point, we should always check to see how close to critical spacing we are. The resolution product $\tau_0 \cdot \phi_0$ must not exceed 1 if we are to have a frame. We have already agreed that a frame is not mandatory for this application, but it is still worthwhile knowing just how far off we are. The product here is 2, so we are in bad but not terrible shape. If it were much larger, we should worry that the gaps between the lattice points are too wide, and we are ignoring an inordinate amount of information. But 2 is something that we can live with as long as we are sure that it is safe to throw away a little information.

This might be a good time to state an obvious fact, just to be safe. We must use the same units for $\tau_0$ and $\phi_0$ when we compute the resolution product. It doesn't matter what units we use, as they will cancel. But we must be consistent. Here, we used the intersample time as the unit, measuring $\tau_0$ in points and $\phi_0$ in cycles per point. We could just as well use actual elapsed time in seconds, or hours, or anything else. Just pick a standard and stick with it. Now let's continue from where we left off.

Since the resolution product exceeds 1, we know that information must be lost where it falls in the gaps between the resolution rectangles centered at each lattice point. But how should these gaps be arranged? If we use Equation (5-22) to choose $\sigma$, the gaps will be balanced between the time and frequency domains. But that is not what we want for this problem. We have already decided that the time-domain gaps can be large. It is more important that we capture plenty of frequency-domain information. Our desired tradeoff is thereby indicated.

We have a rule of thumb (page 160) that the ratio of the lattice point spacing in a particular dimension (time or frequency) to the window radius (delta) in that dimension should not exceed approximately 3 if we are to avoid significant gaps. At critical spacing ($\tau_0 \cdot \phi_0 = 1$) and with optimal rectangles (Equation (5-22)), this ratio is exactly twice the square root of $\pi$, or about 3.54. (It is a nice little exercise for the interested reader to verify this. It is not difficult. See Equation (5-20).) So we make the arbitrary decision to use that ratio for the frequency domain, knowing that the resulting gaps will be large but marginally tolerable. This gives $\Delta_f = 0.02 / 3.54 = 0.00565$.

We can now use Equation (5-20) to determine that $\sigma = 19.9$ and $\Delta_t = 14.09$. Checking for gap size in the time domain, we see that $\tau_0 / \Delta_t = 7.1$. That ratio is considerably in excess of the rule-of-thumb value of 3, so we know that information will be lost in the time domain. But we already agreed to that, so no problem.

The above steps for setting up the Gabor transform are not absolutely firm. There is much latitude for variation. But they do demonstrate a typical thought process and are certainly reasonable. Probably the most important aspect is that the ratio of the lattice point spacing to delta in each dimension should be checked. If it exceeds 3, indicating significant gaps in the windows in that dimension, the implications of these gaps should be considered. If the gaps are a problem, the lattice point spacing in that dimension must be reduced (at the price of increasing the amount of processing that must be done). Often, it is best to distribute the gaps equally by using Equation (5-22) to find $\sigma$. But, as was the case here, that may not always be the best choice. Remain flexible.

Once the Gabor transform has been defined, we must decide how much of it to present to the neural network. In many applications, we will want to use all of the information that results from the parameters so laboriously derived over the last few paragraphs. But for now we will discard most of the information, and blend the rest of it into a few bins.

A fundamental decision is whether or not we want to preserve the phase information, using the actual complex values of the Gabor coefficients, or just use the power. For this application, we choose to ignore the phase to facilitate data reduction. (Phase tends to have little value in many speech applications, though not all.)

If we discard the phase, we have the significant advantage that the powers in the Gabor coefficients can be added. This lets us set up custom resolution rectangles, as shown in Figure 5-14. There is no need to use identical rectangles, or to make them all the same size and shape. We are free to customize them for the problem at hand. In particular, the author chose four rectangles by visually examining the Gabor plots of the four words (in Figures 6-1 through 6-8). The first rectangle runs from times 12 through 16 at frequencies of 7 and 8. The second is from times 1 through 4 at frequencies 9 and 10. The third is times 21 through 26 at frequencies 18 through 21. The last is times 22 through 26 at frequencies 7 and 8. No statistical test was done. This was a purely visual process, based on one random sample. This would, of course, be inexcusable for a serious application. But here it serves to further illustrate the power of neural networks.

Four variables were computed from each word sample by summing the power in each of the four rectangles, then taking the log of the total power. The use of logs is natural for sounds, as that mimics mammalian hearing.

A training set consisting of 50 samples of each of the four words was constructed (200 cases in all). Several real-domain neural networks having different numbers of hidden neurons were trained using the four variables. These networks were tested using an independent collection of 75 of each of the four words, 300 cases in all. A correct-response rate of nearly 97 percent was obtained with a network having 8 hidden neurons. While this rate would be inadequate for practical applications, (100 percent is easily obtained with more care), it is good considering that it is obtained with only four hastily chosen variables and a small training set. The misclassification rate for each of these networks is graphed in Figure 6-9. As commonly happens, the error rate increases when so many hidden neurons are used that idiosyncrasies of the training set are learned, degrading generalization ability.

## Doppler-Shift Radar and the Gabor Transform

Police officers on speed patrol, operators of military vehicles, and scientists mapping the surface of distant planets all use doppler-shift radar. This powerful tool is based on a simple physical principle. When an electromagnetic impulse is bounced off of a reflective object, the amount of time that elapses between the generation of the original impulse and the detection of its reflection is a function of the distance of the reflecting object. Furthermore, if the object is moving toward or away from the emitter/detector, then the frequency of the detected impulse will be different from what it was when it was emitted. Thus, both the distance and the velocity of the object can be estimated.

There are many ways of solving the estimation problem, given knowledge of the emitted and detected impulses. In fact, under fairly reasonable conditions, mathematically optimal solutions exist. We will not cover these traditional algorithms, as they are treated in many standard texts. In fact, the discrete sampling needed for these methods is extremely difficult at radar frequencies. (However, hardware filters can be used to feed

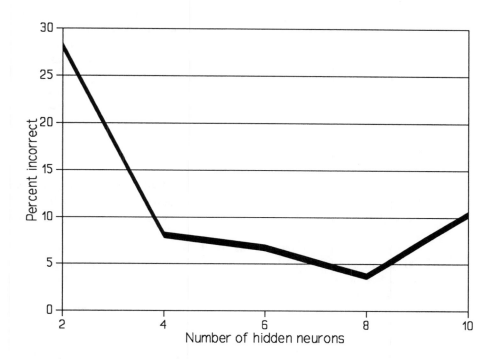

Fig. 6-9: Misclassification of speech classifier.

time, magnitude, and phase information to the network, avoiding the use of sampling and software filters.) We will focus on neural-network solutions to this problem only to demonstrate some general tecniques for doppler problems and to compare real-domain and complex-domain networks. This simple application provides several valuable insights into the behavior of complex-domain neural networks. That is our primary goal.

We will study some minor variations on this general problem. Several different variables will be used, and training and test sets will be constructed differently. But all of these variations will have some common themes. First, the elapsed time to be estimated will vary only slightly. This is because triggering methods or regularly spaced sample bins can be used to obtain the approximate elapsed time. The neural network's task will be to refine the estimate as much as possible. Second, a fairly wide range of detected frequencies will be used. This insures that the algorithm will be tested for effectiveness with large velocities both toward and away from the detector. Actual applications will almost always involve a much narrower frequency range. We use a wide range to violently stress-test the networks. Finally, as few time/frequency localization filters as possible will be used. This economy may not be needed in practice, as parallel processing can often be employed. But it presents the neural networks with a severe test, so that both the strengths and the weaknesses of each model are revealed.

In summary, the methods presented in this section should not be taken as suggested solutions to actual problems. The network's task has been deliberately made more difficult than it needs to be in order to thoroughly test their performance. Real applications would generally use much more straightforward methods. Also, it should be emphasized that the methods described here are *not* mathematically optimal. They cannot compete with traditional solutions under many conditions. This presentation is purely tutorial, to aid the reader in implementing neural solutions to related problems.

## A Naive Approach

The first method tested demonstrates how *not* to solve the problem. It illustrates the fact that phase information may be of dubious importance and is sometimes better ignored. We will not report exact numerical results, but will summarize and comment on the results.

The emitted impulse consists of five cycles of a sine wave having a period of 100 samples. Therefore, the total length of the impulse is 500 samples. The time delay ranges from plus or minus 500 samples, relative to an arbitrary center time. The received frequency varies from a period of 400 to 600 samples. Twenty-one different time delays were used to generate the training set, and 41 time delays covering the same range were used for the validation set. Twenty-one different frequencies went into the training set, and 20 were used for the validation set. The time delays and frequencies for the validation set were chosen to be slightly offset from those of the training set, so that repetition of identical values did not occur. Also, random noise was added to the received signal in the validation set.

A Gabor transform was applied to the received data. Nine complex transform points were used. These correspond to the center time delay as well as to the two extremes. The frequencies were treated similarly. In other words, three filters were constructed. These filters had center frequencies with periods of 400, 500, and 600 samples, respectively. Each of the filters was applied to the received data at a lead of 500 samples, lag of 500 samples, and dead center. As a result, the neural network is presented with 18 real inputs, corresponding to nine complex numbers.

It should be obvious that this is a very naive approach, definitely overkilling the problem. A frivolous amount of expensive preprocessing is being done. The four corners of this rectangle in time/frequency space correspond to the extremes of the training and testing sets, with the remaining five points filling in the gaps nicely. Thus, it should not be surprising that excellent results can be obtained by computing the power of each of the nine Gabor transform points, discarding the phase, and giving the power to an ordinary real-domain neural network. That method outperformed any complex-domain approach. Neural networks having only a very few hidden neurons were able to tolerate quite large amounts of noise before their accuracy in determining the time delay and frequency of the validation set deteriorated. When the phase was included, a source of confusion was introduced. The more complicated representation necessitated more hidden neurons, which, in turn, led to overfitting and poor generalization to the validation set. However, it is interesting to note that even in this "non-test" the complex-domain neural networks consistently outperformed the real-domain neural networks in generalizing. It is just that the power-only networks were the real winners!

There is a very important moral to this story: *We should not be thinking in terms of complex-domain versus real-domain neural networks when the problem does not even need phase information.* The fact that the nine filters' power response alone is sufficient information to solve the problem tells us that we should just use the power in conjunction with a real-domain network. In many cases that will be obvious if only a little thought is applied. But whenever the importance of phase is not clear, it is vital that solutions with and without phase be attempted and the better method selected. If phase is not needed, its inclusion will usually degrade quality.

## A Better Approach

We will now treat in considerable detail another approach. This one is much more economical in that only two Gabor filters are used. As such, it is a more realistic solution to practical problems that have time and/or equipment constraints. More importantly, it presents us with a problem in which phase information is absolutely essential. This lets us thoroughly compare real-domain and complex-domain neural networks. The principles demonstrated in this relatively simple example tend to apply to more elaborate problems and hence deserve careful study.

The emitted impulse here consists of three periods of a sine wave spanning a total of 500 sample points. The received impulse's frequency varies by plus and minus 10 percent. The elapsed time varies over a total range of 160 samples, which is slightly less than one complete cycle of the sine wave. Eleven different elapsed times and eleven different frequencies are used for the training set, giving a total of 121 cases. In particular, the following code fragment is used to generate the received impulses for the training set.

```
centf = 3. / 500. ;              // Center frequency
fdif = 0.1 ;                     // Max deviation fraction

for (istart=1000-250-80 ; istart<=1000-250+80 ; istart+=16) {
  for (ifreq=-500 ; ifreq<=500 ; ifreq+=100) {

    for (i=0 ; i<2000 ; i++)
      x[i] = 0.0 ;

    freq = centf + (centf * fdif * ifreq / 500.0) ;
    for (ipos=0 ; ipos<500 ; ipos++)
      x[istart+ipos] = sin ( ipos * freq * two_pi ) ;
      ...
```

Four validation sets are generated. All four of them are produced from elapsed times and received frequencies randomly generated within the training range. There is an extremely low probability of duplication of exact values that appear in the training set. The net effect is that the neural network will be forced to do extensive interpolation between the relatively widely spaced training points. The first of the four validation sets has no noise added to the received signal, so that set will measure the network's ability to interpolate between training points. The remaining three validation sets have increasing amounts of noise added so that the network's response to noise that did not appear in its training set can be evaluated (in addition to interpolating). The following code fragment shows how the validation-set received impulses were computed. Note that the value returned by rand() ranges from 0 to 32767.

```
for (rep=0 ; rep<5000 ; rep++) {
  istart = 1000 - 250 - 80 + rand() % 161 ;
  ifreq = -500 + rand() % 1001 ;
  freq = centf + (centf * fdif * ifreq / 500.0) ;

  for (i=0 ; i<2000 ; i++)
    x[i] = 0.0 ;

  for (ipos=0 ; ipos<500 ; ipos++)
    x[istart+ipos] = sin ( ipos * freq * two_pi ) ;

  memcpy ( xs , x , 2000*sizeof(double) ) ;
  ... Transform and write to file 1

  for (i=0 ; i<2000 ; i++)
    x[i] = xs[i] + (rand() - rand()) / 65536. ;
  ... Transform and write to file 2

  for (i=0 ; i<2000 ; i++)
    x[i] = xs[i] + (rand() - rand()) / 32768. ;
  ... Transform and write to file 3

  for (i=0 ; i<2000 ; i++)
    x[i] = xs[i] + (rand() - rand()) / 16384. ;
  ... Transform and write to file 4
```

For all training and validation data, the transformation is identical. Two Gabor filters are used, so two complex numbers (four real numbers) are generated for the network input data. One of the filters has a center frequency equal to the minimum range of the data, 10 percent less than the center frequency of 3/500 cycles per sample. The other filter has a center frequency at the upper limit, 10 percent higher than the frequency of the emitted impulse. Both filters are applied at the same time offset. The code used to compute the filters and apply them to the data is now shown.

```
phi0 = centf * (1 + fdif)  -  centf * (1 - fdif) ;
sigma = 170.0 ;
gconst = exp ( 0.25 * log ( 3.141592653589793 )) ;// Pi to the 1/4 power
gconst = 1.0 / (gconst * sqrt ( sigma )) ;          // Actual normalizer for Gaussian
fac = -0.5 / (sigma * sigma) ;                      // Common factor in exponent

for (ipos=1 ; ipos<=500 ; ipos++) {   // Right half of filter
   weight = gconst * exp ( fac * ipos * ipos ) ;
   cr1[ipos-1] = cos ( centf * (1.0-fdif) * ipos * two_pi ) * weight ;
   ci1[ipos-1] = sin ( centf * (1.0-fdif) * ipos * two_pi ) * weight ;
   cr2[ipos-1] = cos ( centf * (1.0+fdif) * ipos * two_pi ) * weight ;
   ci2[ipos-1] = sin ( centf * (1.0+fdif) * ipos * two_pi ) * weight ;
   }

// ---> The code above computed the filters.  The code below applies them.

r1 = gconst * x[1000] ;              // Center real coef is gconst
i1 = 0.0 ;                           // Center imaginary coef is 0

for (ipos=1 ; ipos<=500 ; ipos++) {  // One side of the filter
   r1 += x[1000+ipos] * cr1[ipos-1] ;
   i1 += x[1000+ipos] * ci1[ipos-1] ;
   }

for (ipos=1 ; ipos<=500 ; ipos++) {  // Other side, shown separately
   r1 += x[1000-ipos] * cr1[ipos-1] ; // for clarity
   i1 -= x[1000-ipos] * ci1[ipos-1] ;
   }

r2 = gconst * x[1000] ;
i2 = 0.0 ;

for (ipos=1 ; ipos<=500 ; ipos++) {
   r2 += x[1000+ipos] * cr2[ipos-1] ;
   i2 += x[1000+ipos] * ci2[ipos-1] ;
   }

for (ipos=1 ; ipos<=500 ; ipos++) {
   r2 += x[1000-ipos] * cr2[ipos-1] ;
   i2 -= x[1000-ipos] * ci2[ipos-1] ;
   }
```

It is now time to ponder the nature of the testing. First, observe that since both filters are centered at the same elapsed time, there is no information directly available from different time centers comparable to what we had in the example of the previous section. Any elapsed-time information used by the networks must be gleaned from phase information. This would, of course, be a trivial problem if the frequency of the received signal were fixed in advance. The phase would be related to the elapsed time in a simple way. But when the incoming frequency varies over such a wide range, the relationship between phase and elapsed time becomes far more complicated and must be deduced in the context of the relative outputs of the two filters.

Next, we should see that since the two Gabor filters have center frequencies at the extremes of the range and overlap to some degree, estimating the frequency of the received impulse is almost a trivial problem. Phase has little or no importance. The relative power outputs of the two filters are all that is really important. That sort of problem not only is not helped by a complex-domain neural network operating on phase data, but it is often even hindered by an overabundance of confusing information. Furthermore, experienced neural network researchers know that it is almost always best to train separate networks when multiple outputs are involved. Therefore, we will test only elapsed-time estimation here, as that is the only real test of the competing networks. If frequency estimation is also needed, a real-domain network operating on power alone should be considered.

Some readers may be interested in performing their own tests that are variations on those discussed here. The code listed above will let them easily generate their own customized data sets. For their convenience, a typical control file for MLFN is now listed. Parameters equal to or approximately equal to those shown here were used for all of the training and testing presented in this section. Several dozen TRY restarts were done before training was stopped and results recorded. Any readers who attempt to duplicate the results of this section using just a few tries may be fooled. Like many problems, the error space of this problem is filled with local minima that must be rejected.

```
DOMAIN = COMPLEX-HIDDEN
ERROR TYPE = MEAN SQUARE
MODE = MAPPING
OUTPUT ACTIVATION = LINEAR

INPUTS = 2
OUTPUTS = 1
FIRST HIDDEN = 3
SECOND HIDDEN = 0

CUMULATE TRAINING SET = t2c

ANNEALING INITIALIZATION TEMPERATURES = 3
ANNEALING INITIALIZATION ITERATIONS = 1000
ANNEALING INITIALIZATION SETBACK = 1000
```

```
ANNEALING INITIALIZATION START = 0.15
ANNEALING INITIALIZATION STOP = 0.05
ANNEALING ESCAPE START = 3.0
ANNEALING ESCAPE TEMPS = 1
ANNEALING ESCAPE ITERS = 100

ALLOWABLE ERROR = 0.0
LEARNING ALGORITHM = ANNEALING_CJ
COMPUTE WEIGHTS

SAVE NETWORK = t2c.wts

TEST NETWORK WITH INPUT = t2c
TEST NETWORK WITH INPUT = v2c
TEST NETWORK WITH INPUT = vv2c
TEST NETWORK WITH INPUT = vvv2c
TEST NETWORK WITH INPUT = vvvv2c
```

We can now begin discussing the test results. The most basic experiment tests how well real-domain and complex-domain neural networks are able to interpolate between training points. Recall that training was done with an 11 "elapsed time" by 11 "received frequency" grid, for a total of 121 training cases. The validation set used practically continuous time and frequency parameters within the trained range, with essentially no duplication of training points. No noise was added to this first validation set. We now study real-domain and complex-domain networks having varying numbers of hidden neurons. In order to compare networks having approximately equal numbers of weights, twice as many hidden neurons are used for the real-domain networks as for the complex-domain networks. Experiments were performed with from zero hidden neurons up to a maximum of ten real and five complex. The results are shown in Figure 6-10. Observe that the complex-domain network is superior to the real-domain network in all tests having enough hidden neurons to be of practical use. The real-domain network was better only when there were very few hidden neurons. That reversal-of-fortune phenomenon is very common. The author believes it is due to the fact that there is a critical amount of saturation pattern formation achieved by neuron nonlinearity that is needed to solve some problems. In this case, when the complex-domain network has only a very few hidden neurons to call upon for feature detection, it is severely hindered by a lack of available patterns. The comparable real-domain network has twice as many hidden neurons, so it has the advantage in terms of the number of saturation patterns available. But once the complex-domain network is given enough hidden neurons, it quickly takes over. The author has seen this effect in many diverse situations.

A fairly common (but not universal) weakness of complex-domain neural networks is demonstrated in the next experiment. When the training data is clean but the test data is very noisy, real-domain networks sometimes (but not always!) slightly outperform their complex-domain counterparts. Figure 6-11 shows the mean squared error of the second-noisiest validation set for the same networks used in the previous experiment. It is clear that the real-domain network is superior. But remember that if the noise level of the test data will vary, and we demand the best results possible when the noise is low, our best choice may still be a complex-domain network. The only time the real-domain network is clearly superior is when the test data will always be noisy, so that the price we pay in regard to clean test data is not relevant.

The issue of noisy test data leads us to more closely examine the relative performance of the two network models when presented with varying levels of noise. Recall that four validation sets were constructed, with noise levels ranging from none to a peak-to-peak amplitude twice that of the impulse. The real-domain network having eight hidden neurons achieved the minimum noisy-set error of all networks, and the complex-domain network having four hidden neurons is its counterpart in terms of number of weights. These two networks' errors for all four validation sets are plotted in Figure 6-12. This gives an idea of the noise level at which relative performance is swapped.

There is one more question begging an answer. We have shown that the complex-domain model suffers relative to the real-domain model when the test data is noisy compared to the training data. But what if the training data is noisy? The answer is that the results are very different. We will choose the same models that were used in the previous varying-noise test shown in Figure 6-12. This time we contaminate the training data with noise having a peak-to-peak amplitude equal to that of the pulse. We do this by inserting the following two lines of code before transforming the training signals. Note that the same 121 time and frequency values are still used.

```
for (i=0 ; i<2000 ; i++)
   x[i] = x[i] + (rand() - rand()) / 32768. ;
```

A real-domain neural network having eight hidden neurons and a complex-domain network having four hidden neurons were both trained with this data. They were then tested with the same four validation sets that were used in the previous experiment. The results are shown in Figure 6-13. Notice that the complex-domain network is now the winner across the board. It should not be too surprising that using a noisy training set universally increased the error for the real-domain network. Nor should it be surprising that the noisy training set increased the complex-domain network's error for the clean test set. But what is utterly fascinating is that the error for noisy test sets not only decreased for the complex-domain network, but decreased dramatically. There is clearly a lesson here, although the author wishes that it were more clear. It is well known that using training sets supplemented with noisy training cases can often slightly improve performance on noisy test sets. But that effect is not usually seen when the training data is itself contaminated, rather than being supplemented with noisy cases. And such a great level of improvement is highly unusual.

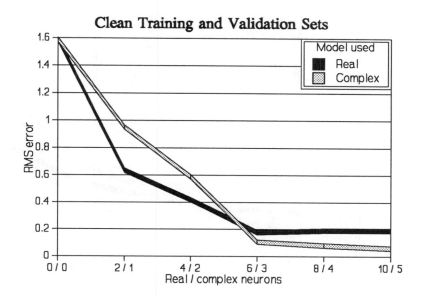

Fig. 6-10: Error with clean validation data.

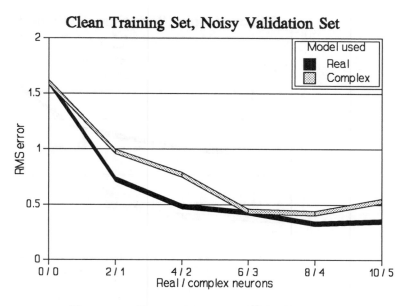

Fig. 6-11: Error with noisy validation data.

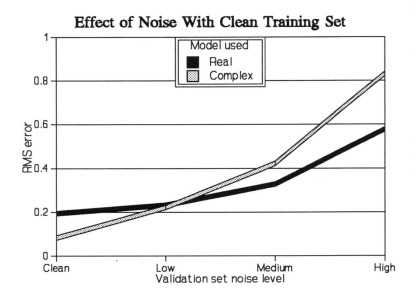

Fig. 6-12: Noise effect with clean training.

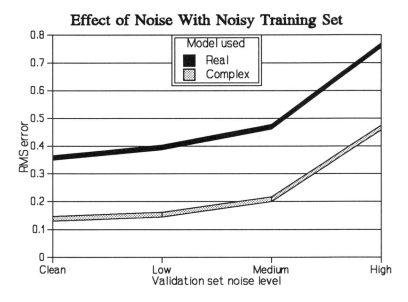

Fig. 6-13: Noise effect with noisy training.

## Phase-Defined Events and Wavelets

Sometimes an event is defined not by its shape or frequency content, but by its position with respect to another event. The electrophysiology of the heart is characterized by many such phenomena. The position of artifacts relative to the sinus cycle can be immensely important. Wavelets, with their ability to zero in on patterns at different scales, are particularly adept at analyzing positional relationships. In this section, we will study a simple problem in which the correct response will be determined by the position of a small but well-defined event relative to the phase of a dominant signal. We will formulate the problem in such a way that the straightforward neural network solutions documented here will provide a very difficult test of the networks' power. This will let us see the power of the combined wavelet/network approach under exceptionally adverse conditions. It will also demonstrate the performance difference for real-domain and complex-domain neural networks. This section will conclude with a brief discussion of how this sort of problem should be approached if we want to make the neural network's life a lot easier.

Let us suppose that the signal we are analyzing primarily consists of a strong periodic wave having moderately varying frequency. Under normal conditions, this waveform is accompanied by a small blip at the bottom of each trough. Occasionally, the machinery goes awry, and this is characterized by the blip moving 180 degrees along the phase of the main wave. When there is a problem, the blip will be at or near the top of each cycle. These two conditions are displayed in Figures 6-14 and 6-15, respectively.

Suppose further that when the machinery is misbehaving, we can glean a little information by the exact position of the spike. It does not necessarily happen at exactly the crest of the wave. It may occur a little before or after the crest. We would like to be able to ascertain its exact position relative to dead center atop the crest.

This would be a trivially easy problem if the frequency of the dominant waveform were constant. And it would be almost as easy if we had a means of locking the sampler into the phase of the dominant varying-frequency wave. But in real life, there will often be complications that prevent a solution that easy. For example, there may be multiple frequencies dominating. Therefore, we take a worst-case approach and train the neural networks with all possible sampled phases. We generate dominant components having periods of from 18 to 22 samples. The position of the spike in the "bad" data ranges from about plus or minus 10 percent of the total period, with the center of that range being as directly atop the peak as possible. Both the training set and the validation set are contaminated with moderate noise.

Our chosen Morlet wavelet uses three scales, each having two voices. The shape parameter is the usual 7.54, and the basic time-domain lattice spacing is six samples. The lowest scale and the two finer scales with centers at up to plus and minus 12 samples are presented to the network. This is a total of 18 complex inputs (the lowest scale, the medium at −12, 0, and +12, and the smallest at −12, −6, 0, +6, and +12). Two voices at each scale and center gives us good but not excellent coverage in the frequency domain.

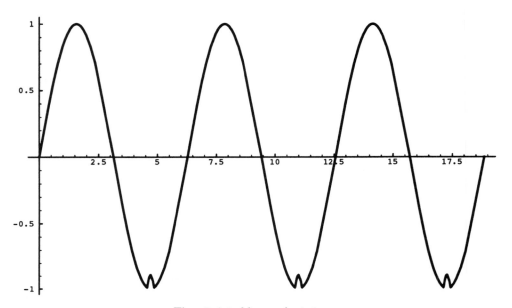

Fig. 6-14: Normal status.

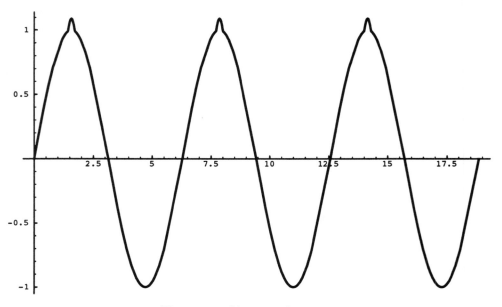

Fig. 6-15: Abnormal status.

There are two separate problems to be solved. The first is a binary decision: Is the condition normal? The second problem involves only the abnormal cases. We must estimate the position of the blip relative to the center of the top crest. For both problems, the training sets consists of several hundred samples at a variety of frequencies, phases, and spike positions. The validation sets consist of three thousand independent samples. The validation-set results for some real-domain and complex-domain neural networks are presented in Figures 6-16 (the good/bad decision) and 6-17 (the position estimation). Note that we see the same pattern that was observed in the previous section. For very few hidden neurons, the real-domain network is superior. But as soon as a reasonable number of neurons are used, the complex-domain network stands out.

## A Better Approach

There is one aspect of this problem that must be considered carefully. Superficially, it might seem that the smallest-scale wavelet provides practically all of the information needed for locating the blip. But that is not the case, for *we do not know the phase of the dominant waveform*. And it is not the position of the blip in the sample that we want. Rather, it is the position of the blip relative to the phase of the dominant waveform. That phase is only available through the largest-scale wavelet's phase and perhaps somewhat from the medium-scale wavelets. *This makes the network's job extraordinarily difficult.* It must learn a tremendous variety of patterns. This approach was deliberately chosen for this test, as it lets us "stress-test" the networks. Naturally, in real-life problems, the last thing we want to do is stress our network! We all know that if we treat our neural networks well, they will treat us well. So, how do we go about making its job easier? The answer is simple. Use two or more specialized networks instead of just one general network.

We must devise some sort of testing criterion that will segregate the possible samples into groups that exhibit as much similarity as possible within each group. That criterion will be applied to each sample, whether it be from the training set or an actual case being applied to the trained network. The result of the test determines which of several networks receives the sample. The bane of the networks is variation that is irrelevant to the decision. The more that can be removed, the better.

In the simple example just studied, the irrelevant variation is the phase of the dominant waveform. Thus, the obvious criterion would be that phase. We would filter the sample with a pair of quadrature filters across a range of frequencies. For the frequency achieving highest output, we would compute the phase of the sample. That phase would be used to determine which neural network gets the sample. Perhaps we could simply use the phase of the largest-scale wavelet. Even using two networks, one for positive phase and one for negative, would significantly improve performance. Using more would be even better.

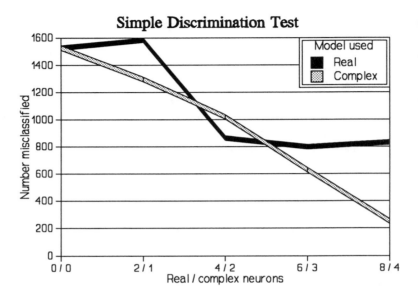

Fig. 6-16: Testing for normal/bad position.

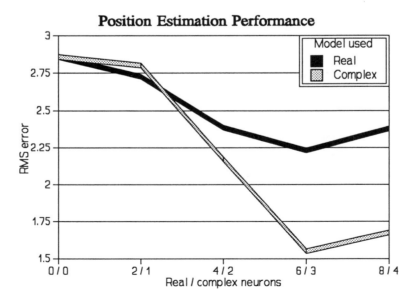

Fig. 6-17: Finding the exact position.

Many readers will immediately wonder why, if we go to that trouble, we even need a neural network. Why not just follow the phase determination with a simple correlation filter or some other primitive but powerful test and dispense with the neural network entirely? There are at least two reasons for avoiding that more straightforward approach. One is that neural networks are generally more robust than traditional techniques. Unusual noise patterns can easily distort the direct phase estimate, thereby invalidating the entire procedure. Neural networks are usually better able to withstand wild data points and other difficulties that are encountered in practice. The other reason is that this example was quite modest in its complexity. It would be simple to compute a phase. But what if the frequency varies so fast that significant sidebands are generated? What if there are two dominant frequencies, and only one of them is significant in conjunction with the blip? The possible complications are endless. The beauty of neural networks is that *we do not need to understand the underlying phenomenon and model it.* We can just throw the data at the network, and it will handle the dirty work.

Alert and skeptical readers will now ask what good a segregation criterion will do if the signal is so complicated that we cannot easily come up with a suitable test like the phase mentioned above. The answer is that *it doesn't really matter terribly!* Suppose that our segregation criterion is a coin flip: Heads, and the sample goes to one network; tails, and it goes to the other. The fact of the matter is that as long as we have enough training data, the performance will still be as good as if we just used one network for every sample. And if the criterion is good, even just a little good, performance will only improve. So we do the best that we can. If the sample is known to contain two frequencies, and we do not know in advance whether the higher or the lower is the important one, we could lock onto both and use four or more networks based on both phases. Or perhaps we could locate the minimum or maximum data point in the sample and base the network choice on the position of that extreme point relative to the extent of the sample. Or maybe there is some other event that is easily detected by a correlation filter, so that we could base the decision on the position of that event. The idea is that if we can find some aspect of the sample that is measurable, we may be able to use that information to break up the problem into several subproblems, training separate networks for each. Even if it turns out that our choice was poor, performance will most likely still be as good as it would have been with just one network as long as we can procure enough training data.

# 7

## Image Processing
## in the Frequency Domain

So far, the applications in this text have focused on signal processing. Most of the single-dimension techniques already discussed have straightforward generalizations into two dimensions. The basic Fourier transform has an immensely useful 2-D counterpart that can be computed very easily. And the Gabor and Morlet wavelet transforms studied in the previous chapter have fairly simple 2-D versions. This chapter will discuss these topics.

## The Fourier Transform in Two Dimensions

We have already studied the Fourier transform in one dimension. (The reader may wish to review the material starting on page 81.) It is often useful for time series analysis. It does have the weakness that it fuses all of the information in the entire series into a set of numbers that has nothing to do with time itself. Time-dependent changes in the frequency content of the signal are lost for all practical purposes. That led us to study the Gabor and wavelet transforms. But sometimes such information fusion is not detrimental to an application. If the nature of the series is such that we can assume that its frequency content is effectively constant across its extent, then the ordinary Fourier transform is acceptable. The same holds true for image processing. Later in this chapter we will discuss the two-dimensional extensions of the Gabor and Morlet wavelet transforms. In this section we will present the two-dimensional discrete Fourier transform (DFT).

The one-dimensional DFT is shown in Equation (4-3) on page 83. Its trivial generalization to two dimensions is shown in Equation (7-1), the forward transform, and Equation (7-2), the inverse transform.

$$H(f_x, f_y) = \sum_{x=0}^{n_x-1} \sum_{y=0}^{n_y-1} h(x,y) e^{i\frac{2\pi x f_x}{n_x}} e^{i\frac{2\pi y f_y}{n_y}} \qquad (7\text{-}1)$$

$$h(x, y) = \frac{1}{n_x n_y} \sum_{f_x=0}^{n_x-1} \sum_{f_y=0}^{n_y-1} H(f_x, f_y) e^{-i\frac{2\pi x f_x}{n_x}} e^{-i\frac{2\pi y f_y}{n_y}} \qquad (7\text{-}2)$$

As was the case in one dimension, $f_x$ in that equation takes on $n_x$ different integer values. Each value is a frequency. It is the number of complete cycles per the horizontal extent of the image. The maximum horizontal frequency that can be resolved without aliasing, the Nyquist frequency, is $n_x / 2$. That corresponds to a period of two pixels. So that array indexing is easy, we generally let $f_x$ range from zero through $n_x - 1$ when programming the transform. As was described on page 89, values of $f_x$ beyond the Nyquist limit correspond to negative frequencies. For the remainder of this section, we will use whichever of these two alternative interpretations is most convenient at the time.

The same interpretation applies to the vertical frequencies. The maximum vertical frequency is $n_y / 2$ cycles per the vertical extent, and $f_y$ will take on $n_y$ different integer values.

The two-dimensional DFT looks long and complicated, but really it is not. Each of the two exponential terms is a cosine + $i$ sine wave. One term varies along the $x$ direction at a frequency determined by $f_x$. The other term varies in the $y$ direction at a frequency determined by $f_y$. This is nothing more than two separate transforms, one operating in the horizontal direction, and the other operating in the vertical direction. Image variation in other directions will have vertical and horizontal components that will be picked up according to their relative strengths.

Let us pursue this line of thought a little more so that we can better understand the nature of the transform. Suppose for a moment that $f_x = 0$. The first exponential term in the transform equation will then equal one for all values of $x$, so that it can be ignored. It can be seen that $H(0, f_y)$ is the sum over all columns ($x$ values) of ordinary Fourier transforms. Each transform is the one-dimensional transform of the rows of a column. In other words, $H(0, f_y)$ represents the overall variation in the image due to up-down ripples at a frequency of $f_y$ cycles per vertical extent. It can similarly be seen that $H(f_x, 0)$ accounts for the variation in a horizontal direction.

The above property can be generalized. To keep things simple, assume that the image is square, so that $n_x = n_y = n$. Recall that $e^a e^b = e^{a+b}$. Express the domain variables of the transform in polar coordinates, as shown in Equation (7-3).

$$(f_x, f_y) = (k \cos \theta, k \sin \theta) \qquad (7\text{-}3)$$

Now consider the product of the exponential terms in the transform. Apply the multiplication rule and express the frequencies in polar coordinates. The product term can then be written as shown in Equation (7-4).

$$e^{i\frac{2\pi x f_x}{n}} e^{i\frac{2\pi y f_y}{n}} = e^{\frac{2\pi i}{n}(x\,k\cos\theta\,+\,y\,k\sin\theta)} \qquad (7\text{-}4)$$

That expression tells us something very useful. It tells us that image variation that is at a frequency of $k$ cycles per unit distance and in a direction of $\theta$ will be detected by $H(k\cos\theta, k\sin\theta)$. Mathematically inclined readers who would like a more rigorous derivation are given the following exercise. Write the expression for a wave at a frequency of $k$ cycles per unit distance and in a direction of $\theta$. Use trigonometry to compute its number of cycles per unit distance in the horizontal and vertical directions, then show the effect of each of these variations on the quantity expressed in Equation (7-1). The result will be the same. The author likes the more intuitive approach of considering each transform term as representing a projection on an individual wave traveling in a particular direction at a particular frequency.

In actuality, the situation is slightly more complicated than the simple formula just described. The representation is orthogonal only on a discrete lattice in Cartesian coordinates. Therefore, parameterizing the transform in terms of polar coordinates can be confusing. Also, issues of side lobes (page 92 and Figures 4-3 and 4-4) have been sidestepped. They will be discussed in more detail on page 248. But the approximation is very good, and the intuitive appeal is excellent. In practice, this interpretation is totally serviceable.

There is one subtlety to note in regard to this formula. The angle of image variation, $\theta$, is assumed to be measured *clockwise* from a right-hand direction. This is because rows increase going downward, just the opposite of Cartesian coordinates. Most computer programs, including the one presented in the next section, transform in this top-to-bottom direction. This should be kept in mind if explicit directions are important.

An example may clarify interpretation of the transform coefficients. Suppose that we know enough about the physics of our application to know that we are particularly interested in detecting variation that occurs downward and to the right, at an angle of 30 degrees clockwise from directly right. This variation will typically have a frequency of 6 cycles per the length of a side of the transformed square. We must multiply 6 times the cosine and sine of 30 degrees to get the values of $f_x$ and $f_y$ that will be most sensitive to this variation. In particular, $f_x = 5.2$ and $f_y = 3$. Since the DFT is most often computed using integer values for the frequencies (although it does not have to be), we find that $H(5, 3)$ will be the transform coefficient most sensitive to the variation of interest. Side lobes will cause some of the energy at that frequency to appear in other terms, but most of it will go into that one. Also note that later in this chapter the restriction to square areas will be dropped.

Integer values for the frequencies are not mandatory. It is just that "fast" algorithms are difficult or impossible for arbitrary real values. In the unusual case that an application involves a few highly specific frequencies, Equation (7-1) can be explicitly evaluated for the frequencies of interest using brute force computation. It will be slow. But if only a few coefficients need to be found, the speed may be commensurate with that obtained by using the fast DFT to compute all of the coefficients, then discarding most of them! And the increased accuracy obtained by zeroing in on the exact frequencies may be worth the effort.

When the image is real, there is an important symmetry in its two-dimensional Fourier transform. The transform value at any $(f_x, f_y)$ point is the complex conjugate of the transform value at the point reflected about both axes. This is made explicit in Equation (7-5).

$$H(f_x, f_y) = \bar{H}(-f_x, -f_y) \qquad (7\text{-}5)$$

Many readers will have been alerted to this symmetry in the example given above. It was stated that the variation was to the lower right at an angle of 30 degrees. But that is essentially the same thing as variation to the upper left, at an angle of $30 + 180 = 210$ degrees. Using the same formula leads us to $H(-5, -3)$. The complex conjugate comes about from the sine wave being an odd function and hence undergoing a sign reversal when the direction reverses. Mathematicians would derive Equation (7-5) by flipping the signs of $f_x$ and $f_y$ in Equation (7-1) and seeing what happens. (The cosine component is even, so it is unchanged. The odd sine term flips its sign.)

The implication of this symmetry is that only about $n_x * n_y / 2$ complex terms are needed to completely describe the transform. This quantity should not be surprising. The original block of data consisted of $n_x * n_y$ real numbers. If the transform has half as many complex numbers (each equivalent to a *pair* of real numbers), then information is conserved, and all is well with the world.

In order to minimize the amount of data presented to the neural network, it is important to understand the symmetry in full detail. The author's habit is to compute the transform for all $n_x$ values of $f_x$, corresponding to both positive and negative horizontal frequencies, but only compute the $n_y / 2 + 1$ values of $f_y$ that correspond to nonnegative vertical frequencies. That gives a few too many numbers. We need only about $n_x * n_y / 2$, but we are computing $(n_y / 2 + 1) * n_x$ numbers. Where did the extras come from? Look back at Equation (7-5). When $f_y = 0$, that equation tells us that the $n_x$ terms corresponding to the positive and negative horizontal frequencies contain redundancy. The terms corresponding to negative frequencies are the complex conjugates of the positive-frequency terms. The same is true at the vertical Nyquist frequency, $f_y = n_y / 2$, since the Nyquist frequency has no sign. (Review the section starting on page 88 if that is not clear.) Thus, when $f_y = 0$ or $f_y = n_y / 2$, we can ignore the terms corresponding to negative values of $f_x$. Those are the extras.

Let us pursue this symmetry issue just a little longer. Many readers have undoubtedly taken offense at the quantity of independent transform terms being said to be *about* $n_x * n_y / 2$. We need to know *exactly* how many there are, for that determines the number of input neurons in the network. Also, can we reconcile the transform with an input containing $n_x * n_y$ real numbers, so that conservation of information is strict? We now do so.

For each value of $f_y$ from 1 through $n_y / 2 - 1$, all $n_x$ values of $f_x$ will produce fully complex numbers. This gives us $(n_y / 2 - 1) * n_x$ complex numbers (network inputs) so far. For $f_y = 0$, we have $n_x / 2 + 1$ numbers that are not redundant, but the first (DC) and last (Nyquist) are strictly real. The others are general complex. The same is true when $f_y = n_y / 2$, the vertical Nyquist frequency. Thus, the vertical DC and Nyquist sets contribute a total of $n_x + 2$ complex network inputs. Adding things up gives us a total of $n_x * n_y / 2 + 2$ complex inputs to the neural network. But four of those inputs, the DC and Nyquist terms in the horizontal and vertical directions, are strictly real. Voila.

There is a standard pattern for storing and visualizing the transform. The subroutine presented in the next section uses this pattern, as do many other common programs. It is illustrated in Figure 7-1. Each transform term in that figure is represented by $f_x$, $f_y$. Studying that figure may clarify some of the preceding discussions.

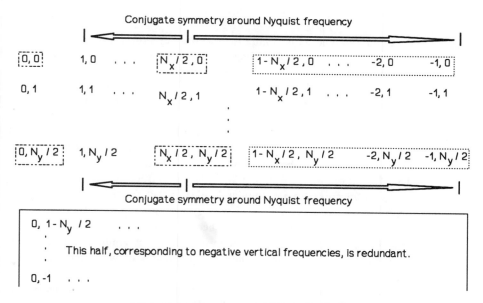

Fig. 7-1: Storage of the 2-D FFT.

To avoid storing redundant information, only terms corresponding to nonnegative vertical frequencies are kept. The terms having negative vertical frequency, noted at the extreme bottom of the figure, are the complex conjugates of those corresponding to positive frequencies, reflected about both axes. The two groups of terms outlined by dotted lines, at the top right and bottom right, are redundant in that they are the complex conjugates of the terms to their left, reflected about the Nyquist term. However, they are still stored to keep things simple. Eliminating their storage would save very little space but would vastly complicate the situation. Finally, the four terms enclosed in dashed boxes are strictly real. This is noted only for interest. In practice, they would be presented to the neural network as if they were complex. It is just that we know in advance that their imaginary components will always be zero.

One final point should be made. This entire discussion has focused on (and will continue to focus on) discarding negative vertical frequencies to avoid redundancy. Actually, there are three other possibilities. One could discard positive vertical frequencies, discard negative horizontal frequencies, or discard positive horizontal frequencies. In other words, the entire transform is a rectangle that is half redundant. We have chosen to keep the top half of the rectangle, shown in Figure 7-1. One could just as well keep the bottom half, left half, or right half. The choice is often a matter of personal preference. The method described here seems to be the most customary, but keeping the right half is also common. They are all correct.

## Data Windows in Two Dimensions

Side lobes are every bit as important a consideration in two dimensions as they are in one dimension. The material starting on page 92 should be reviewed. This section presents a straightforward extension of those results to two dimensions.

It is tempting to think that if the Fourier transform is applied to the original raw image, then we have kept some sort of "purity" by not tampering with the data. In fact, quite the opposite is true. The Fourier transform assumes that the image extends to infinity in all directions. When we transform a rectangular subset of that infinite image, we have implicitly multiplied the image by a rectangular data window function that has the value one inside the selected rectangle and that has the value zero everywhere else. That function resembles Figure 7-2. The net effect of this window function that moves so violently around a rectangular border is that frequency components that do not lie precisely on the integer $f_x$ and $f_y$ values leak into significantly distant transform components. The leakage pattern is graphed as a function of $f_x$ and $f_y$ in Figure 7-3. That function is the amplitude of leakage relative to the amplitude of the center lobe. It has been truncated at 10 percent. Only one quadrant is shown, since the others are symmetric. The worst part about that leakage function is that the height of the side lobes tapers off excruciatingly slowly. That means that serious leakage can occur across very large distances. This situation is virtually always intolerable.

An excellent solution is to generalize the standard one-dimensional Welch data window, shown in Equation (4-7) on page 93. The basic generalization is shown in Equation (7-6).

$$w_{xy} = 1 - \left( \frac{x - 0.5(n_x - 1)}{0.5(n_x + 1)} \right)^2 - \left( \frac{y - 0.5(n_y - 1)}{0.5(n_y + 1)} \right)^2 \qquad (7\text{-}6)$$

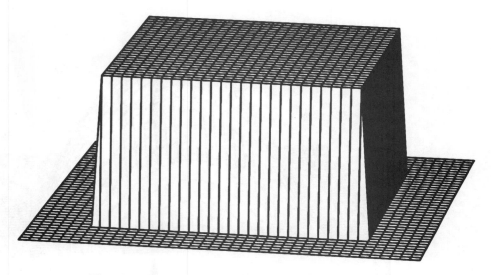

Fig. 7-2: Data window implied by no windowing.

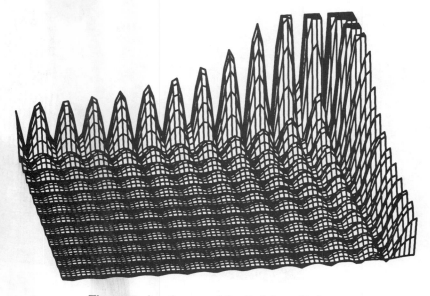

Fig. 7-3: Leakage of "no" data window.

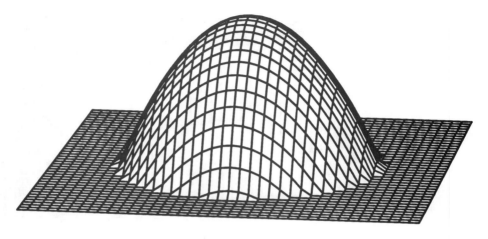

Fig. 7-4: Normal Welch data window.

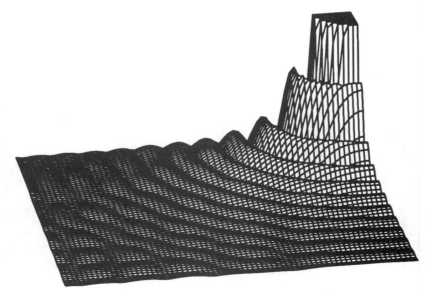

Fig. 7-5: Leakage of Welch data window.

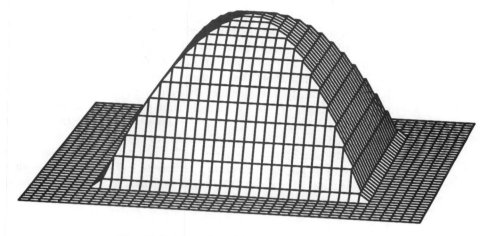

Fig. 7-6: Extended Welch data window.

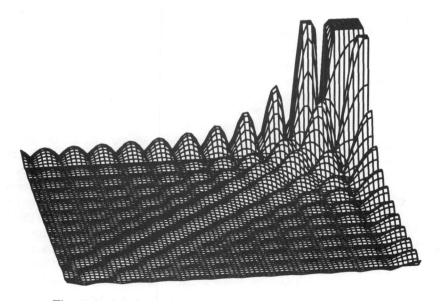

Fig. 7-7: Leakage of extended Welch data window.

Several aspects of that equation should be noted. The rectangle is $n_y$ rows by $n_x$ columns. The $x$ and $y$ indices run from 0 through $n_x - 1$ and $n_y - 1$, respectively. Most importantly, observe that the window as defined in that equation can become negative. We do not let that happen. If a computed value of $w_{xy}$ is negative, set it equal to zero. This window function is graphed in Figure 7-4, and its leakage function is shown in Figure 7-5. Again, the height is truncated at 10 percent of the main lobe's amplitude. Note how much better it is. In particular, observe how quickly the amplitude of the side lobes decreases. That is a vital property of a good data window.

One complaint about the circular Welch window is that it totally ignores the corners, where the $w_{xy}$ values computed by Equation (7-6) are set to zero to avoid being negative. This is actually an exceedingly minor problem. In most cases of practical interest, that data is not terribly important anyway. In the rare case that the reader desperately needs information in the corners, the Welch window can be modified by subtracting only one of the two terms in Equation (7-6), whichever is larger. The resulting window is graphed in Figure 7-6, and its leakage function is shown in Figure 7-7. Note that the small amount of information gained by including the corners is bought at an extremely high price in leakage. This author would never do it, though some do.

## Radial Power

We saw on page 244 that each Fourier coefficient $H(f_x, f_y)$ represents variation having a frequency and direction that can be found by expressing $(f_x, f_y)$ in polar coordinates. This leads to a very natural way of visualizing the transform. Suppose that the matrix of transform values is laid out in such a way that $H(0, 0)$ is in the center, with the first coordinate, $f_x$, having positive values to the right and negative values to the left. The second coordinate's direction would match that of the input. If the tradition of transforming the image from top to bottom is followed, positive values of the second coordinate, $f_y$, would be below the center axis. (In terms of Cartesian coordinates, the transformation is going backwards vertically.) This is shown in Figure 7-9.

The value of this layout is that the position of each coefficient is meaningful. If one draws a line from a coefficient through the origin, that line is the direction of variation represented by that coefficient. Furthermore, the frequency of variation is proportional to the distance separating that coefficient from the origin. For example, look at the image variation portrayed in Figure 7-8. It is slightly counterclockwise from left to right. Therefore, we know that its power will be detected by a Fourier coefficient lying on a line that is slightly counterclockwise from the $X$ axis. Such a line is shown in Figure 7-9. The two black circles show possible Fourier coefficients that will detect this variation. Note that due to the symmetry already discussed, these two coefficients will be complex conjugates of each other.

It should now be apparent that there is an easy way to group sets of coefficients so that they represent similar information. Remember that Fourier coefficients are orthogonal, so their power (squared length) is additive. It makes no sense to add raw Fourier coefficients together. Their sum would be garbage. But it is both legal and

Fig. 7-8: An example of tone variation.

useful to add the *power* of several Fourier coefficients. The main problem of integrating a two-dimensional FFT with a neural network is the large amount of data produced by the FFT. So many network inputs are required that the training set must be gigantic to avoid overfitting. And then training time becomes exorbitant. But if we are willing to discard phase information, working with power alone, the number of inputs can be dramatically reduced. The method for accomplishing this is indicated in Figure 7-10.

Imagine that cells covering the Fourier coefficients are defined as shown in that figure. Any reasonable number of circles and spokes may be used. The power of the coefficients within each cell is summed. That sum represents the power whose direction is within the range determined by the spokes bounding the cell, and whose frequency is within the range determined by the pair of circles bounding the cell. These sums are typically scaled and used for the network inputs.

There is a common variation on this technique. Sometimes we want rotation invariance. We do not know what orientation the image may be in, so the direction of variation is meaningless. Only its frequency is important. For example, we may be photographing terrain features or vehicles from the air. In this case, no spokes would be used. Only the power in each circular band would be sent to the network.

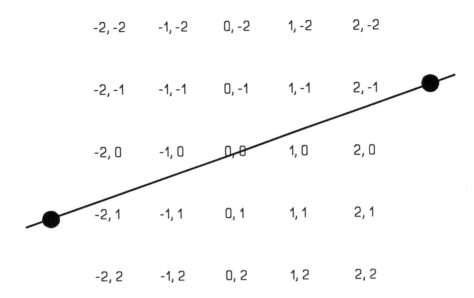

Fig. 7-9: An intuitive Fourier coefficient layout.

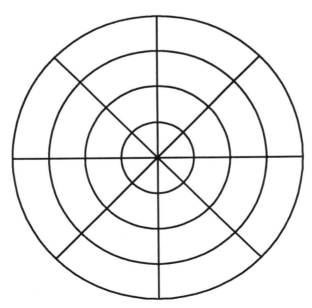

Fig. 7-10: Cells for summing radial power.

## Computing the Transform

It turns out that computation of the two-dimensional DFT is both simple and reasonably fast (compared to some of the transforms to be studied later in this chapter!). Look back at its definition, Equation (7-1). One of the exponential terms, the one involving $x$, is constant throughout the inner summation. We can use the distributive law to pull it outside the summation. This is shown in Equation (7-7).

$$H(f_x, f_y) = \sum_{x=0}^{n_x-1} e^{i\frac{2\pi x f_x}{n_x}} \left( \sum_{y=0}^{n_y-1} h(x,y) e^{i\frac{2\pi y f_y}{n_y}} \right) \tag{7-7}$$

But now look closely at that result. Examine the inner summation that involves $y$. It is an ordinary one-dimensional DFT, operating on the rows of the column determined by $x$. That's easily done. Next, assume that the inner summation (set of DFT's) has been done. Examine the outer summation. That is a set of one-dimensional DFT's again. Each DFT operates on the columns of a row determined by $f_y$. The data being transformed by this outer loop is the transform computed in the inner loop. In other words, we perform a two-dimensional DFT in two steps. First, we do a one-dimensional transform of each column of the image. Then we do a one-dimensional transform of each row of that transform. The computation involves $n_x$ transforms of vectors $n_y$ long, followed by $n_y$ transforms of vectors $n_x$ long.

If one reverses the order of summation in Equation (7-1), the other exponential term can be factored out of the inner summation. Thus, it can be seen that the two-dimensional DFT can just as well be computed by reversing the order of operations. We can first transform all rows of the image, then transform all columns of the result. The end result will be the same either way. In practice, we choose whichever method happens to be more convenient.

In the unusual case that the image is itself complex-valued, we must use the full complex DFT for both passes. However, the image is nearly always real. In that case it is extremely wasteful of both time and memory to convert each image pixel into a complex number having a zero imaginary part and to use the full complex transform. Instead, we should use the real method shown on page 86. Of course, since its result will be complex, we are forced to use the full method for the second pass. But at least we can save time on the first pass by taking advantage of the real image data.

There is one more issue to be resolved. It is nearly always in our best interest to reduce problems with side-lobe leakage by means of a data window. Review page 92 if needed. The program presented here uses the Welch data window described earlier. The distance used to compute the taper is based on the Euclidean distance from the center, so the window is circular.

A complete program for computing the two-dimensional DFT of an image is now given. A discussion of the algorithm follows the code listing.

```
void fft2dr (
   unsigned char *image ,          // Input image, returned intact
   int nrows ,                     // Number of rows in image
   int ncols ,                     // And columns
   int row ,                       // Upper left pixel (origin 0) of
   int col ,                       // the block to be transformed
   int nr ,                        // Size of the transformed
   int nc ,                        // block (must be power of 2!)
   double *work ,                  // Work vector nr long
   double *real ,                  // nr/2+1 (rows) by nc (columns)
   double *imag                    // output matrix of transform
   )

{
   int ir, ic, r, c ;
   double *wr, *wi, rcent, ccent, rden, cden, rdist, cdist, w ;

/*
      Make sure the user did not position the block so that it extends outside the image.
*/

   if ((row < 0)  ||  (row > nrows-nr)  ||  (col < 0)  ||  (col > ncols-nc))
      return ;

   rcent = 0.5 * (nr - 1.0) ;       // Center of block for the
   ccent = 0.5 * (nc - 1.0) ;       // Welch data window
   rden = 0.5 * (nr + 1.0) ;        // Denominators for that window
   cden = 0.5 * (nc + 1.0) ;

/*
Copy each column of the block to the work vector, transform it, then copy the result to the
corresponding column of the output. We split the user-supplied work vector in half, using the first
half for the real part of the transform, and the second half for the imaginary part. Recall that
REAL_FFT wants the even terms of the input vector in the real part, and the odd terms in the
imaginary part. Also recall that REAL_FFT returns the real part of the Nyquist term in wi[0],
which is really zero. The imaginary part of the Nyquist term is also zero.
*/

   wr = work ;                      // User-supplied work vector nr long
   wi = work + nr / 2 ;             // Use second half for imaginary part

   for (ic=0 ; ic<nc ; ic++) {      // Do each column separately
      c = col + ic ;                // Column in image
```

```
cdist = (ic - ccent) / cden ;          // Column dist for Welch window
cdist *= cdist ;                       // Work with squared distance

for (ir=0 ; ir<nr/2 ; ir++) {          // Copy all rows 2 at a time
  r = row + 2 * ir ;                   // Even row in block
  rdist = (2 * ir - rcent) / rden ;// Row distance for Welch window
  rdist *= rdist ;                     // Work with squared distance

  w = 1.0 - rdist - cdist ;            // This is the Welch window
  if (w < 0.0)                         // It must range from 0 to 1
    w = 0.0 ;                          // This cuts off corners though!

  wr[ir] = w * image[r*ncols+c] ;// Even part into real vector
  ++r ;                                // Odd row in block
  rdist = (2 * ir + 1 - rcent) / rden ;// Row distance for Welch window
  rdist *= rdist ;                     // Work with squared distance

  w = 1.0 - rdist - cdist ;            // This is the Welch window
  if (w < 0.0)                         // It must range from 0 to 1
    w = 0.0 ;                          // This cuts off corners though!
  wi[ir] = w * image[r*ncols+c] ;// Odd part into imaginary
  }

real_fft ( nr/2 , wr , wi ) ;          // Transform column ic

for (ir=0 ; ir<nr/2 ; ir++) {          // Because the input was real,
  real[ir*nc+ic] = wr[ir] ;            // there are nr/2+1 unique terms.
  imag[ir*nc+ic] = wi[ir] ;            // The neg freqs are just conjugates
  }                                    // and so are ignored.

real[nr/2*nc+ic] = wi[0] ;             // Returned real part of Nyquist here
imag[nr/2*nc+ic] = imag[ic] = 0.0 ;// These are really zero
} // For all columns

/*
```

At this time, all nc columns of the block have been transformed. They are now in place in the output vectors "real" and "imag". The final step is to transform each of the nr/2+1 rows. The first and last rows are entirely real, so if we were terribly concerned with efficiency, we could use the real transform routine REAL_FFT to do it. But the relatively small savings in time is rarely worth the increase in complexity of the code, so we will use the full complex routine. Of course, the interior rows all need the full complex routine, as they have no special properties.

```
*/
```

```
for (ir=0 ; ir<=nr/2 ; ir++)
  fft ( nc , real+ir*nc , imag+ir*nc ) ;
}
```

The image is declared as unsigned char type, as that is the most common method of storage. The reader can easily change this type if needed. The complete image has a size of nrows rows and ncols columns. We will be transforming a rectangular block whose upper-left corner is at row row and column col. These are origin zero. The size of the block to be transformed is nr rows by nc columns. The image itself will not be modified by this routine. A work vector nr long is needed.

The transform will be output as a matrix strung out into a vector (nr/2+1) * nc long. The first nc elements of this vector correspond to $f_y = 0$. The next nc are for $f_y = 1$, and so on. The final nc elements are for $f_y = $ nr/2, the Nyquist frequency in the vertical direction. Within each group of nc elements, the first represents a horizontal frequency of zero, the next a frequency of one, up to nc/2. Then, just as was the case in one dimension, the next element corresponds to a *negative* frequency of nc/2−1. The last element in the group of nc corresponds to a frequency of −1.

This order can be made more clear by examining Figure 7-1 on page 247. The output starts at the upper-left corner of that pattern, (0,0). It proceeds to the right. Note that for this first row (and also the last, as shown), only the first nc/2+1 values should be presented to the neural network. The remainder of the row is redundant by conjugate symmetry. After the entire first row appears in the output, the next row down is done. This continues until the bottom row in the figure, the vertical Nyquist frequency, is output.

The first thing done by the routine is to verify that the user did not carelessly position the block to be transformed in such a way that it extends outside the image. Then, several constants for the modified Welch data window are computed. Review page 92 if needed.

Each column of the image is now transformed. Recall how real_fft wants its inputs arranged. It wants the even-subscript terms in the real part of the input, and the odd terms in the imaginary part. The user-supplied work vector is split in half for this. The first half is used for the real part, and the second half for the imaginary part. Also recall that there is no explicit room for the Nyquist term in the output. However, it is strictly real, as is the DC (zero frequency) term. Thus, the Nyquist term is stored in the imaginary part of the DC term. That was explained on page 86.

The column transformation would be a few simple lines of code were it not for the Welch data window. The Euclidean distance from the center of the block is computed, and that distance is used to compute the window value. The image data is multiplied by the window, and then the column transforms are performed.

The last step is easy. Each row is transformed. This is done in place, since the column transforms were placed in the output area. The full complex routine is needed.

# The Gabor Transform in Two Dimensions

The single-dimension Gabor transform that we have already studied maps a vector (typically a time series) to a matrix (a time dimension and a frequency dimension). Things get considerably more complicated in image processing. The domain of the two-dimensional Gabor transform is a matrix, the image. The range has four dimensions: row, column, frequency, and orientation. In the case of signal processing, we work with a family of time-localized bandpass filters, each having a different center frequency. We pass these filters along the time series, recording each filter's response at each time location. For image processing, we work with a family of oriented bandpass filters, each of which is characterized by a center frequency *and* an orientation. We pass these filters over the image, recording each filter's response at each spatial location. It is apparent that a lot of data will be generated.

When we use a neural network in conjunction with a Gabor transform for signal processing, we commonly use many frequencies and many time positions as the network input. The goal is to characterize the signal by the time-varying pattern of its frequency content. We will almost never do an analogous deed with an image. While the frequency structure of an image as a function of location in that image can be an effective characterization, the sheer volume of data generated by four dimensions of variation is overwhelming. It is far more likely that we will attempt to individually characterize single areas of an image by their Gabor filter responses for a set of frequencies and orientations. The network's inputs will thus be a two-dimensional matrix, something more manageable.

A common scenario is that a human expert passes a cursor over an image, selecting areas of interest. Each member of a training set is generated by computing the Gabor transform at the one row/column location identified by the cursor. After training the network, it is tested by computing the Gabor transform at many locations throughout the entire image. For each location, the corresponding family of frequency/orientation responses is presented to the network for analysis. We will now develop the two-dimensional Gabor transform in the spatial domain. A frequency-domain method that is often dramatically faster to compute will be presented later in this chapter.

For signal processing, the time-domain window is a Gaussian function normalized to unit energy. That function is shown in Equation (5-18) on page 155. It is easy to generalize the Gaussian to two dimensions so that it can be used to window a two-dimensional function $f(x, y)$. That function may be, for example, an image. In that case, $x$ specifies the column and $y$ specifies the row in the image. In some unusual situations, we might want the window function to scale the $x$ and $y$ dimensions separately. But that is rare, and it significantly complicates things. Also, disparate scaling can often be done in the domain of the function. Therefore, we will assume that the same scale is used in both directions. The two-dimensional Gaussian data window is shown in Equation (7-8).

$$g_\sigma(x,y) = \frac{1}{\sigma\sqrt{\pi}}\, e^{\left(-\frac{x^2+y^2}{2\sigma^2}\right)} \tag{7-8}$$

The complex sinusoid factor in the two-dimensional Gabor transform is defined by a frequency and an orientation of maximum sensitivity. The argument to the complex exponential for a given point is the displacement of that point along the line of orientation. If the orientation angle is theta ($\theta$), then the displacement of the point $(x,y)$ in that direction is $x\cos(\theta) + y\sin(\theta)$. Note that we only need to consider $\theta$ from 0 to 180 degrees. Symmetry makes the other directions redundant.

We are now in a position to define the two-dimensional filter that will be placed at a variety of positions on the image. This filter has a total of three parameters. One of them, the scale parameter $\sigma$, is generally fixed. The remaining parameters are the frequency, $\phi$, and the orientation, $\theta$. These two parameters are typically varied to provide the neural network with an assortment of inputs. The coefficients of the filter are defined by a complex-valued function of two variables, $x$ and $y$. These are the column and row offsets from the center. The filter function is defined in Equation (7-9). Note that the scale parameter, $\sigma$, does not appear in that equation because it is implied in the window function $g(x,y)$.

$$g_{\phi,\theta}(x,y) = g(x,y)\, e^{2\pi i\phi\,(x\cos(\theta)+y\sin(\theta))} \tag{7-9}$$

In the most general case, $g(x,y)$ in Equation (7-9) can be any reasonable window function. But in practice it will nearly always be the Gaussian shown in Equation (7-8). That weight function satisfies the optimality condition of having a resolution rectangle of minimum size, just as was the case for one dimension.

The appearance of these filters can be visualized by thinking of a gentle ripple across a lake. That is the sinusoidal component. The spatial window damps the ripple in such a way that it is unimpeded at a single central point but dwindles into flatness further from that center. If the central, full-amplitude point of the spatial window is placed at a peak in the wave, that is the real component of the filter. If the center is at a zero-crossing, that is the imaginary part. Examples are shown in Figures 7-11 and 7-12.

## Code for the Two-Dimensional Gabor Transform

In this section we present C++ code for computing and applying a Gabor filter. In the signal processing section, the transform routine applied the filter to multiple time placements along the input series. Since we will rarely want to generate so much data in image processing, we take a different route here. The constructor will generate and save the filter coefficients as before. But the transform routine will only apply the filters to one position on the image. It is trivial for the user to move this filter across the entire image, one placement at a time, if that is desired.

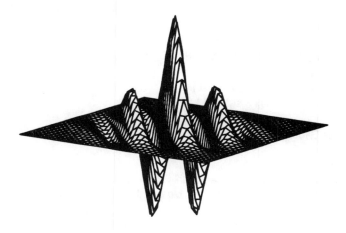

Fig. 7-11: Real part of a 2-D Gabor filter.

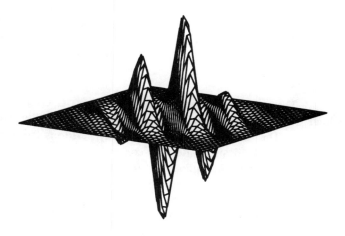

Fig. 7-12: Imaginary part of a 2-D Gabor filter.

Note that this is a spatial-domain implementation of the Gabor transform. If a very large amount of data needs to be processed, the FFT-based algorithm given later in this chapter will be preferable to this approach. Static constants and the class header are now shown.

```
static double pi = 3.141592653589793 ;
static double two_pi = 2. * pi ;
static double root2 = sqrt ( 2.0 ) ;
static double rootpi = sqrt ( pi ) ;

class Gabor2D {

public:
   Gabor2D ( double *sigma , double *delta_t , double *delta_f ,
          int nfreqs , int ndir , double orient , int *halflen , int *ok ) ;
   ~Gabor2D () ;
   void transform ( int nrows , int ncols , double *img ,
               int rcent , int ccent , double *rt , double *it ) ;

private:
   double sig ;               // Sigma
   double dt ;                // Delta_t
   double df ;                // Delta_f
   int hl ;                   // Filter half length (=halflen)
   int nf ;                   // Number of frequencies resolved (=nfreqs)
   int nd ;                   // Number of orientation directions (=ndir)
   int neach ;                // Number of doubles for each filter
   double f0 ;                // Frequency increment = Nyquist / (nfreqs-1)
   double *coefs ;            // FIR coefficients for all frequencies
} ;
```

The constructor computes and returns sigma, delta_t, and delta_f as discussed below. It also allocates memory for holding the FIR filter coefficients that define the transform. The function transform actually performs the transform. It may be called as many times as desired after the constructor has been called. The sample interval defines the time unit throughout this routine.

The following parameters are in the constructor parameter list:

sigma, delta_t, delta_f These are the scale factor and window radii. If sigma is input positive (regardless of the deltas), that value is used. The deltas are then computed according to sigma. If sigma is input negative and delta_t is input positive, that value of delta_t is used, and sigma and delta_f are returned accordingly. If sigma and delta_t are input negative, and delta_f is input positive, that value will be used to compute sigma and delta_t.

nfreqs    This is the number of frequency bands that will be resolved, including a frequency of zero. They will be equally spaced up to the Nyquist limit.

ndir    This is the number of orientation directions, at least 1. If more than 1, they will be equally spaced around a half circle, with the first aligned at orient.

orient    This is the orientation angle in degrees (0–180) of the first filter.

halflen    This is returned as the filter half-length. It is the number of rows and columns in all directions that will be accessed by the transform routine. A very serious runtime error (possibly crashing the system) could occur if the transform is applied closer to an edge than this quantity. Therefore, the transform routine aborts if this is attempted.

ok    This is returned 1 if all went well, and 0 if there was insufficient memory for storing the FIR filter coefficients.

The constructor will need to allocate a work area to hold the FIR filter coefficients. This is 2 * (halflen+1) * (2*halflen+1) * nfreqs * ndir doubles. Each coefficient is complex (2 doubles). There will be nfreqs * ndir filters saved. Only the center row and upper half of each filter will be found. Each row is 2*halflen+1 long, and there are halflen+1 rows saved. An element in the bottom half of the filter is the complex conjugate of the element reflected along both axes, so we save memory by not storing the bottom rows. Code for the constructor and destructor now follows.

```
Gabor2D::Gabor2D (
   double *sigma ,              // Data window scale factor
   double *delta_t ,            // Radius of time-domain window
   double *delta_f ,            // Radius of frequency-domain window
   int nfreqs ,                 // Number of freqs, zero through Nyquist
   int ndir ,                   // Number of orientation directions
   double orient ,              // Angle (0-180 deg.) of first orientation
   int *halflen ,               // Filter half-length (skip this in all direcs)
   int *ok                      // Memory allocation ok?
   )
{
   int ifreq, idir, ix, iy ;
   double freq, direc, weight, fac, gconst, *cptr, cosine, sine, p, dsq ;

   *ok = 0 ;                    // Start out pessimistic
   orient *= pi / 180.0 ;       // Convert degrees to radians
```

```
    nd = ndir ;                          // Number of orientation directions
    nf = nfreqs ;                        // Number of frequencies
    f0 = 0.5 / (nf - 1) ;                // Increment for successive freqs

/*
  Compute (as needed) sigma, delta_t, and delta_f
*/

  if (*sigma <= 0.0) {
    if (*delta_t <= 0.0) {
      if (*delta_f <= 0.0)
        return ; // Error
      else {
        df = *delta_f ;
        *sigma = sig = 1.0 / (two_pi * root2 * df) ;
        *delta_t = dt = sig / root2 ;
        }
      }
    else {
      dt = *delta_t ;
      *sigma = sig = root2 * dt ;
      *delta_f = df = 1.0 / (two_pi * root2 * sig) ;
      }
    }
  else {
    sig = *sigma ;
    *delta_t = dt = sig / root2 ;
    *delta_f = df = 1.0 / (two_pi * root2 * sig) ;
    }

/*
    Compute the filter half-length such that the weight goes to about 1.e-12 times its max by the
    end of the filter.  Each of the nfreqs * ndir filters will be stored with the center row first, then
    the next row up, and so on.  The bottom half is not stored because it can be computed by
    symmetry.  Each row of a filter will have 2*hl+1 coefficients, and there will be hl+1 rows
    stored.  The real part of each coefficient will be followed by the imaginary part.
*/

  *halflen = hl = 1 + 7.4 * sig ;        // Goes to about 1.e-12
  neach = 2 * (hl+1) * (2*hl+1) ;        // Doubles in each filter

  coefs = (double *) malloc ( neach * nf * ndir * sizeof(double) ) ;
  if (coefs == NULL)
    return ;
```

```
    *ok = 1 ;

/*
  Compute the filter coefficients.
*/

    gconst = 1.0 / (sig * rootpi) ;                    // Normalizer for Gaussian

    fac = -0.5 / (sig * sig) ;                         // Common factor in exponent

    cptr = coefs ;                                     // Point to coef area
    for (ifreq=0 ; ifreq<nf ; ifreq++) {               // Covers 0 to Nyquist
      freq = ifreq * f0 * two_pi ;                     // Freq term in complex exp
      for (idir=0 ; idir<ndir ; idir++) {              // Each orientation direction
        direc = orient + idir * pi / ndir ;
        cosine = cos ( direc ) ;
        sine = sin ( direc ) ;
        for (iy=0 ; iy<=hl ; iy++) {                   // Only do top half
          for (ix=-hl ; ix<=hl ; ix++) {               // Do entire row
            p = ix * cosine + iy * sine ;              // Distance along direc
            dsq = ix * ix  +  iy * iy ;                // Distance from center
            weight = gconst * exp ( fac * dsq ) ;      // Data window
            *cptr++ = cos ( p * freq ) * weight ;      // Real part of filter
            *cptr++ = sin ( p * freq ) * weight ;      // And imaginary part
            }
          }
        }
      }
    }


Gabor2D::~Gabor2D ()
{
  if (coefs != NULL)
    free ( coefs ) ;
}
```

The actual transformation routine now follows. It may be called as often as desired once the constructor has been called. The following parameters are in the transform parameter list:

nrows   Number of rows in the image

ncols   Number of columns in the image

img      This is the image. It must be stored as a vector with the first (top) row first, then the second row, and so on.

rcent    This specifies the row (origin 0) where the filter is to be centered. It must be greater than or equal to halflen, the filter half length. It must be strictly less than nrows − halflen. This prevents the filter from spilling outside the image. If a closer approach to the edges is needed, pad the image with zeros.

ccent    This is the column center, as above.

rt       This is the real part of the output transform. It is a vector nfreqs * ndirs long. All orientation directions for zero frequency are first. (This is a little silly, as they are all the same! But it maintains a consistent pattern. Readers needing top efficiency may want to trivially modify this code to avoid the redundancy.) Orientations change fastest, with frequencies then running from 0 through the Nyquist limit of 0.5 cycles per sample.

it       And this is the imaginary part, as above.

```
void Gabor2D::transform (
   int nrows ,                    // Number of rows in image
   int ncols ,                    // And columns
   double *img ,                  // Image
   int rcent ,                    // Row (origin 0) where filter centered
   int ccent ,                    // And column
   double *rt ,                   // Real output
   double *it                     // Imaginary output
   )
{
   int ifreq, idir, row, col ;
   double *cptr, *imgptr_up, *imgptr_dn, rsum, isum ;

/*
       Make sure that we do not overrun the image with the filter.  Recompute img to point to the
       center pixel in the filtered span.
*/

   if ((rcent < hl)  ||  (rcent >= nrows-hl))
      return ;
   if ((ccent < hl)  ||  (ccent >= ncols-hl))
      return ;

   img += rcent * ncols + ccent ;
```

```
/*
      Apply the filters.  Remember that an element in the bottom half of the filter is the complex
      conjugate of the element reflected along both axes.  Therefore we can implement the full
      filter very easily by summing left-to-right for the top half and right-to-left (flipping the sine
      sign) for the bottom half.
*/

    cptr = coefs ;                                    // All coefs here

   for (ifreq=0 ; ifreq<nf ; ifreq++) {               // Do all frequencies
     for (idir=0 ; idir<nd ; idir++) {                // Do all orientations

        rsum = isum = 0.0 ;                           // Will cumulate here

        for (row=0 ; row<=hl ; row++) {               // Only saved top half coefs
          imgptr_up = img - row * ncols - hl ;        // Rows above center row
          imgptr_dn = img + row * ncols + hl ;        // And those below, reversed

          if (row) {                                  // Do both rows at once
            for (col=-hl ; col<=hl ; col++) {         // Do all columns
              rsum += *cptr * *imgptr_up ;            // Row above center
              rsum += *cptr++ * *imgptr_dn ;          // And below
              isum += *cptr * *imgptr_up++ ;          // Ditto for imaginary part
              isum -= *cptr++ * *imgptr_dn-- ;        // Sign flips below center
              }
            }

          else {                                      // Do center row alone
            for (col=-hl ; col<=hl ; col++) {         // All columns
              rsum += *cptr++ * *imgptr_up ;
              isum += *cptr++ * *imgptr_up++ ;
              }
            }
          }

        *rt++ = rsum ;  // Output transform for this freq and dir
        *it++ = isum ;
        }
      }
   }
```

# Morlet Wavelets in Two Dimensions

The mathematics of one-dimensional Morlet wavelets generalizes easily to two dimensions. On the other hand, their application is slightly more complicated than two-dimensional Gabor transforms. There are two significant differences between the Gabor and wavelet transforms. The first is that two-dimensional Gabor filters have only two parameters: orientation and frequency. Wavelet filters have three parameters to deal with. They still have orientation. But rather than having a single frequency parameter, they have a scale parameter that moves in powers of two (usually) and a voice parameter that fills in the gaps between the relatively widely spaced scales. The net effect is the same, and about the same amount of transform data is generated. It is simply generated in a more complicated manner. The Gabor transform uses many values of a frequency parameter. The wavelet transform uses a few scales and a few voices within each scale. In practice, the total number of frequency/scale lattice points will be about the same for both transforms.

The second difference is more severe. The Gabor transform uses the same spatial-domain data window for all frequencies. But wavelet transforms shrink the spatial window as the scale parameter shrinks. The implication is that we can evaluate all frequencies of a Gabor transform at a single center point in space and capture information from the same area of the image. They all use the same data window. However, in most cases we run into trouble if we attempt the same approach with a wavelet transform. We presumably choose the largest scale according to the size of the area of interest in an image. For example, we may be most interested in features covering about 30 by 30 pixels. So we may perhaps choose the largest-scale wavelet to have a period of about 16 pixels. (The exact value is, of course, problem-dependent.) When that wavelet filter is centered on a 30 by 30 block, it will tell us something useful about the energy content of that block at that low rate of variation. But what about the next smaller scale wavelet, having a primary period of 8 pixels? If the effective area covered by the largest wavelet is a 30 by 30 block, this smaller scale will cover only a 15 by 15 block. Information at that scale in the remainder of the 30 by 30 block that surrounds the central 15 by 15 block will be ignored. And things get progressively worse at smaller scales. We will almost always need smaller-scale information at areas surrounding the center of the largest-scale filter. Thus, we are led to compute smaller-scale transform values at a set of additional image centers surrounding the center of the largest-scale filter. This is illustrated in Figure 7-13.

In that figure, the center of the largest-scale wavelet filter is shown as the heavy black dot at the center of the illustration. Its approximate spatial area of influence is delineated by a large circle. For the next-smallest wavelet, we choose nine centers arranged as a three-by-three block. The area of influence of each of these nine filters is shown with dotted lines. Notice that these circles have half the radius of the larger circle.

If we were to go to a still smaller scale, we would need half the spacing again. This would give us a five-by-five block of filters, each having half the radius of the previous scale. It is easiest to lay out this block so that the center and corners correspond to those of the previous filters. With the spacing shown, the result is that the total area

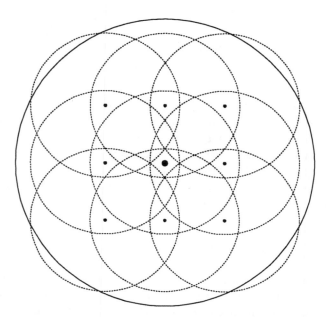

Fig. 7-13: Additional filter centers for coverage.

covered by successively small-scale filters will shrink as the radius of each filter's response area shrinks. We can always compensate for this by making the common block larger as long as the filters within a block at any scale overlap sufficiently to avoid information gaps in the spatial domain. In practice, this is never a problem. We almost never use more than three scales, or we would be overwhelmed with information. Later, when computer code for this transformation is presented, we will discuss how to choose appropriate spacing.

## The Mother Wavelet

The reader should now turn to page 182 and review the one-dimensional version of the Morlet wavelet. A review of the two-dimensional Gabor transform (page 259) would also be beneficial, as the generalization of the Morlet wavelet to two dimensions follows a similar pattern.

The mother wavelet, from which all scales and voices will be derived, has two parameters. One of these, $k$, is the basic shape. It controls the width of the frequency-domain window relative to its center frequency, and it is exactly analogous to the $k$ in Equation (15-3) on page 182. The other parameter is the orientation direction, $\theta$. This has the same interpretation as its analogue in Equation (7-9) on page 260. The Morlet mother wavelet is shown in Equation (7-10). Note that it is normalized to have zero mean and unit energy, just like the one-dimensional version.

$$h_{k,\theta}(x,y) \;=\; \pi^{-1/2}\, e^{\left(-\frac{x^2+y^2}{2}\right)} \cdot$$
$$\left(e^{2\pi i k(x\cos(\theta)+y\sin(\theta))} - e^{-k^2/2}\right)$$

(7-10)

The first factor in that terrible-looking equation is the normalized Gaussian data window. It is identical to Equation (7-8) on page 260 when $\sigma = 1$. The second factor is the sinusoidal component of the wavelet. Its $(x, y)$ term is projected in the $\theta$ direction to determine the complex exponential. As was done in Equation (5-34) on page 182, we subtract a constant function of $k$ before multiplying by the data window. This tiny-but-important correction insures that the mother wavelet integrates to zero. We already know that such centering is vital if high-frequency (small-scale) components are to be protected from corruption by constant offsets and large-scale components.

## Scales and Voices

We derive all needed scales and voices from the mother wavelet in a way that is almost identical to the one-dimensional situation. The only real difference is that since the domain of the function has two dimensions, we scale the height of the derived wavelets by the square of the scale factors used in one dimension. In particular, scaling is done according to Equation (7-11). Child wavelets derived according to that equation retain their normalizations (zero mean and unit energy). And, just like in one dimension, we nearly always let the values of $\xi$ be powers of 2.

$$h_\xi(x,y) \;=\; \frac{1}{\xi}\, h\!\left(\frac{x}{\xi}, \frac{y}{\xi}\right)$$

(7-11)

Multiple voices within a given scale are also a straightforward extension of the one-dimensional case. If there are N voices, each designated with a superscript ranging from 0 through N-1, then these voices are defined in terms of a basic wavelet $h$ by Equation (7-12). It is assumed that $\xi$ for successive scales moves by factors of 2, the usual case.

(7-12)

$$h^j(x,y) \;=\; 2^{-2j/N}\, h(2^{-j/N}x,\, 2^{-j/N}y), \qquad j = 0, \ldots, N-1$$

Mathematically inclined readers will notice that these voices are not normalized to unit energy. That happened in the one-dimensional case also. In fact, the squared norm of voice $j$ is $2^{-2j/N}$. For an intuitive grasp of why this must be so, imagine that we have a huge number of voices. The largest of them will have scaling almost twice the scale of its basic ($j = 0$) voice. In other words, its scale will be almost equal to the scale of the next-largest scale basic wavelet ($j = 0$ with $\xi$ doubled). At the same time, its lattice-point spacing in both of the spatial dimensions will be twice as dense as that of the next-largest scale basic wavelet. There will be four times as many of them per unit area. Therefore, the energy of each must be one-quarter of that of the next-largest scale. An examination of Figure 5-24 on page 180, which shows the same effect in one dimension, may help. This is an admittedly rough justification, but it is the best that can be done without invoking far more mathematics than is appropriate here.

## Code for the 2-D Morlet Wavelet

Two-dimensional Morlet wavelets are often more useful for neural network image processing than two-dimensional Gabor transforms. At the same time, they are more difficult to implement efficiently. Their comparative advantages and disadvantages arise from one factor. For any given frequency, the Gabor transform fuses all of the information from the data window surrounding a single spatial placement. The contents of that data window at that frequency are distilled down into one complex number. The same area is covered for all frequencies, since the data window is the same for all filters; only the periodic component is different. The Morlet wavelet as typically employed does such information fusing only for the largest-scale wavelet. For smaller scales, multiple centers are nearly mandatory unless only small-scale information at the exact center is important. This is illustrated in Figure 7-13 on page 268. The result is that more detail is available from a wavelet transform than from a Gabor transform. This is obviously good in that we are provided with an abundance of information. But that is a two-edged sword. So much information can easily overwhelm a neural network, slowing training and encouraging overfitting. Moreover, the user has more to think about in terms of experimental design and parameter selection. Therefore, this section, which discusses the program in detail, will need to consider issues related to the choice of the number of scales and voices and to the spacing of the supplementary filter centers. It will be helpful to the reader if Figure 7-14 is kept handy throughout the remainder of this section.

If only one scale is used, a call to the transform routine will result in only one filter (for each voice and rotation) being applied to the image at the user's specified center pixel. If two scales are employed, nine filters (for each voice and rotation) at the smaller scale will be applied. One of these nine will be centered at the user's specified center, which is also the center of the larger-scale filter. The centers of the other eight will surround the user's center. This is illustrated in the top of Figure 7-14. Notice that the center placement has one neighbor on each side. This number of neighbors will double with each successively smaller scale. Thus, if there are three scales, the smallest scale will be computed with 25 different filter centers as shown in the bottom of Figure 7-14. In the unlikely event that the user required four scales, there would be four neighbors (two, doubled) around the center placement for the smallest scale. This would result in 81 filter centers. That almost always would be impracticably excessive. Remember that those 81 filter centers are used for each voice and orientation! And that is in addition to the filters at larger scales. We are getting into supercomputer territory.

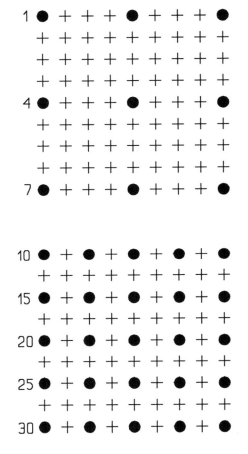

Fig. 7-14: Second- and third-largest scale centers.

When we pass from one scale to the next smaller, the number of neighbors doubles. At the same time, the distance between each center is halved. The net effect is that the size of the square determined by the outermost centers remains constant. This keeps the area covered by the set of centers at each scale approximately the same. It does reduce slightly for smaller scales. This is because the filter half-length is halved for each successively smaller scale as shown in Figure 7-13 on page 268. But this effect is not severe if the spacing between filter centers has been chosen wisely.

When the constructor is called to compute and save the wavelet filter coefficients, the user must specify many parameters. These will now be discussed. Many of them are identical to parameters for the one-dimensional version. In particular, the reader should refer back to the section starting on page 190 for a discussion of the shape, rate, nvoices, samprate, voicefac, and ok parameters.

Several other parameters for the two-dimensional Morlet wavelet are identical to parameters for the two-dimensional Gabor transform. The reader should consult the section starting on page 262 for a description of the ndir and orient parameters.

This leaves us with only a few special parameters to be individually treated. Many of these are similar to parameters already seen. For example, we must tell the constructor how many scales to compute. This is done with the nscales parameter. In the one-dimensional case, this parameter set an upper limit on the number of scales that could be computed. The actual number computed depended on the length of the time series. In the current case, nscales scales will always be computed. It is up to the user to make sure that the specified center is far enough from the edges of the image.

The spacing parameter is also nearly identical to the one-dimensional case. It specifies the number of pixels in each dimension that separate the filter centers at the smallest scale. For example, look at Figure 7-14. The crosshairs (+) represent image pixels. In this illustration, which shows the smaller two of three scales, spacing = 2. There is an easy way to find a good value for this parameter. Call the constructor with spacing set to zero. The constructor will return a value for the sample rate in samprate. Delete the newly created Morlet2D object, as no filter coefficients were computed. Rounding the returned sample rate will give an excellent choice for spacing. If the spacing is made too small, the filters will produce wastefully redundant information, and too little area of the image will go into their computation. If the spacing is too large, the filters will be far apart relative to their spatial radius. The resulting information gaps in the spatial domain will be deleterious. Consider 1.75 times samprate to be an absolute upper limit for spacing.

Alert users may notice what seems to be an anomaly concerning the spacing and filter sizes. Figure 7-13 implies that the square of additional centers thoroughly covers the area encompassed by the largest-scale filter. Yet there may be concern that the half-length of that filter is very large compared to the size of the square. It seems as though the square only covers a small area near the center of the largest filter. This conflict is resolved when it is realized that the half-length is a gross exaggeration of the actual area covered. The outermost coefficients are tiny. The vast majority of image information that contributes to the results for the largest scale resides very near the center. The circles drawn in Figure 7-13 do not represent the half-lengths of the filters. They really are a fraction of the half-length, enclosing the area of the image that makes the bulk of the contribution to results.

The constructor returns the half-length of the largest-scale filter in halflen. This gives the user an idea of how many pixels around the specified center go into computation of the wavelet transform results. Of course, pixels outside of about half of this half-length make very little contribution. Since the window is a bell curve, most information comes from near the center. Another use for this parameter is to help the user understand memory requirements. The constructor will use 2 * (halflen+1) * (2*halflen+1) * ndir * nvoices doubles for storing the filter coefficients.

The user must be informed as to how close to an edge of the image each wavelet transform can be computed. It is illegal for a filter to extend past a border. (If that is truly needed, the user must pad the image.) The border parameter is returned for that purpose. In most cases it will be equal to halflen, the half-length of the largest filter. But if the user specifies an unusually large spacing, the outermost extent of the complete transform may be determined by the second-largest scale filters. The half-length of these filters is half of halflen, but their centers run closer to the edge of the image because there are nine of them. That may push them outside the range of the largest filter.

Finally, the user needs to know how much memory to allocate for the transform outputs. We already know that there will be one filter placement (for each voice and orientation) for the first scale. There will be nine more for the next, 25 more for the next, 81 (heavens!) more for the next, and so on. The constructor returns tcents as the total number of filter centers output. Thus, the real and imaginary outputs will each consist of tcents * nvoices * ndir doubles.

Code for the class header, the constructor, and the destructor now follows. Comments concerning this code will appear after the listing.

```
static double pi = 3.141592653589793 ;
static double two_pi = 2. * pi ;
static double root2 = sqrt ( 2.0 ) ;
static double rootpi = sqrt ( pi ) ;


class Morlet2D {

public:
   Morlet2D ( double shape , double rate , int spacing , int nvoices ,
           int nscales , int ndir , double orient , int *halflen ,
           int *border , double *samprate , double *voicefac ,
           int *tcents , int *ok ) ;
   ~Morlet2D () ;
   void transform ( int nrows , int ncols , double *img ,
             int rcent , int ccent , double *rt , double *it ) ;


private:
   double kparam ;              // "Frequency" shape parameter k
   double srate ;               // Sample rate (samples per unit time)
   int nv ;                     // Number of voices
   int ns ;                     // Number of scales
   int nd ;                     // Number of orientation directions (=ndir)
   int space ;                  // Distance between centers of shortest filter
   int hl ;                     // Longest filter half length (=halflen)
   int bord ;                   // Most distant filter extent (=border)
   int nsmall ;                 // Number of smallest-scale centers each side
```

```
   int neach ;                      // Number of coefs for each filter
   double *coefs ;                  // FIR coefficients for all frequencies
} ;


Morlet2D::Morlet2D (
   double shape ,                   // f/delta, Morlet used 7.54
   double rate ,                    // Multiple of Nyquist, at least 1.0
   int spacing ,                    // Samples between smallest-scale filters
   int nvoices ,                    // Number of voices, at least 1
   int nscales ,                    // Number of scales, at least 1
   int ndir ,                       // Number of orientation directions
   double orient ,                  // Angle (0-180) of first orientation
   int *halflen ,                   // Half-length of longest filter
   int *border ,                    // Furthest extent from center
   double *samprate ,               // Samples per unit time
   double *voicefac ,               // Factor by which period is * for voices
   int *tcents ,                    // Number of transform centers output
   int *ok                          // Memory allocation ok?
   )
{
   int i, idir, ix, iy, iv ;
   double direc, weight, window_radius, fac, vfac, con, gconst, *cptr ;
   double cosine, sine, p, dsq, vgc, dsqmult, vf, fx, fy ;

   orient *= pi / 180.0 ;  // Convert degrees to radians

/*
      Verify that parameters are legal.  If the shape-determining frequency is very low, normaliza-
      tion is poor unless the sample rate is well above the Nyquist limit.  (This is a crude check,
      but it inspires more care than no check at all.)  The specified sample rate must be at least
      1.0 times that limit.  The number of voices and scales must be at least 1.
*/

   *ok = 0 ;               // Start out pessimistic
   coefs = NULL ;          // So destructor doesn't do bad free if failure here
   if ((shape < 6.2)  &&  (rate < 2.0))
      return ;
   if (rate < 1.0)
      return ;
   if (nvoices < 1)
      return ;
   if (nscales < 1)
      return ;
```

```
/*
    The user specified the shape as a multiple of the frequency-dimension window radius. Use
    that to find k. Multiply k by 2 pi right now to avoid having to do it later for trig functions. The
    sample rate is the Nyquist frequency (twice the shape frequency) times the user's multiple
    (which must be at least 1.0).
*/

    window_radius = 1.0 / (root2 * two_pi) ;
    kparam = window_radius * shape * two_pi ;
    *samprate = srate = window_radius * shape * 2.0 * rate ;
    *voicefac = pow ( 2.0 , 1.0 / (double) nvoices ) ;


/*
    Save other user parameters in private storage area. Compute the filter half-length such that
    the weight goes to about 1.e-5 times its max by the end of the filter. If multiple voices are
    used, remember that we must accommodate the largest voice. For each voice, we will store
    the coefficients for the largest scale member of that family. Smaller scale members will be
    derived by decimation. Thus, the number of coefficients needed for that longest filter is the
    half-length of the shortest filter times the largest scale (2 ** (ns-1)). Note that 2 ** (ns-1) / 2
    is also the number of smallest-scale filters that will be applied in all directions (radius)
    around the user's center.
*/

    nd = ndir ;                      // Number of orientation directions
    ns = nscales ;                   // Number of scales
    nv = nvoices ;                   // Number of voices
    space = spacing ;                // Dist between centers of shortest filter

    i = ns ;                         // Number of scales
    nsmall = 1 ;                     // Compute 2 ** (ns-1)
    while (--i)                      // which is twice the number of smallest-scale
      nsmall *= 2 ;                  // filters on each side of the user's center

    hl = 1 + 4.8 * srate * pow ( 2.0 , (double) (nv-1) / (double) nv ) ;
    *halflen = hl = hl * nsmall ; // Half-length of longest filter

    if (space <= 0)                  // Abort if user just wants to examine parms
      return ;


/*
    Normally, the extent of the image needed (border) will be the half-length of the longest filter.
    But if the user chooses to sacrifice frame quality by specifying a large 'spacing' parameter,
    this max extent may be determined by the extent of the second-largest scale filter, because
```

```
                its center is out at the limit.  Also, we are now done using nsmall to hold 2 ** (ns-1), so
                divide it by 2 to get the true value of nsmall, which will be 0 if ns=1.
   */

      nsmall /= 2 ;

      if ((space * nsmall + hl / 2)  >  hl)    // Extent of second-largest filter
        *border = bord = space * nsmall + hl / 2 ;
      else                                        // Extent of largest
        *border = bord = hl ;

   /*
                Compute tcents, the number of transform centers that will be output by the transform routine.
                The user will have to allocate rt and it for tcents * nvoices * ndir doubles.  The largest-scale
                wavelet is evaluated at one center.  The next scale is evaluated at the center, plus one on
                each side.  The next has the center plus two on each side, and so on.
   */

      *tcents = 1 ;                    // Largest-scale wavelet at center only
      i = ns ;                         // Number of scales
      idir = 1 ;                       // Number of centers on each side
      while (--i) {                    // Sum for each additional scale
        idir *= 2 ;                    // Side count doubles for each smaller scale
        *tcents += (idir+1) * (idir+1) ;   // Both sides plus center
        }

   /*
                Each of the nvoices * ndir filters will be stored with the center row first, then the next row up,
                and so on.  The bottom half is not stored because it can be computed by symmetry.  Each
                row of a filter will have 2*hl+1 coefficients, and there will be hl+1 rows stored.  The real part
                of each coefficient will be followed by the imaginary part.
   */

      neach = 2 * (hl+1) * (2*hl+1) ;        // Doubles in each filter

      coefs = (double *) malloc ( neach * nv * nd * sizeof(double) ) ;
      if (coefs == NULL)
        return ;
      *ok = 1 ;
```

```
/*
    Compute the filter coefficients.  The sample rate for a voice is the smallest-scale sample
    rate, srate, times the scale of the largest-scale member (since we compute and save only its
    coefs) times the rate factor for that voice.  The unit time change per point is the reciprocal of
    that sample rate.
*/

    fac = 1.0 / (srate * pow ( 2.0 , (double) (ns-1))) ;// Largest-scale rate
    gconst = fac / rootpi ;
    con = exp ( -0.5 * kparam * kparam ) ;                // Centering constant

    for (iv=0 ; iv<nv ; iv++) {                            // For all voices
        vfac = pow ( 2.0 , -(double) iv / (double) nv ) ; // Voice factor
        vgc = vfac * vfac * gconst ;
        vf = vfac * fac ;
        dsqmult = vf * vf ;
        cptr = coefs + iv * neach * nd ;                  // Point to this voice's coef area

        for (idir=0 ; idir<nd ; idir++) {                 // Each orientation direction
            direc = orient + idir * pi / nd ;
            cosine = cos ( direc ) ;
            sine = sin ( direc ) ;

            for (iy=0 ; iy<=hl ; iy++) {                  // Only do top half
                fy = iy ;                                 // Convert to fpt
                for (ix=-hl ; ix<=hl ; ix++) {            // Do entire row
                    fx = ix ;                             // Convert to fpt
                    p = vf * (fx * cosine + fy * sine) ;  // Scaled dist along direc
                    dsq = dsqmult * (fx * fx + fy * fy) ; // Distance from center
                    weight = vgc * exp ( -0.5 * dsq ) ;   // Window
                    *cptr++ = (cos ( p * kparam ) - con) * weight ; // Real part
                    *cptr++ = sin ( p * kparam ) * weight ;    // And imaginary part
                    } // For ix
                } // For iy
            } // For idir
        } // For iv
}


Morlet2D::~Morlet2D ()
{
    if (coefs != NULL)
        free ( coefs ) ;
}
```

The first thing done by the constructor is to convert the initial orientation angle from degrees (easier for the user) to radians (easier for computers). It then verifies that all parameters are reasonable. If not, it returns immediately. The parameter ok will be left at 0. Hopefully, the caller will always check this parameter before trying to use the new object. Also, the private coefs area will be left NULL so that the destructor will not have problems.

The next step is to compute, save, and pass to the user several important parameters. The half-length of the largest-scale filter is computed in such a way that the coefficients at the extreme end of the filter are about 1.e-5 times their maximum value (which they have at the center of the filter). This is less conservative than was done for the one-dimensional version. Larger filters simply require too much time to compute and too much space to store. Also, less precision is usually required for image processing compared to signal processing in practical applications. The user can easily change this by increasing the factor of 4.8. A value of 6.1 extends the filter until the coefficients are down to 1.e-12 times the center value. This should be far more than adequate for any situation.

At this point, the constructor checks the user's supplied spacing parameter, now stored in the private space. If it is not positive, the user just wanted to interrogate the constructor about computed parameters, especially samprate for estimating a good value for spacing. In this case, it returns.

In most cases, the longest filter (the largest scale) will determine how far from the center the complete wavelet transform will extend. But if the user specified an unusually large spacing, the second-longest filter, which is centered at points other than the overall center, may extend even further. The constructor checks for this, passing the appropriate value to the user.

The next step is to compute, for the user's convenience, the total number of filter centers that will be output. That is done by a straightforward summation at each scale. This value, tcents, lets the user compute the amount of memory to allocate for the output vectors.

Now it needs to allocate memory for the filter coefficients. Each of the nvoices voices and ndir orientations will be stored separately. We only need to store the coefficients for the largest-scale set of filters (all voices and orientations), as the other filters will be derived by decimation by powers of two. The mathematically complete filter for any voice and orientation is 2 * halflen + 1 square. But it has the same sort of symmetry that we saw for the two-dimensional Fourier transform and the Gabor transform. If one reflects about both axes, the coefficients on either side of that reflection are complex conjugates of one another. Therefore, we only store the middle row and all rows above it. In fact, if we were really greedy, we could even avoid storing half of the center row! (Reader exercise: why?) But that adds a lot of complexity for very little savings. So, each filter will require space for halflen + 1 rows, with each row containing 2 * halflen + 1 columns. Each of these filter elements is complex (a real part, followed by an imaginary part). The constructor allocates memory for all filters.

The final step is to compute the filter coefficients. This is very similar to what was done in the one-dimensional situation. The only real difference is in normalization.

The normalization factor for unit energy, expressed in Equation (7-10) on page 270, is the square root of pi rather than the fourth root. Similarly, the scaling factor for moving from the smallest scale to the largest is the square of what it was in the one-dimensional case. That is expressed in Equation (7-11) on page 270. Finally, Equation (7-12) on page 270 tells us that the voice scaling factors are also the square of their one-dimensional counterparts. The periodic component is computed by projecting the $(x, y)$ position along the orientation direction, exactly as was done for the two-dimensional Gabor transform.

After the constructor has been called, the transform routine may be called as many times as desired. Its parameters are very straightforward. It is given a pointer to an image, img, that is nrows by ncols in overall size. The image is stored starting with the upper left pixel, working from left to right. The top row is followed by the second row, and so forth. The transform will be centered at row rcent and column ccent. These are origin 0. They must be greater than or equal to border. The row center, rcent, must also be strictly less than nrows - border, and similarly for ccent. This prevents the filter from spilling outside the image. If a closer approach to the edges is needed, pad the image with zeros.

The entire transform will be output in two vectors. One vector, rt, holds the real parts. The other vector, it, holds the corresponding imaginary parts. The order of the outputs in these vectors is a little complicated. Each vector is tcents * nvoices * ndir doubles long. Results for all voices and orientations at the single center of the largest scale appear first. These are followed by the 9 sets of results for the 3-by-3 block at the next scale. Then come the 25 sets of results for the next scale, and so forth. All block centers appear starting at the upper-left corner, working across first, then down. This was illustrated in Figure 7-14. The small numbers next to the leftmost filter center in each row refer to the relative position (origin 0, for the largest scale) of the results in the output array.

For each filter center, the basic voice appears first, then increasing period voices follow. For each voice, the direction at orient appears first, followed by the other orientations in counterclockwise rotation. In other words, for the output array, the orientation changes fastest, then voice, then column center (left to right), then row center (top to bottom), and finally, scale, running from largest to smallest, changes most slowly. Code for the transform routine is now given.

```
void Morlet2D::transform (
    int nrows ,                    // Number of rows in image
    int ncols ,                    // And columns
    double *img ,                  // Image
    int rcent ,                    // Row (origin 0) where filter centered
    int ccent ,                    // And column
    double *rt ,                   // Real output
    double *it                     // Imaginary output
    )
```

```
{
    int iscale, ivoice, idir, ipos, row, col, decim, roff, coff, maxoff, off ;
    int offspace ;
    double *cptr, *imgptr_up, *imgptr_dn, rsum, isum, scafac, *cent ;

/*
    Make sure that we do not overrun the image with the filter.
    Recompute img to point to the center pixel in the filtered span.
*/

    if ((rcent < bord)  ||  (rcent >= nrows-bord))
        return ;
    if ((ccent < bord)  ||  (ccent >= ncols-bord))
        return ;

    img += rcent * ncols + ccent ; // Point to user's center

/*
    Apply the filters.  Remember that an element in the bottom half of the filter is the complex
    conjugate of the element reflected along both axes.  Therefore we can implement the full
    filter very easily by summing left-to-right for the top half and right-to-left (flipping the sine
    sign) for the bottom half.
*/

    decim = 1 ;                          // Filter coefs decimated for smaller scales
    scafac = 1.0 ;                       // Scaling factor
    maxoff = nsmall * space ;            // Max offset of centers
    offspace = 2 * maxoff ;              // Space between offset centers

    for (iscale=0 ; iscale<ns ; iscale++) {              // Start with longest filter
        off = iscale  ?  maxoff : 0 ;                    // Largest scale has only one center
        for (roff=-off ; roff<=off ; roff+=offspace) {   // Centers in square
            for (coff=-off ; coff<=off ; coff+=offspace) {   // around user's cent
                cptr = coefs ;                           // All coefs here
                cent = img + roff * ncols + coff ;       // Offset center
                for (ivoice=0 ; ivoice<nv ; ivoice++) {  // Do all voices
                    for (idir=0 ; idir<nd ; idir++) {    // Do all orientations
                        rsum = isum = 0.0 ;              // Will cumulate here
                        for (row=0 ; row<=hl/decim ; row++) {    // Only saved top half
                            imgptr_up = cent - row * ncols - hl / decim ; // Rows above
                            imgptr_dn = cent + row * ncols + hl / decim ; // Below
```

```
        if (row) {                              // Do both rows at once
          for (col=-hl ; col<=hl ; col+=decim) {  // All columns
            rsum += *cptr * *imgptr_up ;        // Row above cent
            rsum += *cptr * *imgptr_dn ;        // And below
            isum += *(cptr+1) * *imgptr_up++ ;  // Imaginary part
            isum -= *(cptr+1) * *imgptr_dn-- ;  // Sign flips
            cptr += 2 * decim ;                 // Skip real, imag
            }
          }
        else {                                  // Do center row alone
          for (col=-hl ; col<=hl ; col+=decim) {  // All columns
            rsum += *cptr * *imgptr_up ;
            isum += *(cptr+1) * *imgptr_up++ ;
            cptr += 2 * decim ;                 // Skip real, imag
            } // For all cols
          } // Else center row
        cptr -= 2 * (decim-1) ;                 // May skip too many
        if (row != hl/decim)
          cptr += 2 * (2 * hl + 1) * (decim-1) ;  // Decim rows
        } // For all rows

      *rt++ = scafac * rsum ;                   // Output for this voice and dir
      *it++ = scafac * isum ;
      } // For all directions
     } // For all voices
    } // For all column center offsets
   } // For all row center offsets
  decim *= 2 ;                                  // Decimate coefs for smaller scales
  scafac *= 2.0 ;                               // Scaling factor
  offspace /= 2 ;                               // Space between offset centers
  } // For all scales
}
```

The first thing done by this routine is to catch the careless user who places the transform center so close to the edge of the image that a filter would run past the border. Once this is done, the img pointer is reset to point to the user's specified transform cent.

The filters are applied starting with the largest scale and working toward the smallest. For each scale, decim is the decimation ratio for the filter coefficients. It is initialized to one, since the largest-scale filter is what is stored, then decim is doubled for each successive scale. The coefficient in Equation (7-11) on page 270 is scafac. It is initialized to one because we explicitly compute the coefficients for that scale, then it is doubled for each successive scale. We could have saved a little space by using decim here, as they are always equal. But this makes the code a little clearer. The maximum offset of any supplementary filter center (the distance in pixels to the sides of the square) is maxoff. This is equal to the distance between each center at the smallest scale, times the number of centers at that scale out from the overall center. The space between centers at any scale will be kept in offspace. This is initialized to twice what its first value will be, because it will be halved in the scale loop before it is used the first time.

The row and column offsets are done in nested loops within the scale loop. For the first scale, the largest, the offset from the true center will be zero, as only one center is done at that scale. The filter coefficients for all voices and orientations are stored contiguously in the coefs array, so their pointer is initialized to the start of that array each time a new filter center is done.

Remember that only the coefficients for the center row and all rows above it (positive $y$) are stored. Thus, for each filter, we work outward from the center row of the image. One pointer, imgptr_up, is used for rows above the center row. Another, imgptr_dn, is used for rows below the center. Also, recall that the coefficients below the center row are the conjugate of those above the center row, *reflected along both axes*. The easiest way to implement this is to sum the upper row of the image from left to right, just like the coefficients are stored, but sum the lower row from right to left. When cumulating the imaginary part of the filter's output, flip the sign of the imaginary part for the lower row to take care of the conjugate property. Naturally, the center row must be processed alone.

The most confusing part of the whole routine comes at the end of the loops for an individual filter. After each column is processed, the filter coefficient pointer is incremented by 2 * decim. This is because each coefficient has a real and an imaginary part, and we are decimating by a factor of decim. But the first and last coefficients of each row are always used, no matter what the decimation ratio. After the last coefficient for a row, the first coefficient of the next row appears. By adding 2 * decim after the last coefficient, we have gone too far. We passed the first coefficient of the next row. We must back up by decim - 1 complex coefficients. (If you do not understand this, draw a picture for a small filter.)

Also, we need to decimate the rows as well as the columns. But for the same reason as above, we must not skip rows after the last (top) row is done, because the next filter is coming right up. So, except for the last time, we increment the coefficient pointer so as to skip decim - 1 rows.

## Application Notes

In this section, we will discuss several miscellaneous topics concerning application of the Morlet wavelet transform just described.

Remember that the programs for implementing the two-dimensional Gabor and wavelet transforms are quite different from the one-dimensional versions. The latter routines perform a complete transformation along the entire time series, repeatedly placing the filters at appropriate positions in the time dimension. The two-dimensional versions place the transform center at just one point (though the wavelet version uses supplementary centers). If the user wants to transform an entire image, it is necessary to call the transform routine once for each desired placement in the spatial domain. This is in deference to the most probable use for each model. There is no loss of efficiency when this is done with the Gabor transform. The user simply calls the transform routine in a nested loop, moving the center across the image in a grid pattern.

Unfortunately, that is not the case for the wavelet transform. In order to follow the power-of-two scaling pattern in the spatial domain, the centers would be spaced at a distance equal to twice the spacing at the second-largest scale. But this results in the outermost supplementary centers being processed more than once. That can be a serious waste of time. The author has not yet seen a practical application in which such action is appropriate, so he is not too concerned. It is nearly always the case in real life that a prescreening criterion is employed so that massive evaluation of the transform over a whole grid is not needed. The transform is selectively placed over areas of interest. On the other hand, for any readers who need uniform grid transformations, there is a fairly simple method that can be used. Call the constructor several times, once for each scale. Specify nscales = 1 each time and vary the rate in powers of two. This will provide separate Morlet2D objects for each scale. Then pass these across the image, varying their spacing in powers of two. This method is only slightly more complicated than doing it with one object, and it provides full efficiency. *An even better alternative may be to use the fast general algorithm discussed later in this chapter.*

When one uses the Gabor transform, the center frequency of each filter is explicitly known. For wavelets, we are working in terms of scale rather than frequency. Sometimes we need to know the center frequency that corresponds to a particular scale. The reader may recall from the one-dimensional Morlet wavelet that the period of maximum response is equal to twice the rate specified to the constructor. The same is true in two dimensions. If the constructor is called with rate = 2.5, then the center frequency of the smallest-scale wavelet will be 1 / 5 cycles per pixel. In other words, that wavelet will be maximally sensitive to image components varying with a period of five pixels. The next-largest wavelet will respond to energy having a period of ten pixels. The next will respond at 20 pixels, and so forth.

Image components having intermediate periods will be covered by the voices. The previously mentioned periods of 5, 10, and 20 are for voice 0, the basic voice, at each scale. Suppose that we specify four voices. The constructor will return voicefac = 1.189, the fourth root of two. Voice one of each of the three scales will be that factor times the basic period for each scale: 5.95, 11.89, and 23.78, respectively. Voice two for each is obtained by multiplying by that factor again, and voice three by multiplying one last time.

## Comparing the Methods

We have discussed three methods for extracting the frequency information from an area of an image. The first was the discrete Fourier transform, the second was the Gabor transform, and the third was the Morlet wavelet transform. They are quite different approaches, and each has its own advantages and disadvantages in any application. In this section we will compare and contrast these three approaches.

The Gabor transform is the most straightforward in interpretation. Each Gabor filter has only two simple parameters: frequency and orientation. In many applications, just a few judiciously chosen filters are all that is needed to extract the important information from the image. If the user knows in advance that only a few bands and orientations are all that is needed to characterize the data, the Gabor transform is often the method of choice due to its inherent simplicity. If one adheres to the eminently worthwhile KISS principle (Keep It Simple, Stupid), then the Gabor transform stands out.

At the other end of the simplicity scale, we have wavelets. The fact that their spatial-domain windows shrink at smaller scales means that we must worry about keeping the area of interest covered. The version presented earlier in this chapter accomplishes this by using arrays of filters at smaller scales. Considerably more data is generated than is the case for Gabor filters. Also, this data is more complicated to interpret due to the unusual layout of the filter centers. On the other hand, in some situations this is a very valuable property. Both the DFT and the Gabor filter fuse all of the information in the spatial window being examined. Only the wavelet approach espoused here is able to pick out different details in different parts of the covered area. In some applications, this provides crucial details that would not be obvious from the other methods. In other applications this is a needless waste of resources. The user must decide whether or not such information is desired. If it is expected that the high-frequency information will be fairly constant across the area of interest, then this approach is overkill.

Sometimes the user does not know in advance that only certain frequencies and orientations are of interest. This may be due to a simple lack of knowledge, or it may in fact be the case that nearly all information is important. In either case, great care must be exercised if the Gabor or Morlet wavelet transform is used. Unless the parameters are chosen most carefully, a frame will not be attained. Furthermore, it is unavoidable that considerably more variables will be generated than are theoretically needed to represent the spatial-domain data. This is because both of those transforms have a lot of redundancy when they are performed thoroughly enough to be complete. In such cases, the discrete Fourier transform may be the best choice. It is ideal from an information and efficiency point of view. If no spatial-domain data window is applied, the DFT is the holy grail of transforms: It is based on an orthogonal basis. That means two things. First, it means that every last bit of information in the data is captured in the transform. Second, it means that not one extra bit of redundant information appears. We have it all, but nothing more to overburden the network. The DFT presented at the start of this chapter employs a modified Welch data window, so a small amount of information is actually lost. However, it is a very small amount, negligible in nearly all cases. And a data window is important to avoid side-lobe problems. The DFT generates a lot of data, but if full information is needed, this is probably the best approach.

The DFT has two shortcomings. The worst is that it automatically generates a very large number of transform values, each encompassing a small window in the frequency domain. This degree of detail is not often needed in practical applications, and the sheer volume of data can easily overwhelm neural networks by slowing training and causing overfitting. A smaller shortcoming is that the DFT transform coefficients do not lend themselves to straightforward interpretation. When one considers a coefficient at some particular location in the DFT output matrix, it is ridiculous to talk about that coefficient representing horizontal variation at some rate and vertical variation at some other rate. One wants to invoke some trigonometry to pinpoint a direction and a frequency in that direction. That extra step is small, and it is easy to do. But it adds to a certain mysticism that surrounds DFT coefficients. Most people are instinctively more comfortable with transforms in which frequency (or scale) and direction are explicit.

Finally, we may be concerned with orientation bandwidth. If only a few specific orientations are known to be important, this is not a serious problem. But if all orientations are possible and significant, we must be sure that none of them is missed by the transform. The Morlet wavelet is simple to handle. If enough orientations are chosen to handle the shape of the mother wavelet, that will be sufficient at all scales and voices. But the Gabor transform is a problem. As the frequency increases, the width of the frequency window remains constant. Thus, the relative width of the window shrinks, and the width of the orientation window shrinks along with it. To ensure thorough coverage, we need more orientations at higher frequencies. If we are trying to minimize the amount of information presented to the neural network, we can compute a sufficient number of orientations to handle the highest frequencies desired, then discard some orientations at lower frequencies. The DFT implicitly deals with this need by computing more coefficients per degree of arc for large frequencies. All three transforms can handle this problem, but the considerations for each are very different and must be considered.

## Fast Image Filtering

This section extends to images the fast signal filtering discussed on page 200. That section should be reviewed before proceeding into this image processing discussion. Also, a review of the two-dimensional fast Fourier transform presented on page 243 would not hurt. Careful study of the coefficient layout shown in Figure 7-1 on page 247 and the discussion of radial power on page 252 would be particularly helpful.

The central idea here is exactly the same as for fast filtering in one dimension. The Gabor and Morlet wavelet methods of this chapter so far compute a spatial-domain filter and convolve that filter with the image at selected areas. We will now show how one can use the FFT to transform the image, apply the appropriate filter(s) in the frequency domain, then transform back to the spatial domain. If we have numerous large filters that must be applied to many areas of the image, the time savings by using this method can be substantial.

As we did for fast signal filtering, we will take this opportunity to introduce a new shape and scaling for the filter. This shape is very similar to those discussed earlier in this chapter, but it is generally more appropriate for two-dimensional problems. Readers who want to use the traditional symmetric Gaussian shape of the Gabor and Morlet wavelet filters will have no trouble modifying these algorithms. Be aware that when we change the shape to anything other than a Gaussian, we lose the valuable property of optimal (Heisenberg limit) simultaneous location in space and time. But for image processing, it is usually a small price and a price worth paying.

The traditional Gabor and Morlet wavelet filters have frequency-domain responses that are circularly symmetric. The problem with this shape should be apparent when one examines Figure 7-10 on page 254. Notice how the angular width for each frequency remains constant when the wedges are defined as shown in that figure. This is nearly always what we want. Now imagine that instead of wedges, we have the circular filters of the Gabor transform. For the lower frequencies in a window, a relatively wide band of angular variation is passed. For the higher frequencies, the angular variation that is passed is much less. This is particularly pronounced at low frequencies. For many applications this is unacceptable. And even if this effect can be tolerated, it complicates the life of the person designing the filters. What we would often prefer is filters having the shapes shown in Figure 7-15.

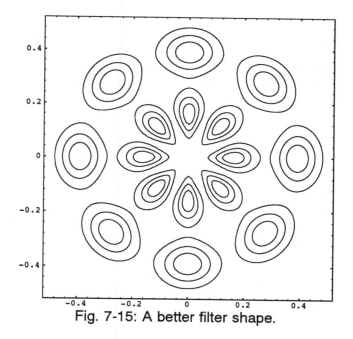

Fig. 7-15: A better filter shape.

The frequency-domain filters illustrated in that figure demonstrate how the angular width decreases as the frequency decreases. However, there are two ways in which the filters shown are not representative of filters that we would employ in practice. For clarity, orientations around the full circle are shown. But remember that the two-dimensional Fourier transform has conjugate symmetry across the origin. The results of applying a filter in any of the orientations shown there would be the complex conjugate of that from using the 180-degree-opposite filter (across the origin). Hence, half of the filters in that illustration are redundant. In practice we pick any two adjacent quadrants and limit our choice of orientations to those quadrants.

The other way in which the illustrated filters are not representative of actual use is that they do not overlap. It is only in the rarest of applications that we know in advance that we will be interested in a few disjoint areas of the frequency domain. We will nearly always want our filters to overlap to a significant degree in both the frequency and orientation dimensions. It is simply more clear to illustrate the filters as was done here.

The frequency-domain representation of that wedge-shaped filter is easily defined. We use the traditional Gaussian function to taper the response. But rather than parameterizing it in terms of vertical and horizontal frequencies, we do so with a linear frequency and a direction. This is shown in Equation (7-13).

$$h(f, \theta) = e^{a(f-f_0)^2 + b(\theta-\theta_0)^2} \qquad (7\text{-}13)$$

In that equation, the frequency and angle of maximum response are $f_0$ and $\theta_0$, respectively. The widths of the response in these coordinates are determined by the constants $a$ and $b$. A convenient way of specifying the filter width is in terms of the distance from the center frequency or angle at which the amplitude response of the filter drops to one-half of its center response. In Equation (7-14), $w$ represents that distance and $k$ represents the corresponding constant, $a$ or $b$.

$$k = \frac{\log(0.5)}{w^2} \qquad (7\text{-}14)$$

The reason that half amplitude is convenient is that this is often an effective degree of overlap for the filters. For example, suppose that we are interested in frequencies ranging from 0.1 to 0.2 cycles per unit time (0.5 is the Nyquist frequency). We decide to divide that range into five linearly spaced bands. The center frequencies are then 0.1, 0.125, 0.15, 0.175, and 0.2. Letting $w = 0.025$, Equation (7-14) tells us that $a = -1109$. Similarly, suppose that we want to present the neural network with information from all possible orientations. We choose to divide the half-circle into four directions separated by 45 degrees. The center orientations would then be 0, 45, 90, and 135 degrees. The program given later will use radians, so we should point out that these angles can also be expressed as 0, $\pi/4$, $\pi/2$, and $3\pi/4$ radians. Continuing to use radian measure, Equation (7-14) tells us that $b = -1.1237$.

It should be noted that the use of 0.5 in Equation (7-14) gives us a fairly substantial overlap. The amplitude response of one filter is down one-half at the center of the next adjacent filter. On the other hand, we nearly always want that much overlap. Remember that these figures assume that only one parameter (frequency or orientation) is changed; the other is at its center. Those areas of the frequency domain corresponding to variation of both parameters are a bit thinner. Also, experience indicates that neural networks often perform better when activation levels change gradually as data characteristics change. If the neuron count is critical, less overlap may be used. But that should be a last resort.

## Implementing the Filter

Recall how we implemented the fast filter for a signal. We observed that the frequency-domain filter function for the sum of the in-phase and (pure imaginary) in-quadrature filters canceled below the Nyquist frequency and reinforced above it (where the negative frequencies live). Essentially the same thing happens in two dimensions. The only difference is that now negative frequencies have a real meaning in terms of orientation. We do not apply the filter function only in negative frequency areas. Rather, we apply the filter function in those areas that correspond to the negative (opposite) orientation direction. This may actually involve positive frequencies. Figure 7-16 should make this more clear. But before studying that figure, let us briefly summarize the fast filtering process.

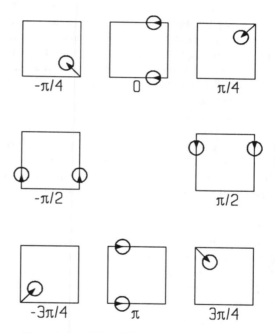

Fig. 7-16: The filter response area.

Our actions in fast filtering are actually very straightforward. First we transform the image using the FFT. The transform has conjugate symmetry, but multiplication by the asymmetric filter will destroy that symmetry. Thus, we must extend the transform to the full image size. Then we multiply the transform by the filter function. Finally, we transform back to the spatial domain. The real part of that transform is the in-phase filter, and the imaginary part is the in-quadrature filter. Now it is time to examine Figure 7-16 to gain insight into the application of the filter function. Note that this discussion applies to filters of nearly any shape. This includes the symmetric Gaussian of the Gabor and

Morlet wavelet, as well as the modified shape defined by Equation (7-13). This figure simply illustrates the center frequency and orientation by means of an arrow and the rough extent of the filter by a circle around the arrow's tip.

This would be a good time to turn back to Figure 7-1 on page 247. That matrix, including the redundant bottom half, corresponds to each box in Figure 7-16. This orientation was chosen because it corresponds to the traditional storage of the FFT coefficients, which also happens to be the storage method used in the program to be presented later. The upper-left coefficient is the grand DC term. Horizontal frequencies increase toward the right until the horizontal Nyquist limit is hit in the middle of the box. The rightmost coefficient corresponds to a horizontal frequency of −1. However, for the purposes of this illustration, we will assume that the resolution is so fine that the right edge of each box gets us back to a horizontal frequency of zero. (Remember that the transform is periodic in all directions. Each box is just a member of an infinite tiling of identical boxes.) Similarly, the vertical frequency increases downward until the vertical Nyquist frequency is encountered halfway down. Again, in the actual storage, the bottom row will correspond to a vertical frequency of −1. But we will visualize the box as just another tile and call that bottom row a vertical frequency of zero. Thus, all four corners correspond to horizontal and vertical frequency origins and may be used interchangeably. The grand Nyquist point is in the center of the box.

Start by considering a filter having a center frequency of about half the Nyquist limit and an orientation direction of 0 radians. This means that the filter is maximally sensitive to variation that moves directly to the right (and left, expressed as the conjugate of that to the right). This situation appears in the top center box. As was discussed at the beginning of this section, we center the filter at the point in the transform space that corresponds to that frequency in the *opposite* direction. The response area of the filter extends into adjacent tiles above and below. In terms of the elements of this particular tile, the response wraps around to the other side of this tile as shown.

Wraparound does not always occur. Look at the upper-right box, which illustrates an orientation of $\pi/4$ radians. Unless the angular width is very large, the effect of this filter in the frequency domain will be entirely contained within a single contiguous area of the transform matrix. We will not take the time to examine the other six illustrations in this figure. However, most readers would benefit from studying them, as they demonstrate an important part of the workings of the program that will appear soon.

Deep mathematical detail is not a goal of this text. However, for the benefit of readers who would like a little more in the way of exactitude than just the heuristic illustration given above, we will now briefly sketch the mathematics involved in this filter. By definition, the in-phase filter is real and even (symmetric with respect to the origin). That implies that its Fourier transform is also real and even. Also by definition, the in-quadrature filter is the imaginary $i$ times a function that is real and odd (antisymmetric with respect to the origin). That implies that the function's Fourier transform is $i$ times a function that is pure imaginary and odd. Multiplication by $i$ causes that transform to be pure real. The same window functions are used for both the in-phase and in-quadrature filters in the spatial domain, so the magnitudes of their transforms are identical. However, the squaring of $i$ for the in-quadrature filter's transform causes it to have the opposite sign as the in-phase filter's transform on one side of the origin, causing them to cancel. On the opposite side of the origin, they will be equal, so they reinforce. Proofs of these claims can be found in nearly any text that covers two-dimensional Fourier transforms. One example is [Lim, 1990].

## Decimation

Frequency-domain decimation of a signal is easy. We simply collapse out the interior coefficients in the transform, the ones that correspond to frequencies above the new Nyquist point. That was illustrated in Figure 5-48 on page 213. It is slightly trickier in two dimensions, but not too bad. Look at Figure 7-17, which demonstrates decimating a square image by a factor of two in both dimensions.

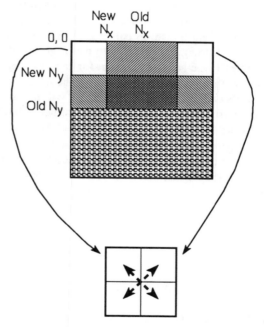

Fig. 7-17: Frequency-domain image decimation.

The complicated top portion of the figure is the Fourier transform of the original image. As usual, the grand DC point is at the upper-left corner. The transform has conjugate symmetry, so the redundant bottom half can be ignored. In order to decimate in the vertical direction, we remove all coefficients from the new vertical Nyquist frequency through the old one. This corresponds to the horizontal band extending across the entire width of the figure about one-quarter of the way down. We also are implicitly removing the corresponding band across the redundant lower half of the transform. That is not shown as it is not relevant to the computation.

In order to decimate horizontally, we remove a vertical band extending for the entire height of the transform. The lower half of this band is in the redundant part, so it is ignored here. But the upper part of the band includes both positive and negative horizontal frequencies.

The two small squares that remain are the unique low-frequency components. They are combined as shown to define the upper half of the transform of the decimated image. The lower half of this transform can be derived from the upper half by conjugate symmetry.

## Code for the Transform

We now present code for implementing fast image filtering. It assumes that the number of rows and columns are powers of two. Readers who substitute a mixed-radix transform for the power-of-two transform called here may relax this restriction to just being even.

The constructor is called with the original image as input. It performs the two-dimensional FFT and saves it. Except for conjugate symmetry in the first and last rows, the transform that we save is unique. The redundant bottom half is ignored. The constructor also allocates a work vector that will be used by itself and by the filter. Only half of that work vector is used by the constructor, but the filter needs the entire array.

As the transform is being computed, we also multiply each coefficient by the constant $2 / (nr * nc)$. The factor of two is the same factor that we encountered in one dimension. It comes from the in-phase and in-quadrature filters reinforcing in the negative-orientation half-plane. The other factors are a necessary part of the inverse transform. If we apply them now, it saves doing it every time the filter is called. The class declarations, constructor, and destructor now follow.

```
class FastFilterImage {

  public:
    FastFilterImage ( int nrows , int ncols , double *image ) ;
    ~FastFilterImage () ;
    void filter ( double freq , double f_width , double angle ,
                  double a_width , int rdec , int cdec ,
                  double *real , double *imag ) ;


  private:
    int nr, nc ;                    // Size of image
    double *xr, *xi ;               // Transform saved here
    double *work ;                  // Used by constructor (nr) and filter (2nr)
} ;
```

```
/*
   Constructor
*/

FastFilterImage::FastFilterImage (
   int nrows ,                        // Number of rows in image
   int ncols ,                        // And columns
   double *img                        // Image
   )
{
   int n, ic, ir, r ;
   double *wr, *wi, wt ;

   nr = nc = 0 ;                      // Cheap insurance
   xr = xi = work = NULL ;            // So destructor does not err

   if (nrows <= 0 || ncols <= 0)      // Protect from careless user
      return ;

   n = (nrows / 2 + 1) * ncols ;      // Size of (almost) unique transform
   xr = (double *) malloc ( 2 * n * sizeof(double) ) ;
   if (xr == NULL)
      return ;

   work = (double *) malloc ( 2 * nrows * sizeof(double) ) ;
   if (work == NULL) {
      free ( xr ) ;
      xr = NULL ;
      return ;
      }

   xi = xr + n ;                      // Saves a call to malloc

   nr = nrows ;
   nc = ncols ;
```

```
/*
   Copy each column of the block to the work vector, transform it, then copy the result to the
   corresponding column of the output. We split the temporary work vector in half, using the
   first half for the real part of the transform, and the second half for the imaginary part. Recall
   that REAL_FFT wants the even terms of the input vector in the real part, and the odd terms
   in the imaginary part. Also recall that REAL_FFT returns the real part of the Nyquist term in
   wi[0], which is really zero. The imaginary part of the Nyquist term is also zero.
```

Also include the factor of two that results from the in-phase and in-quadrature filters
reinforcing, and 1/(nr*nc) for the two inverse Fourier transforms that will be done for the
filtering.  Doing it once now saves doing it every time 'filter' is called later.
*/

```
wr = work ;          // Work vector
wi = work + nr / 2 ;  // Use another part for imaginary
wt = 2.0 / ((double) nr * (double) nc) ; // Reinforcement, inverse transform

for (ic=0 ; ic<nc ; ic++) {          // Do each column separately

  for (ir=0 ; ir<nr/2 ; ir++) {       // Copy all rows 2 at a time
    r = 2 * ir ;                      // Even row in block
    wr[ir] = wt * img[r*nc+ic] ;      // Even part into real vector
    ++r ;                             // Odd row in block
    wi[ir] = wt * img[r*nc+ic] ;      // Odd part into imaginary
    }

  real_fft ( nr/2 , wr , wi ) ;       // Transform column ic

  for (ir=0 ; ir<nr/2 ; ir++) {       // Because the input was real,
    xr[ir*nc+ic] = wr[ir] ;           // there are nr/2+1 unique terms.
    xi[ir*nc+ic] = wi[ir] ;           // The neg freqs are just conjugates
    }                                 // and so are ignored.
  xr[nr/2*nc+ic] = wi[0] ;            // Returned real part of Nyquist here
  xi[nr/2*nc+ic] = xi[ic] = 0.0 ;     // These are really zero
  } // For all columns
```

/*

At this time, all nc columns of the block have been transformed.  They are now in place in
the private vectors "xr" and "xi".  The final step is to transform each of the nr/2+1 rows.  The
first and last rows are entirely real, so if we were terribly concerned with efficiency, we could
use the real transform routine REAL_FFT to do it. But the relatively small savings in time is
rarely worth the increase in complexity of the code, so we will use the full complex routine.
Of course, the interior rows all need the full complex routine, as they have no special properties.
*/

```
  for (ir=0 ; ir<=nr/2 ; ir++)
    fft ( nc , xr+ir*nc , xi+ir*nc ) ;
}
```

```
/*
   Destructor
*/

FastFilterImage::~FastFilterImage ()
{
   if (xr != NULL)
      free ( xr ) ;
   if (work != NULL)
      free ( work ) ;
}
```

The real work is done by the filter routine, which appears soon. However, we will start out with a pair of simple local routines that filter calls. They both concern the frequency-domain filter function. The first routine, weight_init, uses Equation (7-14) to compute and save fwt and awt. It also makes sure that the orientation angle lies in the interval from $-\pi$ to $\pi$, which is the range of the C library routine atan2. This initialization routine appears now.

```
static double fcent, acent, fwt, awt ;
static double pi = 3.141592653589793 ;
static double two_pi = 2.0 * pi ;
static double half_pi = 0.5 * pi ;

static void weight_init (
   double freq ,                    // Center frequency, 0-.5
   double f_width ,                 // Half-amplitude frequency difference
   double angle ,                   // CW Direction in radians
   double a_width ,                 // Half-amplitude angle difference
   )
{
   fcent = freq ;
   if (angle <= pi)
      acent = angle ;
   else
      acent = angle - two_pi ;
   fwt = log ( 0.5 ) / (f_width * f_width) ;
   awt = log ( 0.5 ) / (a_width * a_width) ;
}
```

The second local routine is used to compute the weight function at any frequency and orientation. Since the matrix form of the Fourier transform implies horizontal and vertical frequencies, that is the form of this routine's calling parameters. Readers who want to use the traditional Gaussian function of the Gabor transform and Morlet wavelet need to modify this routine, and in addition, make trivial changes to the weight_init routine and to the calling parameters of the filter routine. The weight routine starts by computing the linear frequency and then subtracting the center frequency. It also computes the angle and its difference from the center angle. The only tricky part is explicitly returning zero rather than evaluating Equation (7-13) when the angle error exceed 90 degrees. This saves a little computation time, since in all cases of practical interest the function would be essentially zero there anyway. Moreover, in case the user carelessly specified an excessive filter width, it enforces a quadrature pair of filters by restricting response to a half-plane.

```
static double weight ( double xfreq , double yfreq )
{
  double fdist, angle, adist ;

  fdist = sqrt ( xfreq * xfreq + yfreq * yfreq ) - fcent ;
  angle = atan2 ( -yfreq , -xfreq ) ;
  adist = fabs ( angle - acent ) ;

  if (adist > pi)                    // Wrapping around the discontinuity
    adist = two_pi - adist ;         // would cause error
  if (adist > half_pi)               // Keep response on half-plane
    return 0.0 ;                     // so we get a quadrature pair
  return exp ( fwt * fdist * fdist  +  awt * adist * adist ) ;
}
```

We now arrive at the most complicated routine here, the one that actually generates the filtered output. Logically and mathematically, it is not at all difficult. It simply applies the filter function, Equation (7-13), in the manner shown in Figure 7-16 on page 289. It simultaneously decimates the output using the method shown in Figure 7-17 on page 291. What makes this routine confusing is keeping track of subscripts, especially in regard to handling coefficients in the bottom half of the transform matrix. We will try to explain it thoroughly, starting with the calling parameters.

freq　　This is the center frequency of the filter, normalized as cycles per unit sample time. Thus, the Nyquist limit is 0.5.

f_width　The filter width in the absolute frequency dimension is specified as the difference at which the filter's amplitude response drops to one-half of its maximum.

angle    Since we follow the convention of computing the column transform from top to bottom, this is the *clockwise* angle from +$X$ of the direction of maximum sensitivity (in radians).

a_width    The filter width (in radians) in the orientation dimension is the difference at which the amplitude response is cut in half.

rdec    The row decimation ratio must be a power of two. Use 1 for no decimation.

cdec    The column decimation also must be a power of two.

real    This is the real part of the filtered output, which is the in-phase filter component.

imag    This is the imaginary (in-quadrature) output.

The first action taken by filter is to call weight_init to initialize the filter function constants. Then it computes the new image size (after decimation, if any). The remainder of the program is divided into four sections: process the upper half of the transform (up to but not including the vertical Nyquist frequency), process the lower half, process the vertical Nyquist frequency, and finally, perform the inverse transform. Let us take these one at a time.

The upper half of the transform is the easiest, as that is the part that was stored for us by the constructor. For each row, process the columns from the outside in, working toward the middle. The nonnegative frequencies are trivial. We just copy the stored transform to the output, applying the weight function at that time. The first (horizontal DC) column has no corresponding negative frequency. For all others, we also do a direct weighted copy. But now we must compute the subscripts so that we access the right side of the rows. Speed fanatics may want to use pointers rather than explicitly computing subscripts each time. However, most optimizing compilers do a very good job of avoiding the redundant computation that explicitly appears in the code. Doing it this way makes the code much more clear and easy to translate to other languages, while having little or no impact on speed in most cases.

When the row processing gets to the center column, the horizontal Nyquist frequency, that case is handled separately. If column decimation is occurring, we simply set that term to zero. Purists may want to do a quadrature addition of the two Nyquist components. However, that is superfluous in practical situations. If we are decimating, we are assuming that the filter function has dropped to essentially zero by then. If it has not, we have more worries than simply discarding the Nyquist power! If we are not decimating, we just copy the Nyquist term. Note that the horizontal frequency at this new Nyquist point is the original (0.5) divided by the decimation ratio.

Processing the bottom half of the transform is similar to that above, but it is complicated by the fact that we cannot do a simple copy. We must use conjugate symmetry to procure each term from the opposite side of the top half. Also, all vertical frequencies are now negative. The first column, the horizontal DC term, must be handled separately, as it is not retrieved by symmetry. It is a direct copy from the same column in the top half. From then on, operation is exactly the same as in the top half, except for getting all terms by symmetry.

The vertical Nyquist frequency row is handled separately. If no row decimation is being done, we do a weighted direct copy, much as was the case for the top half. If row decimation is used, we set this entire row equal to zero for exactly the same reason as for a decimated horizontal Nyquist frequency.

The final step is to invert the transform, bringing us back to the spatial domain. This is a straightforward application of the usual algorithm. Each column is transformed first, with conjugation being done as the column is copied. Then the rows are transformed, with conjugation again done. A tiny amount of time could be saved by using an FFT routine that implicitly performs an inverse transform. However, in the grand scheme of things, the time taken by conjugation is minuscule. This approach avoids the need for an invertible routine. Readers who have an invertible FFT may want to substitute it for the one here and skip the two conjugations.

```
void FastFilterImage::filter (
   double freq ,               // Center frequency, 0-.5
   double f_width ,            // Half-amplitude frequency difference
   double angle ,              // CW Direction in radians
   double a_width ,            // Half-amplitude angle difference
   int rdec ,                  // Row decimation ratio (power of 2)
   int cdec ,                  // And column
   double *real ,              // Real output (in-phase)
   double *imag                // Imaginary output (in-quadrature)
   )
{
   int irow, icol, newnr, newnc ;
   double xfreq, yfreq, wt, rnyq, inyq, *wr, *wi ;
   double root2 = sqrt ( 2.0 ) ;
   double rootp5 = sqrt ( 0.5 ) ;

   weight_init ( freq , f_width , angle , a_width ) ;

   newnr = nr / rdec ;
   newnc = nc / cdec ;

/*
   First, do the upper half, which is nothing more than a weighted copy of the stored transform.
   Do the DC and all positive vertical frequencies below the Nyquist. For each, do all horizon-
   tal frequencies.
*/
```

```
      for (irow=0 ; irow<newnr/2 ; irow++) {          // Nonneg Y freqs below Nyquist
         yfreq = (double) irow / (double) nr ;         // Vertical frequency
         for (icol=0 ; icol<newnc/2 ; icol++) {        // All X freqs except Nyquist
            xfreq = (double) icol / (double) nc ;       // Positive Horizontal freq
            wt = weight ( xfreq , yfreq ) ;
            real[irow*newnc+icol] = wt * xr[irow*nc+icol] ;
            imag[irow*newnc+icol] = wt * xi[irow*nc+icol] ;
            if (! icol)                                  // DC has no corresponding negative
               continue ;                                // So skip the negative frequency stuff below
            wt = weight ( -xfreq , yfreq ) ;
            real[irow*newnc+newnc-icol] = wt * xr[irow*nc+nc-icol] ;
            imag[irow*newnc+newnc-icol] = wt * xi[irow*nc+nc-icol] ;
            }
         // X Nyquist is assumed zero if decimating, else just conjugate
         if (cdec > 1)
            rnyq = inyq = 0.0 ;
         else {
            rnyq = xr[irow*nc+nc/2] ;
            inyq = xi[irow*nc+nc/2] ;
            }
         wt = weight ( 0.5 / (double) cdec , yfreq ) ;
         real[irow*newnc+newnc/2] = wt * rnyq ;
         imag[irow*newnc+newnc/2] = wt * inyq ;
         }

/*
      Now do the lower half.  This is negative vertical frequencies.  We get the values by using the
      conjugate of the stored transform at the opposite diagonal position (negative of the frequen-
      cy).
*/

      for (irow=1 ; irow<newnr/2 ; irow++) {          // Negative Y freqs
         yfreq = - (double) irow / (double) nr ;       // Vertical frequency
         for (icol=0 ; icol<newnc/2 ; icol++) {        // All X freqs except Nyquist
            xfreq = (double) icol / (double) nc ;       // Nonnegative Horizontal freq
            wt = weight ( xfreq , yfreq ) ;
            if (! icol) {                                // DC is a simple weighted copy
               real[(newnr-irow)*newnc] = wt * xr[irow*nc] ;
               imag[(newnr-irow)*newnc] = -wt * xi[irow*nc] ;
               continue ;
               }
            // Get positive X frequencies from right side of source
            real[(newnr-irow)*newnc+icol] = wt * xr[irow*nc+nc-icol] ;
            imag[(newnr-irow)*newnc+icol] = -wt * xi[irow*nc+nc-icol] ;
```

```
        // Get negative X frequencies from left side of source
        wt = weight ( -xfreq , yfreq ) ;
        real[(newnr-irow)*newnc+newnc-icol] = wt * xr[irow*nc+icol] ;
        imag[(newnr-irow)*newnc+newnc-icol] = -wt * xi[irow*nc+icol] ;
        }
     // X Nyquist is assumed zero if decimating, else just conjugate
     if (cdec > 1)
        rnyq = inyq = 0.0 ;
     else {
        rnyq = xr[irow*nc+nc/2] ;
        inyq = -xi[irow*nc+nc/2] ;
        }
     wt = weight ( 0.5 / (double) cdec , yfreq ) ;
     real[(newnr-irow)*newnc+newnc/2] = wt * rnyq ;
     imag[(newnr-irow)*newnc+newnc/2] = wt * inyq ;
     }

/*
   Now do the vertical (Y) Nyquist frequency
*/

  if (rdec == 1) {
     for (icol=0 ; icol<newnc/2 ; icol++) {   // All X freqs except Nyquist
        xfreq = (double) icol / (double) nc ; // Positive Horizontal freq
        wt = weight ( xfreq , 0.5 ) ;
        real[newnr/2*newnc+icol] = wt * xr[nr/2*nc+icol] ;
        imag[newnr/2*newnc+icol] = wt * xi[nr/2*nc+icol] ;
        if (! icol)        // DC has no corresponding negative
           continue ;      // So skip the negative frequency stuff below
        wt = weight ( -xfreq , 0.5 ) ;
        real[newnr/2*newnc+newnc-icol] = wt * xr[nr/2*nc+nc-icol] ;
        imag[newnr/2*newnc+newnc-icol] = wt * xi[nr/2*nc+nc-icol] ;
        }
     // X Nyquist is assumed zero if decimating, else just conjugate
     if (cdec > 1)
        rnyq = inyq = 0.0 ;
     else {
        rnyq = xr[nr/2*nc+nc/2] ;
        inyq = xi[nr/2*nc+nc/2] ;
        }
     wt = weight ( 0.5 / (double) cdec , 0.5 ) ;
     real[newnr/2*newnc+newnc/2] = wt * rnyq ;
     imag[newnr/2*newnc+newnc/2] = wt * inyq ;
     }
```

```
    else {  // Case of row decimation
      for (icol=0 ; icol<newnc ; icol++)
        real[newnr/2*newnc+icol] = imag[newnr/2*newnc+icol] = 0.0 ;
      }

/*
      Do the inverse Fourier transform.  Note that the division by n squared was done in the
      constructor.  So we can avoid that now.
*/

    wr = work ;
    wi = work + newnr ;

    for (icol=0 ; icol<newnc ; icol++) {

      for (irow=0 ; irow<newnr ; irow++) {
        wr[irow] = real[irow*newnc+icol] ;
        wi[irow] = -imag[irow*newnc+icol] ;
        }

      fft ( newnr , wr , wi ) ;

      for (irow=0 ; irow<newnr ; irow++) {
        real[irow*newnc+icol] = wr[irow] ;
        imag[irow*newnc+icol] = wi[irow] ;
        }
      }

    for (irow=0 ; irow<newnr ; irow++) {
      fft ( newnc , real+irow*newnc , imag+irow*newnc ) ;
      for (icol=0 ; icol<newnc ; icol++)
        imag[irow*newnc+icol] = -imag[irow*newnc+icol] ;
      }
    }
```

Astute readers may have noticed one logical dissimilarity between our presentations of signal and image filtering. When a signal was filtered, we explicitly zeroed all transform coefficients except those that were in the negative frequency area. The image-processing algorithm given here computes the entire filtered transform, calling the weight function for every term. The reason for this discrepancy is that for a signal, orientation is not an issue. We know in advance that only the terms corresponding to negative frequencies will be kept. But for image processing, the terms that are retained depend on the orientation. We do not in general have advance knowledge of which terms will be kept and which will be discarded. The weight routine does explicitly zero terms on one side of a half-plane, although we usually do not know where that half-plane will be located.

Sometimes, though, we may have useful advance information. In this case, specialized routines may be used to speed time-critical applications. For example, we may want to write a special filter function that is only used when the filter response lies entirely in the right half of the transform matrix. In this case, we only need to apply the filter function to the right half. Much more importantly, when we invert the transform, we only need to apply the column transforms to the columns in the right half. That can be a tremendous savings. If we need a special routine that processes either the top or bottom half exclusively, we may want to reverse the order of the inversion. First, transform rows, doing only the nonzero half. Then do the columns. Remember that a two-dimensional FFT can be done in either order; the results will be the same both ways. Since these are special applications, code is not given here. However, the modifications should be very easy. The filter code given above would simply have a few parts removed or reordered.

## Zero Padding

On page 201 we discussed the need for padding the end of a signal with zeros to prevent wraparound error. The same principle applies in two dimensions. Unfortunately, zero padding is much more expensive in two dimensions than in one. Both memory and time requirements grow rapidly as rows and columns of zeros are added. Therefore, we must be as judicious as possible. Complicating the matter even more is the fact that the unusual wedge shape of the frequency-domain filter function makes it difficult to find a universally applicable function for estimating the extent of the filter in the spatial domain. Generally, the easiest and safest approach is to start by padding one or more sample images with a relatively large number of zeros. Then apply each of the filters that will be used. Carefully examine the right and bottom borders of the filtered output. They should ideally be entirely zero, but wraparound from the top and left sides will generate nonzero terms. *For every filter*, see how far inward the seriously nonzero terms extend. That indicates how many zeros are needed in the final application.

There is a more rigorous, if slightly more complicated way of determining the extent of wraparound. That is to examine the actual impulse responses of the filters. There are at least two ways of doing this. The fastest method (in terms of computation speed) is to generate a large matrix consisting of the frequency-domain filter function, then compute the inverse Fourier transform of this matrix. With this approach, the impulse response will be returned in the usual 2DFFT order. The center of the filter will be in the upper-left corner and will extend halfway into the matrix in both directions. The other three quadrants of the impulse response will lie in the remaining three matrix quadrants, with the centermost coefficients occupying the outer corners.

A much simpler approach, which can be done with the program supplied here, is to explicitly filter an impulse. Generate a large matrix that is entirely 0 except for a single 1 somewhere near the center. Call the constructor with this image, then examine all filtered outputs. The center of each impulse response function will be in the same location as the 1 in the original image, so in some sense it will be easier to handle. The disadvantage of this method is that time is unnecessarily wasted transforming the original impulse matrix. Of course, in most applications, that is done off-line, so it is no real problem.

Since the impulse responses are conjugate symmetric, we only need to examine a half-plane in order to determine their practical extent. This operation can be automated by writing a small routine that computes the absolute value of the (complex) impulse response at an ever increasing distance from the center. The user can then choose a distance (number of zeros to append) at which the magnitude of the coefficients is sufficiently small.

## An Image-Processing Methodology

By now, many readers are probably wondering how a book that is supposedly focused on neural networks has apparently managed to stray so far from that topic. This section will pull together image processing, filtering techniques, and neural networks. We will discuss in general terms a methodology that is extremely powerful and that is often used in practical applications.

The following situation is a common occurrence. Large images arrive. Perhaps each is a chest X-ray that may contain suspicious lesions. Or maybe it is an infrared scan of enemy terrain in which hostile vehicles are potentially hiding. Or it might be a satellite image that contains areas of interest that can be characterized by specific textures. In all of these cases, we will often have the same goal and will follow the same procedure to reach that goal.

1. An expert, usually human, is employed to identify and collect training samples from small areas of an existing assortment of images.
2. A set of feature variables is defined that will be used to identify the regions of interest.
3. Those features are computed for each training set case.
4. Normalizations appropriate for neural network inputs are performed.
5. A neural network is trained and validated with an independent collection of known samples.
6. We can now locate these same regions of interest on unknown images.

Let us consider some of the decisions that are involved in the above sequence. It is implicitly assumed here that some features based on spatial frequency at particular orientations are important. In most cases, other variables, such as raw tone or perhaps even special texture characteristics, are also useful. Those variables are ignored in this section, as they are treated elsewhere.

The first question to be answered is whether the final classifier will be called upon to scan entire images, looking in great detail for areas of interest, or whether, instead, the classifier will just be presented with occasional potential targets for analysis. In the former case, the fast filtering methods of the previous section are mandatory unless computer time is of no importance. In the latter case, the spatial-domain methods that appeared first in this chapter are often more appropriate.

A consideration that is vital but often overlooked is that the implicit filter size must be commensurate with the object size. For the spatial-domain filters, we automatically know the spatial size of each filter, since the filter is designed and implemented in that domain. But for the frequency-domain fast filters, it is imperative that we use one of the methods described on page 302 to compute the spatial-domain extent of the filter's impulse response. Ideally, the area in which the filter has its strongest response should be about the same size as the object being sought. If the filter's extent is much smaller than the object, that is usually not a problem. At worst, we may want to compute variables for several spatial positions. That was done in the Morlet wavelet algorithm presented on page 271. At best, we may be able to get away with just one common center for all filters. On the other hand, if the spatial extent of a filter is much larger than the object, we are in big trouble. The variables will probably be unduly influenced by the contents of the object's neighborhood. When that happens, it is time to rethink the approach. Chances are very good that the center frequency of the offending filter is too low to be a useful feature anyway.

In many image-processing applications, we will want to work with power, discarding phase information. The sum of the squares of the real and imaginary parts of the filtered output, usually with some additional transformations, will be the foundation variables. It is certainly possible to work with the full complex values, in which case a complex-domain neural network would probably be indicated. Some applications may benefit greatly from this, as phase information is occasionally of significant importance. However, there is a high price to pay. The phase normalizations discussed in the context of signal processing are difficult or impossible to effectively implement in two dimensions. Some special techniques are available, but they are beyond the scope of this text. The best hope is to use extra-large training and validation sets in order to handle the variety of phase relationships that will have to be learned.

There is one procedure for handling complex filter outputs that is useful in some situations. Divide the training set into subsets based on the phase of one or more of the *lowest* frequency filter outputs, then train a separate network with each subset. When it comes time to classify an unknown case, choose the network based on the same phase decision. The decision rule may be very simple, such as using one network if the phase of the lowest-frequency horizontal filter is in one half-plane, and using a second network otherwise. Or it may be very complicated. We may look at a horizontal *and* a vertical filter, and divide the phase of each into four quadrants. This implies the use of 16 neural networks! At first thought, it may seem that this approach would require a prohibitively large training collection. But the truth of the matter is that the number of training examples needed to train a single large network to handle all 16 phase conditions will often greatly exceed the number needed to train 16 smaller specialized networks. If phase is important, do not neglect this technique. It can be immensely valuable.

As always, proper transformation and scaling is crucial to neural network performance. We dare not blithely present the raw filter outputs to the network. If we are using power, a square root or log transform is often an appropriate first step. If we are using the full complex outputs, avoid the temptation to separately transform the real and imaginary parts. Instead, convert to polar coordinates, transform the magnitude (leaving the phase unchanged), then transform back to the complex plane. In any case, the median of each variable should be computed for the training set and subtracted from all training and unknown cases to center the data. Scaling per the standard deviation or interquartile range is also recommended. (Again, scale complex values per the magnitude, not separately!) If any readers are getting sick of hearing this same story over and over, remember that it is being repeated for a reason: It is vital.

The issue of decimation of low-frequency outputs involves the resolution of the image relative to the size of the anticipated targets. If we are scanning an image for large irregular regions having a uniform texture, and if the effective spatial extent of the filters is much smaller than the regions of interest, and if we do not need an exact boundary located, we may want to decimate. In this case we would classify an unknown image only at the pixels corresponding to the largest decimation. At the other extreme, we may be looking for a target occupying only several dozen pixels. In this case we would nearly always keep full resolution and send each pixel's filter outputs to the network.

We usually want to collect each training case in such a way that the pixel whose filtered outputs are presented to the network is in the exact center of the object being sought. This facilitates locating the center of the detected objects in the unknown image. If the exact same target is repeated in the training set with slightly different offsets, this may slightly increase the likelihood of detecting that target. But the price paid will be less accurate location of the centroid of the target. Unless economics dictates otherwise, use each training target only once, exactly centered.

This section has presented a powerful method for target detection in images based on oriented periodic features. A good experimental platform for implementing this method can be constructed as follows. Write a program that displays the application's images on a video screen. Give the user a mouse-driven pointer for locating areas of interest on the image. A deluxe implementation will allow the user to interactively design filters, with the frequency, orientation, and extent of the filter in the spatial domain displayable as an overlay on the screen. Let the user build training sets (including a reject category) by clicking mouse buttons. Another menu selection causes a displayed image to be classified, with the results written to the screen. The user can then immediately see errors and append corrective cases to the training set. A few iterations with an expert are often sufficient to construct an extremely powerful classifier system.

# 8

# Moment-Based Image Features

Many applications involve segmenting an image, locating regions of interest, and then determining the identity of the regions. Medical laboratories might stain cells, isolate them based on unique coloration, then make a diagnosis based on the shape of those cells. Defense applications might locate target vehicles based on infrared return or laser radar range, then attempt to assign them to vehicle classes based on their profile. A sandpaper manufacturer might use photomicrographs of their abrasive coating to isolate individual particles, then ascertain their quality based on the shapes of those particles. The segmentation process will not be discussed in this chapter. Both neural and nonneural segmentation is possible. This chapter presents a powerful family of shape descriptors based on moments. These descriptors are ideal for use as neural network inputs.

## The Utility of Shape Moments

There are two principal approaches to describing the shape of a two-dimensional closed figure. (Three-dimensional figures and unconnected regions are special cases dealt with later.) One approach is to follow the perimeter of the figure, basing the shape descriptors on that path. Fourier methods for perimeter description were discussed in the section *Complex Data in the Time Domain*, starting on page 105. Perimeter methods extract a lot of fine detail about the shape of the object. When this detail is meaningful, that is good. But often a lot of that detail is contributed by noise. The effects of noise contamination of the object's outline can be reduced by basing shape descriptors on the entire mass of the figure, rather than just on its perimeter. That is what moments do.

Shape moments do have a few disadvantages that should be mentioned. They are not sensitive to fine detail. Only low-order moments are easily computed with any reliability, and those low-order moments are primarily sensitive to gross shape features. Computation of the high-order moments that capture more detail is almost always too numerically unstable for practical applications. The basic fact of the matter is that fewer numbers are available than with other methods. On the other hand, many applications are satisfactorily served by relatively few measures of shape, so moments are just fine. The choice depends on how much detail is significant and how much noise is present.

Another disadvantage of moments for shape description is that rotation normalization can sometimes be difficult. Some applications do not want rotation invariance. The rotational position of the object is fixed or perhaps is even useful information. But many applications need the shape descriptors to be independent of the rotational position of the figure. Certainly, cells under a microscope cannot be asked to all line up in some standard way! Later in this chapter, a method for removing the effect of rotation will be given. However, it will be seen that it is prone to error. That problem must be carefully approached on a case-by-case basis. If it is any consolation, remember that Fourier perimeter descriptors are also difficult to normalize against rotation. The positive real energy criterion discussed on page 123 is not exactly foolproof.

Moments do have several good points (in addition to being widely effective at their job of describing the shape of an object). Their meanings are quite intuitive. This is always a good trait when one needs to convince a boss or a review committee that the right approach is being taken. They are easy to program and not terribly slow to compute. Finally, they have several straightforward generalizations. Unlike Fourier descriptors, they are trivially adaptable to describing figures that are not connected. And if the figure contains a third dimension, such as range or tone information, that additional information can be included in the shape measure with little additional work.

## Binary Shape Moments

The most basic moment variables are based on binary figures. Every pixel in the image either belongs to the figure, or it does not. This formulation will be generalized later. For now, a binary definition of the figure will keep things a little simpler. That is also the most common situation.

In their most general form, moments are defined using integrals across continuous coordinates. Let $f(x, y)$ be the membership function of the figure. This function takes the value 1 if the point $(x, y)$ is a member of the figure. Otherwise its value is 0. The $(p, q)$ moment of the figure, where $p$ and $q$ are nonnegative integers, is given by Equation (8-1). That equation should look familiar to statisticians, who use something very like it to define the moments of probability densities.

$$M(p, q) = \int_{-\infty}^{\infty} \int_{-\infty}^{\infty} x^p y^q f(x, y) \, dx \, dy \qquad (8\text{-}1)$$

Since we virtually never deal with continuous figure definitions, we need a version of that equation that is applicable to discrete samples. The figure will be assumed to be composed of a finite number of pixels, each of which is centered on a point of an equispaced discrete lattice. The horizontal position on this lattice will be defined by an $x$ coordinate, with $x$ increasing to the right. The vertical position is defined by $y$, which increases upward. The definition of moments using discrete data is given in Equation (8-2).

$$M(p, q) = \sum_x \sum_y x^p y^q f(x, y) \tag{8-2}$$

That equation shows that the moments are computed using the $(x, y)$ coordinates of the pixels that make up the figure. Pixels that are not part of the figure are ignored. One simply passes through the figure, pixel by pixel. For each pixel, sum $x^p y^q$. It seems easy.

Alas, there are some small but worrisome problems with a direct application of Equation (8-2). The most obvious worry is that the moments are extremely dependent on the location of the origin, $(0, 0)$, with respect to the figure. For any fixed origin, moving the figure will impact the computed moments. That is not good. If we want to use moments to represent shape, they should not be affected by location.

A more subtle but equally serious problem is in regard to numerical stability. We don't have it. Suppose that the origin is at the lower-left corner of the figure. Pixels near that corner will contribute relatively small $x^p y^q$ products to the sum that makes up a moment. But if $p$ and $q$ are anything but tiny, the pixels near the upper-right corner will contribute gigantic products to the sum. Floating-point sums will be combining terms that are so different in magnitude that precision could easily be lost. The contributions of pixels near the lower-left corner may be totally ignored if not enough mantissa bits are used in the implementation.

The problem of location affecting moments can be totally eliminated, and the problem of numerical instability can be helped, by one simple change in the previous formula. Rather than working with an arbitrary coordinate system, decree that the origin is at the physical centroid of the figure. Do this by subtracting from each coordinate the mean of the figure in that coordinate. The mean (center) of the object in the horizontal direction, $\mu_x$, is the sum of the $x$ coordinates of all pixels in the object, divided by the number of pixels in the object. We leave it as a very simple exercise for the reader to verify Equation (8-3).

$$\mu_x = \frac{M(1, 0)}{M(0, 0)}$$
$$\mu_y = \frac{M(0, 1)}{M(0, 0)} \tag{8-3}$$

When we force the origin to be at the center of the figure, the moments are called *central moments*. These are, for all practical purposes, the only moments ever actually used. It is obvious that central moments are invariant to change of location. There can still be numerical difficulty related to the magnitude of contributions near the center versus those near the periphery. But this is not nearly so much of a problem for two reasons. One reason is that the distance separating a corner from the centroid is generally about half the distance separating two corners, so the worst-case scenario tends to be much less troublesome. The other reason is that the pixels that are most likely to get short shrift from loss of floating-point precision are those near the center, and they usually have no great importance in determining the figure's shape anyway. The most important determiners of shape are the peripheral pixels, and those are the ones that numerically dominate the sum.

The central moments are defined in Equation (8-4). Note that rather than using $M$ for their labels, we use $U$. However, for most of the remainder of this chapter, we will use $M$ rather than $U$ to refer to all moments, while understanding that in practice they are actually central moments. That offense is standard in the literature, so the tradition will be maintained. There should be no confusion, as noncentral moments are virtually never used for anything.

$$U(p, q) = \sum_x \sum_y (x - \mu_x)^p (y - \mu_y)^q f(x, y) \qquad (8\text{-}4)$$

The literature contains elaborate formulae for expressing the central moments in terms of the raw, noncentral moments. They are of academic interest only, since intelligent, caring programmers would *never* compute central moments by first computing raw moments, then applying those formulae. The effect of doing so would be to subtract two huge numbers to get a small number as a result, a practice that is abhorrent. The only correct way to compute central moments is first to compute the centroid, then directly compute the central moments using Equation (8-4).

## Scale Invariance

In most applications we do not want the size of the figure to have an impact on its shape as defined by moments. There can be exceptions. If we are photographing an object at a fixed distance, and its size is a truly useful bit of information, there is a chance that we may want to let size affect the moments. But usually we do not. The figure is often produced by an object at unknown range from the camera, and so its size can vary. And even if size is an important descriptor, we may want to let it exert its influence in a way other than affecting moments. For example, we may want to use a "size" variable, and let the moments be independent of the size. That is almost always the best approach.

Compensating for size changes is easy. Simply rescale the figure to unit area. Then the moments will be the same at any magnification. Since the figure that we are working with is defined on a discrete lattice of pixels, we cannot directly adjust its size. But we can investigate what would happen to the various moments if we did so.

If the height and width of the figure are multiplied by some scale factor, then its area will be multiplied by the square of that scale factor. The area of the figure is $M(0, 0)$. (That assumes that the unit of area is the square pixel. $M(0, 0)$ is the number of pixels in the object.) So now we see how to scale the figure so that its area is 1. The scale factor is the reciprocal of the square root of the number of pixels in the figure, as shown in Equation (8-5). If we scale the figure by multiplying its dimensions by that scale factor, the area of the scaled figure will be 1.

$$\lambda = \frac{1}{\sqrt{M(0,\,0)}} \tag{8-5}$$

We now know the factor by which the dimensions of the figure must be multiplied in order to reduce it to unit area. We also know that we cannot literally adjust its size, because we are not in possession of the actual figure. All we have is its projection on a discrete lattice of pixels. Thus, we must be content to compute the effect that a hypothetical size change in the continuous domain would have on the moments. Look back at Equation (8-1). Readers who have a good grasp of calculus should be able to verify Equation (8-6).

$$\begin{aligned} M_\lambda(p,\,q) &= \int_{-\infty}^{\infty} \int_{-\infty}^{\infty} x^p y^q f(x/\lambda,\,y/\lambda)\,dx\,dy \\ &= \lambda^{2+p+q}\,M(p,\,q) \end{aligned} \tag{8-6}$$

Thus, to compute moments that are independent of the size of the object, first compute the ordinary central moments using Equation (8-4). Then normalize them according to Equation (8-7).

$$V(p,\,q) = \frac{U(p,\,q)}{\sqrt{M(0,\,0)}^{(p+q+2)}} \tag{8-7}$$

A subroutine for computing moments up to the third order is now given. Size normalization is also done. Note that the three lowest-order moments are not returned, as their values are fixed at constants by the normalizations.

```
void moments (
   int n ,                        // Number of points
   double *xpts ,                 // X coordinates
   double *ypts ,                 // Y coordinates
   double *mom                    // Moments: 20, 11, 02, 30, 21, 12, 03
   )
{
   int i ;
   double xm, ym, x, y, xsq, ysq, factor ;
   xm = ym = 0.0 ;
   for (i=0 ; i<n ; i++) {
      xm += xpts[i] ;
```

```
        ym += ypts[i] ;
        }
    xm /= (double) n ;                        // This is the centroid
    ym /= (double) n ;

    for (i=0 ; i<7 ; i++)
        mom[i] = 0.0 ;

    for (i=0 ; i<n ; i++) {                    // Cumulate all moments
        x = xpts[i] - xm ;                     // Make them central moments
        y = ypts[i] - ym ;
        mom[0] += (xsq = x * x) ;              // M20
        mom[1] += x * y ;                      // M11
        mom[2] += (ysq = y * y) ;              // M02
        mom[3] += xsq * x ;                    // M30
        mom[4] += xsq * y ;                    // M21
        mom[5] += x * ysq ;                    // M12
        mom[6] += y * ysq ;                    // M03
        }

    factor = 1.0 / ((double) n * (double) n) ;        // For size normalization

    mom[0] *= factor ;                         // Second-order moments
    mom[1] *= factor ;
    mom[2] *= factor ;

    factor /= sqrt ( (double) n ) ;

    mom[3] *= factor ;                         // Third-order moments
    mom[4] *= factor ;
    mom[5] *= factor ;
    mom[6] *= factor ;
}
```

## Relation to Mathematical Statistics

The physical moments just defined are closely related to the statistical moments of a random variable following a uniform distribution over a domain defined by the shape of the object. The condition $M(0, 0) = 1$ insures that we have a probability density function. The means of $x$ and $y$ are $M(1, 0) / M(0, 0)$ and $M(0, 1) / M(0, 0)$, respectively. $V(2, 0)$ and $V(0, 2)$ are constant multiples of the variances. $V(1, 1)$ is the same multiple of the covariance. These insights, while not directly important to the subject at hand, should be of interest to some readers.

## Rotation Invariance and Standard Moments

We must now deal with the leading problem in any shape description system. In many applications, the figure may appear at any rotational position. We may be in an airplane, photographing a vehicle on the ground. Or we may be on the ground, photographing an airplane that can assume nearly any orientation. Or we may be examining white blood cells under a microscope. The possibilities are endless. We need to be able to compute shape descriptors that remain the same regardless of the orientation of the figure. Unfortunately, that can be terribly difficult in many cases.

There is nothing intrinsically defective about moments that makes it difficult to normalize them against rotation. The ultimate cause of the problem is that some shapes exhibit symmetries that make them similar in several rotational positions. Whatever our shape description system, we need a way to rotate an arbitrarily placed figure into some sort of standard position. The criterion by which we choose this standard position generally depends on the shape measures. The Fourier shape descriptors already described rely on phase relationships. Moment descriptors rely on other criteria. But regardless of the method used, the underlying problem does not go away. It only manifests itself in different forms.

There are two very different approaches to rotation normalization. In a later section we will discuss moment-based shape descriptors that are automatically independent of rotation. These are sometimes called *moment invariants*. No standard position is needed, for those variables are simply not affected by rotation. The advantage of this approach is that errors introduced by an incorrect choice of standard placement are not a consideration. The disadvantage is that those variables can sometimes be more difficult to use effectively. Actual moments, based on a standardized orientation, can be used as raw material for more sophisticated shape descriptors. Also, we can use as many standardized moments as we like, going easily to high orders if that is desired. Moment invariants, on the other hand, are difficult to define for high orders. Restricting ourselves to moment invariants can sometimes lead to failure to detect important aspects of shape. The general rule is this: *If the expected shapes make it easy to find a standard orientation, use the standard moments described in this section. Otherwise, use the moment invariants described later.*

In this section we will discuss how to find a standard rotational position. By rotating the figure into this position, all of the moments will be independent of the original position. Therefore, all variables computed from the moments will also be independent of rotation. Some very sensitive and useful variables can be employed. But we must be absolutely sure that the standard position chosen is the correct one. If not, all of our shape variables will be wrong. The shapes likely to be encountered in the application will determine which of these two approaches is best. Perhaps it is appropriate to feed the fruits of both methods to the neural network.

When moments are used to describe a shape, the easiest way to define a standard position for the figure is to find its major axis (the direction along which most of its spread occurs) and rotate it until that axis is horizontal. There are two orientations that satisfy that requirement, 180 degrees from each other. That little ambiguity can be resolved in any number of ways. The most popular way is to require that the heaviest part of the figure lies on its left side. In statistical parlance, we say that the object is skewed to the right. For example, the egg shown in Figure 8-1 is in this standard position.

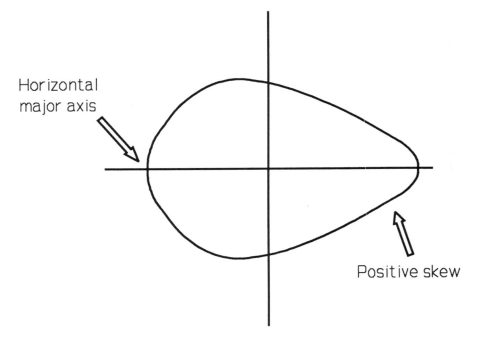

Fig. 8-1: An egg in standard position.

These conditions can also be stated in terms of their effect on the central moments. When a figure satisfies the rotation conditions just stated, its moments will have the following properties:

- $M(1, 1) = 0$
    *This will be true when its principal axes are aligned with the horizontal and vertical coordinate axes.*

- $M(2, 0) > M(0, 2)$
    *The major axis (its longer principal axis) is horizontal.*

- $M(3, 0) > 0$
    *The bulk of the object lies on its left side, with its tail skewing out to the right.*

Note that the above conditions may not always be able to be met. And sometimes when they are met, they will not be unique. In most cases, imposition of those three conditions will result in a single, unique orientation. The predominant difficulty is encountered when they are met, but just barely. In that case the standard orientation is subject to random error, and we can expect stability problems with the shape variables that are subsequently computed. Let us briefly explore some of the conditions under which these conditions are less than ideal.

The first condition can always be satisfied by at least (and in general, exactly) four rotational positions. These are when the two principal axes of the figure are horizontal and vertical. However, some symmetries will result in more than four positions that have $M(1, 1) = 0$.

The second condition states that the length of the (rotated) horizontal axis exceeds that of the vertical axis. The figure is lying down as opposed to standing up. But what if the object is a square or a circle? Even if it is some highly irregular, unique, and interesting shape, its principal axes may still be of equal length. We cannot make it lie down. This presents a most serious problem. If the principal axes are of essentially equal length, we have no way of finding a standard position using the above criteria. There are other exotic possibilities, most of which rely on higher-order moments. But the bottom line is that if our application can be expected to encounter figures that do not have a dominant spreading direction, the methods discussed here are probably not appropriate. In most such cases, a method that is custom designed for the expected shapes is the best solution.

The third condition is designed to eliminate the final ambiguity. Suppose that we have found the major principal axis of the figure and rotated that axis to a horizontal direction. We could rotate it another 180 degrees, and it would still be horizontal. So we hope that the figure, in this prone position, does not have left–right symmetry. If it does, the third horizontal moment will be 0. But if that moment is not 0, we choose whichever of the two possibilities causes the bulk of the mass of the figure to lie to the left of the centroid, with the tail pointing to the right.

The first step in standardizing the figure's position is to find the direction of its major axis. For statisticians, we need its first principal component. For physicists, we need the axis of maximum moment of inertia. For all of us, we need the direction in which the figure is most spread out. There are many formulae for finding the orientation of this axis. Each has its own ambiguities that must be resolved in tricky ways. The author's favorite method is to start out with the second-order moment matrix, which is practically a covariance matrix:

$$V = \begin{bmatrix} V(2, 0) & V(1, 1) \\ V(1, 1) & V(0, 2) \end{bmatrix} \tag{8-8}$$

The elements of that matrix are computed with Equation (8-7). This matrix has two eigenvalues, $v_{max}$ and $v_{min}$, and two corresponding eigenvectors, $e_{max}$ and $e_{min}$. These quantities can be explicitly calculated. To simplify the equations, let us use the following abbreviations: $V(2, 0) = a$, $V(1, 1) = b$, $V(0, 2) = c$. We will frequently need the value shown in Equation (8-9). Then the eigenvalues and (not normalized) eigenvectors can be computed as shown in Equations (8-10) and (8-11), respectively. Finally, the counterclockwise angle of the major axis from the $x$ axis can be found from Equation (8-12).

$$r = \sqrt{(a-c)^2 + 4b^2} \tag{8-9}$$

$$v_{max} = \frac{1}{2}(a+c+r)$$
$$v_{min} = \frac{1}{2}(a+c-r) \tag{8-10}$$

$$e_{max} = (e_{x,max}, e_{y,max}) = (a-c+r, 2b)$$
$$e_{min} = (e_{x,min}, e_{y,min}] = (a-c-r, 2b) \tag{8-11}$$

$$\theta = \tan^{-1}\left(\frac{e_{y,max}}{e_{x,max}}\right) \tag{8-12}$$

An important warning about those computations is in order. Recall that our goal is to locate the major principal axis of the figure. If the major and minor axes are the same length due to any number of symmetries, the major axis is undefined. That is manifested in those equations in an easily detected way. And it had better be detected, or Equation (8-12) will be undefined. Readers having a background in linear geometry know that the eigenvectors are ill-defined when the eigenvalues are nearly equal, and not defined at all when the eigenvalues are exactly equal. This intuitively agrees with the observation that we will have great difficulty identifying the major principal axis if its length is about the same as its smaller sibling. Look back at Equation (8-10) to see how the eigenvalues are calculated. They are identical except that $r$, which is never negative, is added in one and subtracted in the other. Therefore, we must look at the value of $r$ relative to $a + c$. If it is small, then the style of rotation normalization put forth in this chapter is not appropriate and should not be attempted. Luckily, $r$ has plenty of opportunity to avoid being small. It is the sum of two nonegative numbers. If either $a$ is reasonably different from $c$, or if $b$ is reasonably nonzero, we are fine. In most applications, this method works out well.

How do we define "small," and what do we do if $r$ is small? Unfortunately, the reader must be dealt a nonanswer here. Numerical difficulties with floating-point operations do not become an issue until $r$ is vastly smaller than what is reasonable from a practical point of view. Much of the decision is dependent on subtle aspects of the

problem.  The crux of the matter is that when *r* is small, it is easy for noise in the shape
to corrupt the estimate of the direction of the major axis.  As a result, the higher moments
are incorrect, so any shape variables computed with them are rubbish.  If the shapes are
very clean, much smaller values of *r* can be tolerated than if the shapes are noisy.  The
best approach is to analyze carefully a representative sample of the shapes that are likely
to be encountered in the application.  Look at the estimated major axis and see if it is
consistent with what is expected.  If not, the only alternative is to look for some other
feature to use as a reference point for rotation.  That is such a broad topic that it is far
out of the scope of this text.  Good luck.  The problem will probably be very difficult.
Try to find shape descriptors that do not require rotation normalization.  A later section
in this chapter will describe some possibilities.

   We now present a subroutine for computing the eigenvalues and vectors of the
second-order moment matrix.  With one exception, this code is a straightforward
implementation of the above equations.  The only exception involves a subtle point in the
eigenvector computation.  Notice that 2 * b always appears as one of the coordinates in
both eigenvectors.  If the principal axes happen to be nearly vertical and horizontal, this
quantity will be tiny.  For one of the vectors, the other coordinate will have to be even
tinier to cause the vector to point nearly directly out an axis.  This is not good from a
numerical point of view.  The solution is to use Equation (8-11) to compute only the
longer eigenvector, which will be numerically stable.  Then derive the other eigenvector
from the fact that the vectors are orthogonal.

```
void eigen (
   double *moments ,           // Moments: 20, 11, 02
   double *valmax ,            // Larger eigenvalue
   double *valmin ,            // And smaller
   double vectmax[] ,          // (x, y) eigenvector corresponding to valmax
   double vectmin[]            // And valmin
   )
{
   double a, b, c, r, t1, t2 ;

   a = moments[0] ;   // m20
   b = moments[1] ;   // m11
   c = moments[2] ;   // m02

   r = sqrt ( (a - c) * (a - c)  +  4.0 * b * b ) ;        // Equation (8-9)

   *valmax = 0.5 * (a + c + r) ;     // Equation (8-10)
   *valmin = 0.5 * (a + c - r) ;

   t1 = a - c + r ;                  // If the axes are vert / horiz
   t2 = a - c - r ;                  // one of these will be tiny, along with b
```

```
  if (fabs(t1) >= fabs(t2) ) {        // Avoid using the tiny one
     vectmax[0] = t1 ;                // Use the larger
     vectmax[1] = 2.0 * b ;           // And get the other
     vectmin[0] = 2.0 * b ;           // by orthogonality
     vectmin[1] = -t1 ;
     }
  else {
     vectmax[0] = 2.0 * b ;
     vectmax[1] = -t2 ;
     vectmin[0] = t2 ;
     vectmin[1] = 2.0 * b ;
     }
  }
```

At this point, the first step in rotation normalization is complete. We have found the orientation of the major principal axis via Equation (8-12). Next, we will be rotating the hypothetical continuous figure through a *clockwise* angle of $\theta$ to bring the major axis to a horizontal position. All we need now is to know the effect of this rotation on the moments. That is expressed in Equation (8-13).

$$
M_\theta(p, q) = \sum_{r=0}^{p} \sum_{s=0}^{q} (-1)^{q-s} \binom{p}{r} \binom{q}{s} \cdot
$$

$$
(\cos \theta)^{p-r+s} (\sin \theta)^{q-s+r} M(p+q-r-s, r+s) \qquad (8\text{-}13)
$$

We are just about finished. The major axis is horizontal. The last property to verify is that it has positive skewness along the $x$ axis. Check the sign of $V(3, 0)$. If that sign is negative, we must rotate an additional 180 degrees. One could simply use Equation (8-13) on the already rotated moments, with $\theta = 180$. But the most numerically stable way is to add 180 to the original $\theta$ and apply that equation to the original moments.

It must be understood that rotation normalization is not always called for. In some applications the rotational position of an object is an important piece of information. For example, we may be in a military vehicle, searching for enemy vehicles with a radar sensor. Unfortunately, trees reflect radar. However, there should be little confusion if the shape of a sensed object is considered. Trees will, in most cases, have a major axis that is vertical, while most vehicles will have a horizontal major axis. In such cases, we certainly do not want to apply rotation normalization right up front. We may still want to do it before computing some other shape variables, but it should not be an automatic first step in all applications.

The following subroutine computes the moments that result from a *clockwise* rotation through an angle of theta. It replaces the moments in the calling parameter vector. This implementation is slightly inefficient in that it copies the moments into a matrix and uses loops so that Equation (8-13) can be easily used. If top speed were a priority, those loops could be opened up and the moments accessed directly. But that is rarely needed, as the computation of the original moments is invariably the slowest operation. Also, this method makes it easy for the reader to generalize the program to higher-order moments if desired. Also note that a small special power routine is needed. This routine guarantees that 0 to the 0 power, which is technically undefined, actually evaluates to 1.

```
void rotmom (
   double *moments ,                    // In/Out 7 moments:20,11,02,30,21,12,03
   double theta                         // Clockwise rotation
   )
{
   double m[4][4] ;

   m[0][0] = 1.0 ;                       // Size was normalized to 1
   m[1][0] = m[0][1] = 0.0 ;             // Central moments have mean 0
   m[2][0] = moments[0] ;                // Copy moments to matrix
   m[1][1] = moments[1] ;
   m[0][2] = moments[2] ;
   m[3][0] = moments[3] ;
   m[2][1] = moments[4] ;
   m[1][2] = moments[5] ;
   m[0][3] = moments[6] ;

   moments[0] = rotmom1 ( 2 , 0 , m , theta ) ;        // M20
   moments[1] = rotmom1 ( 1 , 1 , m , theta ) ;        // M11
   moments[2] = rotmom1 ( 0 , 2 , m , theta ) ;        // M02
   moments[3] = rotmom1 ( 3 , 0 , m , theta ) ;        // M30
   moments[4] = rotmom1 ( 2 , 1 , m , theta ) ;        // M21
   moments[5] = rotmom1 ( 1 , 2 , m , theta ) ;        // M12
   moments[6] = rotmom1 ( 0 , 3 , m , theta ) ;        // M03
}


double rotmom1 ( int p , int q , double m[4][4] , double theta )
{
   int r, s, sgn ;
   double sum = 0.0 ;
   double ct = cos ( theta ) ;
   double st = sin ( theta ) ;
```

```
    for (r=0 ; r<=p ; r++) {                    // Equation (8-13)
      for (s=0 ; s<=q ; s++) {
        sgn = ((q - s) & 1)  ?  -1.0  :  1.0 ;
        sum += sgn * binom ( p , r ) * binom ( q , s ) *
             zpow ( ct , p-r+s ) * zpow ( st , q-s+r ) * m[p+q-r-s][r+s] ;
        }
      }
    return sum ;
}


double zpow ( double base , int power )
{
  double result = 1.0 ;

  while (power--)
    result *= base ;
  return result ;
}


int binom ( int n , int m )
{
  int prod = 1 ;
  int den = 1 ;

  while (m) {
    prod *= n-- ;
    den *= m-- ;
    }

  return prod / den ;
}
```

## Other Normalizations

There are several other normalizations that are less common but that can be immensely useful. These are *reflection* and *aspect* normalization. If used, these are normally applied after the previously described normalizations (location, scale, and rotation) have been done.

In some applications we may not know in advance which side of an object we will be seeing. Perhaps flat widgets tumble out of a chute onto a conveyor belt where they are photographed for quality control. Defects will be found by examining their shape moments. It places an unnecessary burden on the neural network to make it work with the shapes of both sides of the widget. It would be nice to have some means of insuring that the widgets always land on the conveyor belt heads-up. Alas, that is usually too much to ask. The next best thing is to force their shapes to be what they would be if they were heads-up. That means that we need a way of telling whether a widget has landed tails-up, then modifying this deviate's moment set to mimic what it would have been had it landed right. The former is usually easy, the latter is trivial.

We eliminated the 180-degree rotation ambiguity by forcing the horizontal skewness to be positive. That assumed the object was not right–left symmetric, and that $M(3, 0)$ could pick up on the lack of symmetry. Otherwise, we were in trouble. We can similarly use skewness in the $y$ direction for reflection normalization. The problem is not so bad here (and it wasn't usually that bad there, either). If the object is top–bottom symmetric, it doesn't matter which side it lands on; they both look the same. So all that we have to worry about is that $M(0, 3)$ responds to any lack of symmetry in the figure. That is a very reasonable assumption. If the reader's application would benefit from using a higher-order odd $y$ moment, that is fine. (Naturally, the even moments are invariant to reflection.) All we need to do to insure uniformity with regard to reflection is to verify that $M(0, 3)$ is not negative. If it is, we flip the figure about the $x$ axis. The result is that all moments that depend on an odd power of $y$ change sign. This is expressed in Equation (8-14).

$$M'(p, q) = (-1)^q M(p, q) \qquad \text{(8-14)}$$

Another normalization that seems to be very valuable, especially for extremely irregular figures, is *aspect ratio* normalization. This technique was developed by Anthony Reeves and presented in the classic paper [Reeves *et al.*, 1988]. It is based on the idea that the performance of a shape-description system is enhanced if the variables are commensurate. Suppose that an object is very elongated. Moments along the long dimension will greatly exceed those along the short dimension. This can lead to everything from numerical difficulties to unbalanced weighting of importance for features derived from moments. It would be better if the major and minor principal axes of the figure were equal. We can achieve such an effect by squashing the object along its long dimension, which is assumed to be the $x$ axis by virtue of having already performed rotation normalization. In the process of squashing the object horizontally, we let its height increase so that its area, $M(0, 0)$, remains equal to 1. If a single dimension of a figure is multiplied by a constant, the figure's area is multiplied by that same constant. So, to keep the area constant while changing the aspect ratio, all we do is multiply the $x$ coordinates by a constant and divide the $y$ coordinates by the same constant. Let us now find that constant.

The first step is to determine the effect on the moments of multiplying the figure's dimension by a constant. It is left as an exercise for readers versed in calculus to prove that when a figure is expanded in a particular direction by a constant multiplier, the moments in that direction are multiplied by that constant raised to the power of 1 plus the order of the moment in that direction. Moments in a direction perpendicular to the expanded direction are not affected. Readers who are not able to prove this will benefit from attempting to find an intuitive feel for it. When we combine the two reciprocal changes, we find that the effect on the moments is expressed in Equation (8-15). The condition that the spread in both directions be equal is expressed in Equation (8-16). Substituting the former into the latter and solving for $a$ gives us the multiplier that will do the job, as shown in Equation (8-17). Equation (8-15) can then be used with the computed $a$ to find all moments after normalization for aspect.

$$M_a(p, q) = a^{p+1} (1/a)^{q+1} M(p, q)$$
$$= a^{p-q} M(p, q) \tag{8-15}$$

$$M_a(2, 0) = M_a(0, 2) \tag{8-16}$$

$$a = \left( \frac{M(0, 2)}{M(2, 0)} \right)^{\frac{1}{4}} \tag{8-17}$$

Of course, in the vast majority of applications, the aspect ratio of the original figure is a valuable shape descriptor. The above normalization throws it out. Therefore, when aspect normalization is performed, the original aspect ratio must be provided to the neural network in some way. The author's usual method is to compute the logarithm of $a$ and perhaps apply a linear transform (that is problem-dependent) to the log so that the final variable is roughly centered at 0 and attains moderate absolute values throughout the range of expected shapes. This is discussed in more detail in a later section.

## Raw Moment Variables

The most straightforward use of moments as shape descriptors is to apply a simple transformation to each individual moment and use these as inputs to the neural network. The moments are nearly always first normalized against change in location and scale, as described earlier. That means that $M(0, 0) = 1$, and $M(0, 1) = M(1, 0) = 0$. Those three moments are fixed by the normalization, so they are worthless to the network. For many applications, we will also apply rotation normalization. If so, then $M(1, 1) = 0$; so it, too, should be ignored. If the aspect normalization discussed in the previous section was performed, then $M(2, 0) = M(0, 2)$, so only one of them is needed. The logarithm of the scaling constant $a$ takes the place of the other. Finally, reflection normalization does not result in the loss of any moment. But it does guarantee that $M(0, 3)$ cannot be negative, so that may impact the transformation selected. Rotation normalization has the same effect on $M(3, 0)$.

How many moments do we need, and which ones are best? That is, in general, a problem-dependent issue, so no absolutely firm answer is possible. However, a few guidelines can be given. Let us start with a definition. The *order* of a particular moment, $M(p, q)$, is $p + q$. There are $(n + 1)$ $(n + 2)$ / 2 moments of order $n$. It is traditional, but by no means mandatory, that all moments up to a specified order be used. Only those that are fixed at constant values by normalization, as discussed at the beginning of this section, are excluded. Most applications go up to order three, and a few applications that exhibit low noise and high feature definition may go up to order four. Beyond that, the moments become so susceptible to tiny changes in shape that their utility is impaired. As a general rule, if the application could actually benefit from still higher moments, it could probably benefit even more from an alternative, specialized set of shape descriptors. Although moments are very good at shape description, they are not the best solution when extremely fine detail is important. They are at their best when handling approximate overall shape in the presence of significant noise.

There is one mandatory step in preparing moments for presentation to a neural network and one optional step. It is important that the network's inputs be roughly centered so that the mean of each input is approximately 0 across the range of expected figures. Also, the absolute value of the input data should lie within a moderate range. For example, the author usually strives to have a large fraction of the cases generate inputs whose absolute value exceeds two, while few cases ever exceed ten. This is easily done with a linear transformation. Compute the median of the training set and subtract it to center the data. Compute the interquartile range of the data, divide by that quantity, and multiply by two or three. The additive and multiplicative constants computed this way or some other way for each moment variable should be saved. These same constants will be used for the unknown data when the trained network is ultimately put to use.

The above step, or a close relative, should be considered mandatory. An optional step is to apply a nonlinear transformation to the moments before the linear transformation just described is done. This is usually a compressing transform. The author favors taking the logarithm of moments that can never be negative. (Those are the ones for which both $p$ and $q$ are even.) Negative moments are handled by using cube roots or even higher odd roots.

It is time for a few specific examples. The functions presented now are good general-purpose moment transformations that are useful when a wide variety of shapes will be encountered. Applications that deal with a restricted set of shapes would do well to linearly rescale or otherwise modify these functions to suit the narrower range of moments likely to be produced.

Let us start with the basic height and width measures when only the usual location and scale normalization is done. These are $M(2, 0)$ and $M(0, 2)$. For a perfect circle, these moments will have the value $1 / (4\pi)$, which is about 0.0796. If the figure is very narrow in one dimension (skinny or flat), the moment in that direction will approach 0. These moments have no theoretical upper limit, increasing as the longest radius increases. However, in practice, discretization limits their actual maximum, since the total area of the figure is fixed at 1 by the size normalization. A good transformation maps the moment of a perfect circle to 0. Smaller moments would map to negative numbers, with the limit being around −10 or so. Larger moments would map to positive numbers, with the possibly large values being tempered by compression. An excellent all-purpose transformation is given in Equation (8-18).

$$nm20 = 7.21 \log \left( \frac{m(2, 0) + 0.02}{0.099577} \right)$$

$$nm02 = 7.21 \log \left( \frac{m(0, 2) + 0.02}{0.099577} \right)$$

(8-18)

If rotation normalization has been done, the preceding transformation loses much of its value, for now $M(2, 0) >= M(0, 2)$. That forces one transformed value to be always positive and the other to be always negative (or 0). The resulting reduction in range is not a disaster, for it can be somewhat offset by judicious linear mapping. But it is not good. If aspect normalization has been done, a generally good idea, then things are even worse. Now $M(2, 0) = M(0, 2)$. In either case, a better approach is to use Equations (8-21) and (8-23) described in the section *Moment Invariants and Their Relatives* starting on page 327.

Most other moments contain at least one odd index and are not restricted in any way by the normalizations presented in this text. The exceptions will be considered later. These odd moments can in general be positive or negative. Their absolute value has no theoretical upper limit, but it rarely gets large in practice. The author has had success with the transformation expressed for one of these moments in Equation (8-19).

$$nm11 = 20 \sqrt[3]{m(1, 1)}$$

(8-19)

The cube root is widely applicable for transforming moments. The factor of 20 was chosen empirically and should be optimized for each moment and each application. Equation (8-19) is a good foundation for transforming all moments that contain at least one odd power.

When we restrict ourselves to third-order moments, there are no other totally even moments. But if we advance to fourth order, $M(2, 2)$ appears. It, and all moments that are even in both dimensions, can never be negative. If the figure has been normalized against rotation, with $M(3, 0)$ being used to resolve the 180-degree ambiguity, that moment will never be negative. If $M(0, 3)$ was used to normalize against reflection, that moment also will never be negative. In such cases, the transformation shown in Equation (8-20) is generally effective.

$$f(x) = K_1 \log \left( \frac{x + \epsilon}{K_2 + \epsilon} \right)$$

(8-20)

There is a simple way to find good values for the three constants in that equation. Compute the moment variable for a representative sample of figures. The median is the value of $K_2$. That forces the transformation to map the median to 0. Next, look at the maximum value in the sample. Ignore $\varepsilon$, as in most cases it will be tiny compared to the maximum. Compute $K_1$ so that the maximum moment in the sample maps to 10. Finally, compute $\varepsilon$ so that 0 maps to $-10$.

We conclude this section with a short code fragment for computing all of the transformed moments discussed here. This is the code used to compute the variables in the large table later in this chapter. Note that it uses a special routine for computing the cube root of a possibly negative number.

```
momvars[0] = 7.21 * log ( (m[0] + 0.02) / 0.0995775 ) ;   // m20
momvars[1] = 20.0 * root3 ( m[1] ) ;                       // m11
momvars[2] = 7.21 * log ( (m[2] + 0.02) / 0.0995775 ) ;   // m02
momvars[3] = 20.0 * root3 ( m[3] ) ;                       // m30
momvars[4] = 20.0 * root3 ( m[4] ) ;                       // m21
momvars[5] = 20.0 * root3 ( m[5] ) ;                       // m12
momvars[6] = 20.0 * root3 ( m[6] ) ;                       // m03


double root3 ( double x )
{
  if (x > 0.0)
    return pow ( x , 0.33333333333 ) ;
  else if (x < 0.0)
    return -pow ( -x , 0.33333333333 ) ;
  else
    return 0.0 ;
}
```

## Measuring Orientation

Sometimes the orientation of a figure is an important piece of information. The orientation is, of course, reflected in the raw moments. But there are at least three reasons why we might not want to get at it that way. One reason is that orientation impacts moments in obscure ways involving ratios and trigonometric functions. While neural networks are certainly capable of handling this, we may be able to lighten their load by simplifying their data. Another reason is that we may not even be interested in looking at a lot of moments. We may be interested in orientation only, or orientation in conjunction with just a few related variables. The third reason why we may want to avoid inferring important orientation information from raw moments is that the figure's orientation may impact other moment features in ways that are confusing. The interpretation of moments, whether by humans or neural networks, is inevitably easier if the figure has been normalized against rotation. But then the orientation information has been lost. The bottom line is that we often need a way to extract valuable orientation information from the moments, then apply rotation normalization, then do all further work with the moments after rotation normalization.

The most straightforward measure of orientation is the major axis inclination, θ, computed with Equation (8-12). Like all neural network input variables, it should be scaled to suit the problem. Its mean should be around 0, and its range should be primarily single-digit numbers. In many situations, that is easily done. When this angle is input to a neural network, the author usually multiplies its measure in radians by 6.3662. That causes it to range from −10 to 10. That multiplication is done in the examples presented later in this chapter.

Unfortunately, there can be a very serious problem with such a simplistic approach. The angle exhibits a dramatic discontinuity at the ends of its range. Figures that are very tall and thin can easily flip-flop between 90 degrees and −90 degrees depending on noise or minor variations in the exact shape. Neural networks hate that. Fortunately, there is a very simple fix. Use the inclination angle to define *two* variables: cos 2θ and sin 2θ. These functions are continuous as θ runs from −90 to 90, then immediately wraps back to −90. The cosine variable can be thought of as measuring "horizontalness." It achieves its maximum of 1 when the major axis is horizontal. At inclinations of plus or minus 45 degrees, it is at its central value of 0; and when the major axis is vertical, it attains its minimum of −1. The sine variable measures direction of inclination. It has positive values when the major axis is inclined counterclockwise from directly right, attaining its maximum of 1 when the inclination is exactly 45 degrees. It is negative for clockwise inclinations, reaching its minimum of −1 at −45 degrees. Together, these two variables provide a complete description of the inclination of the major axis, and they do it in such a way that there is no discontinuity. When these variables are presented to a neural network, it is the author's habit to multiply them by 10 so that their range is from −10 to 10. That is done in the examples later in this chapter.

This fix is not needed if the expected angles lie within a range that is significantly narrower than 180 degrees. In that case, just subtract the mean angle from all measured values to center it in the range without danger of wrapping around the ends. But if the data can be expected to cover the whole gamut, this two-variable approach is mandatory.

We now present a trivial code fragment for computing the angle of inclination and its trigonometric relatives. This code was used to compute the shape variables tabled later in this chapter. It refers to the eigenvector corresponding to the larger eigenvalue. Code for computing that was given on page 317.

```
if (fabs (vectmax[0]) > 0.0)
   theta = atan ( vectmax[1] / vectmax[0] ) ;
else
   theta = PI / 2.0 ;
thetavar = 6.3662 * theta ;                    // Theta
cosvar = 10.0 * cos ( 2.0 * theta ) ;   // Cos
sinvar = 10.0 * sin ( 2.0 * theta ) ;          // Sin
```

## Moment Invariants and Their Relatives

This section approaches the problem of rotation invariance from a direction that is radically different from the approach taken so far. Up until now we have sought a means of rotating the figure into some sort of standard position, so that when we compute the moments, they will not be affected by the original orientation of the figure. Now we will examine a collection of variables that are intrinsically unaffected by rotation. The figure does not have to be placed in any particular position before computing these variables, because they will be the same no matter what the orientation of the figure.

One very easily computed variable is the aspect ratio of the figure. We already mentioned that variable in conjunction with aspect normalization, where it was based on the scaling constant $a$. An equivalent, and perhaps more intuitive way of computing the aspect ratio, is by computing the ratio of the eigenvalues of the second-order moment matrix. These eigenvalues are easily found via Equation (8-10). They are often needed for other purposes, and so it is easy to find their ratio as long as they are handy. Before presenting this ratio to a neural network, it is best to take its square root or log, then center and scale it as usual. A good general-purpose transformation is given in Equation (8-21).

$$ASPECT = 2 \log \left( \frac{\dfrac{v_{MAX}}{v_{MIN}} - 0.99999}{M - 1} \right) \tag{8-21}$$

The constant $M$ in that equation should be found by collecting a representative sample of figures. $M$ is equal to the approximate median of the ratio $v_{max} / v_{min}$. The multiplier of 2 should be modified, if necessary, to produce a reasonable range of transformed values. A wide range of shapes can be accommodated by setting $M = 2$. That is the value used in the examples presented later in this chapter.

Although this section focuses on variables that are immune to rotation, this is a good place to mention a variable that is closely related to the aspect ratio just defined but is useful for applications when random, unknown orientation does not occur. A general form of that variable is shown in Equation (8-22).

$$FIXED\ ASPECT = K_1 \log \left( \frac{M(0, 2)}{M(2, 0)} \right) - K_2 \tag{8-22}$$

In that equation, the moments are *before* any rotation or aspect normalization is done. This variable responds to the height of the figure relative to its width. It is particularly effective at distinguishing trees from freight trains. The two constants would be chosen based on a representative sample so that the transformed variable is centered and well scaled. The author uses $K_1 = 2$ and $K_2 = 0$ when a wide range of shapes must be accommodated. Those values are used in the examples presented later in this chapter.

We have just discussed means for measuring the *relative* lengths of the principal axes. We should also be interested in their *absolute* lengths. For a perfect circle, $M(2, 0) + M(0, 2) = 1/(2\pi)$. All other shapes will have larger values of this sum. Despite the fact that the area of the figure is standardized at 1, this sum has no theoretical upper limit. A shape that is smashed so flat that its major axis is extremely long will have a correspondingly large sum. On the other hand, discrete digitization of the figure prevents the actual sum from ever becoming gigantic.

Several aspects of this sum are interesting. The sum is invariant under rotation. It is also equal to the sum of the eigenvalues computed in Equation (8-10) on page 316. Finally, extreme values of the aspect ratio produce large values of this sum, but the converse does not have to be true. Consider an exceedingly narrow strand of hair wrapped into a loop. The lengths of the principal axes can easily be equal, but the radius spanned by that unit-area hair is much larger than the radius of a unit circle. This leads to unusually large values of all moments. A transformation for this variable that is generally effective is shown in Equation (8-23).

$$SPREAD = K_1 \log \left( \frac{V_{\max} + V_{\min} - 0.5/\pi + \epsilon}{K_2 - 0.5/\pi + \epsilon} \right) \qquad (8\text{-}23)$$

The way to find good values for the three constants in that equation is similar to a method discussed earlier. Compute the moment sum for a representative sample of figures. The median sum is the value of $K_2$. That forces the transformation to map the median to 0. Now look at the maximum value in the sample. Ignore $\epsilon$. Compute $K_1$ so that the maximum sum in the sample maps to 10. Finally, compute $\epsilon$ so that the theoretical minimum, $1/(2\pi)$, maps to a moderately large negative number. The author often uses $K_1 = 10$, $K_2 = 0.25$, and $\epsilon = 0.001$. Those values are used in the examples presented later in this chapter.

We now show the code used to compute these three variables in the shape variables tabled later. The first variable refers to the eigenvalues computed on page 317.

```
aspect = 2.0 * log ( valmax / valmin - 0.99999 )
fixed = 2.0 * log ( m[0] / m[2] ) ;
spread = 10.0 * log ( (m[0] + m[2] - 0.158155) / 0.091845) ;
```

A famous set of rotation-independent moment-based shape descriptors was proposed in [Hu, 1962]. He used the concept of algebraic invariants to produce what he logically called *moment invariants*. They become extremely complicated for high orders, so he restricted them to third order. They are based on moments that have been normalized against location shift and scale, as defined by Equation (8-7). Rotation and aspect normalization are not done.

This method of compensating for unknown and variable orientation has an obvious advantage over the previously discussed method. There can be no ambiguity, because no special indicator of standard position is needed. We don't care if symmetries or just plain bad luck cause the principal axes to be so similar in length that the major axis cannot be accurately found. In some applications, that is a major worry that is taken away by these variables. On the other hand, there may be a high price to pay. The plain truth is that a lot of empirical evidence indicates that when the major axis is well defined, rotation-normalized moments simply perform better than the moment invariants described in this section. These variables should, in general, be reserved for situations in which it is known that a serious proportion of the expected figures will fail to have a dominant major axis. Please note that the author does not wish to issue a blanket condemnation of these variables. It is quite conceivable, perhaps even likely, that some applications exist for which they are excellent. It is just that the author and several of his colleagues have been less than impressed with their power compared to rotation- and aspect-normalized moments.

There are ten moments of order three, but location and scale normalization fix three of these at constant values. Rotation invariance removes one more degree of freedom, so we are left with six independent dimensions to quantify. Hu actually defines seven variables. The first six span the full dimension, except that they are also immune to reflection. The seventh variable's magnitude is not independent of the first six, but its sign changes under reflection. Therefore, if reflection invariance is desired, only the first six variables should be used. Otherwise, the sign of the seventh variable should also be made available to the neural network. Hu's seven moment invariants are defined in Equations (8-24) through (8-30).

$$h_1 = V(2, 0) + V(0, 2) \tag{8-24}$$

$$h_2 = [V(2, 0) - V(0, 2)]^2 + 4[V(1, 1)]^2 \tag{8-25}$$

$$h_3 = [V(3, 0) - 3V(1, 2)]^2 + [V(0, 3) - 3V(2, 1)]^2 \tag{8-26}$$

$$h_4 = [V(3, 0) + V(1, 2)]^2 + [V(0, 3) - V(2, 1)]^2 \tag{8-27}$$

$$\begin{aligned} h_5 = {} & [V(3, 0) - 3V(1, 2)][V(3, 0) + V(1, 2)] \\ & ([V(3, 0) + V(1, 2)]^2 - 3[V(0, 3) + V(2, 1)]^2) \\ & + [3V(2, 1) - V(0, 3)][V(0, 3) + V(2, 1)] \\ & (3[V(3, 0) + V(1, 2)]^2 - [V(0, 3) + V(2, 1)]^2) \end{aligned} \tag{8-28}$$

$$h_6 = [V(2,0) - V(0,2)] \tag{8-29}$$
$$([V(3,0) + V(1,2)]^2 - [V(0,3) + V(2,1)]^2)$$
$$+ 4V(1,1)[V(3,0) + V(1,2)][V(0,3) + V(2,1)]$$

$$h_7 = [3V(2,1) - V(0,3)][v(3,0) + V(1,2)]$$
$$([V(3,0) + V(1,2)]^2 - 3[V(0,3) + V(2,1)]^2) \tag{8-30}$$
$$+ [3V(1,2) - V(3,0)][V(0,3) + V(2,1)]$$
$$(3[V(3,0) + V(1,2)]^2 - [V(0,3) + V(2,1)]^2)$$

These variables tend to be very small and to have an enormous dynamic range. Thus, they are unsuitable for direct presentation to most neural networks. The author has developed a set of transformations that seem to have broad applicability. If a reader's application has a limited set of shapes, these transformations should be modified accordingly. The transformations are shown in Equations (8-31) through (8-37).

$$nh_1 = 10 \log (10h_1 - 1.2) \tag{8-31}$$

$$nh_2 = 3 \log \left( \frac{h_2 + 0.0001}{0.0101} \right) \tag{8-32}$$

$$nh_3 = 8 \log \left( \frac{h_2 + 0.0005}{0.0025} \right) \tag{8-33}$$

$$nh_4 = 2 \log \left( \frac{h_4 + 0.0000001}{0.0000101} \right) \tag{8-34}$$

$$nh_5 = 1000 \sqrt[3]{h_5} \tag{8-35}$$

$$nh_6 = 200 \sqrt[3]{h_6} \tag{8-36}$$

$$nh_7 = 500 \sqrt[3]{h_7} \tag{8-37}$$

We conclude this section with code that can be used to compute the raw Hu variables, and code for transforming them to values suitable for presentation to a neural network. The moment vector used throughout this code was computed on page 311. This is the code used for computing the shape variables tabled in the next section.

```
huvars[0] = m[0] + m[2] ;
huvars[1] = (m[0] - m[2]) * (m[0] - m[2])  +  4.0 * m[1] * m[1] ;
huvars[2]= (m[3] - 3.0 * m[5]) * (m[3] - 3.0 * m[5]) +
          (m[6] - 3.0 * m[4]) * (m[6] - 3.0 * m[4]) ;
huvars[3]= (m[3] + m[5]) * (m[3] + m[5])  +  (m[6] + m[4]) * (m[6] + m[4]) ;
huvars[4]= (m[3] - 3.0 * m[5]) * (m[3] + m[5]) *
   ((m[3] + m[5]) * (m[3] + m[5])  -  3.0 * (m[6] + m[4]) * (m[6] + m[4]))
   + (3.0 * m[4] - m[6]) * (m[6] + m[4]) *
   (3.0 * (m[3] + m[5]) * (m[3] + m[5])  -  (m[6] + m[4]) * (m[6] + m[4])) ;
huvars[5] = (m[0] - m[2]) * ((m[3] + m[5]) *
   (m[3] + m[5]) - (m[6] + m[4]) * (m[6] + m[4]))
   + 4.0 * m[1] * (m[3] + m[5]) * (m[6] + m[4]) ;
huvars[6] = (3 * m[4] - m[6]) * (m[3] + m[5]) *
   ((m[3] + m[5]) * (m[3] + m[5])  -  3.0 * (m[6] + m[4]) * (m[6] + m[4]))
   + (3.0 * m[5] - m[3]) * (m[6] + m[4]) *
   (3.0 * (m[3] + m[5]) * (m[3] + m[5])  -  (m[6] + m[4]) * (m[6] + m[4])) ;


huvars[0] = 10.0 * log ( 10.0 * (huvars[0] - 0.12) ) ;
huvars[1] = 3.0 * log ( (huvars[1] + 0.0001) / .0101 ) ;
huvars[2] = 8.0 * log ( (huvars[2] + 0.0005) / .0025 ) ;
huvars[3] = 2.0 * log ( (huvars[3] + 0.0000001) / .0000101 ) ;
huvars[4] = 1000.0 * root3 ( huvars[4] ) ;
huvars[5] = 200.0 * root3 ( huvars[5] ) ;
huvars[6] = 500.0 * root3 ( huvars[6] ) ;
```

## Examples of Shape Variables

So far, we have discussed a wide variety of shape variables. It is time to see them in action. We will study eight different shapes, most of which are chosen for their pedagogical value. They include a circle, three moderately flattened ellipses at different orientations, an unsharpened pencil, an egg, a truck, and an airplane. These figures are all digitized at a resolution of approximately 40 to 50 pixels in each direction. The circle and the pencil (two wide horizontal rows of pixels) are not shown because their shape is obvious. The other six profiles are shown in Figures 8-2 through 8-7. Shape variables for these eight profiles are listed on page 333.

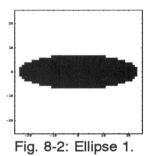
Fig. 8-2: Ellipse 1.

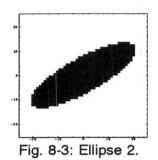
Fig. 8-3: Ellipse 2.

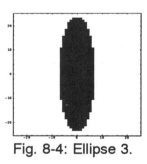
Fig. 8-4: Ellipse 3.

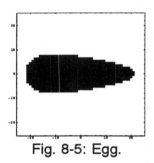
Fig. 8-5: Egg.

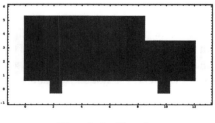

Fig. 8-6: Truck.

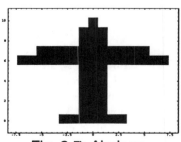
Fig. 8-7: Airplane.

| | Circle | Ellipse 1 | Ellipse 2 | Ellipse 3 |
|---|---|---|---|---|
| nm20 | 0.00 | 7.72 | 5.83 | -6.00 |
| nm11 | 0.00 | 0.00 | 9.40 | 0.00 |
| nm02 | 0.00 | -6.00 | 0.31 | 7.72 |
| nm30 | 0.00 | 0.00 | 0.00 | 0.00 |
| nm21 | 0.00 | 0.00 | 0.00 | 0.00 |
| nm12 | 0.00 | 0.00 | 0.00 | 0.00 |
| nm03 | 0.00 | 0.00 | 0.00 | 0.00 |
| nh1 | -9.38 | 5.53 | 5.15 | 5.53 |
| nh2 | -13.85 | 5.40 | 5.21 | 5.40 |
| nh3 | -12.88 | -12.88 | -12.88 | -12.88 |
| nh4 | -9.23 | -9.23 | -9.23 | -9.23 |
| nh5 | 0.00 | 0.00 | 0.00 | 0.00 |
| nh6 | 0.00 | 0.00 | 0.00 | 0.00 |
| nh7 | 0.00 | 0.00 | 0.00 | 0.00 |
| theta | -.- | 0.00 | 3.33 | 10.00 |
| cos | -.- | 10.00 | 5.00 | -10.00 |
| sin | -.- | 0.00 | 8.67 | 0.00 |
| aspect | -23.03 | 4.72 | 4.60 | 4.72 |
| fixed aspect | 0.00 | 4.90 | 1.77 | -4.90 |
| spread | -45.17 | 3.90 | 3.42 | 3.90 |

| | Pencil | Egg | Truck | Airplane |
|---|---|---|---|---|
| nm20 | 21.08 | 6.59 | 6.19 | 3.54 |
| nm11 | 0.00 | 0.00 | -5.31 | 0.00 |
| nm02 | -10.62 | -5.06 | -4.25 | 3.43 |
| nm30 | 0.00 | 5.88 | 5.37 | 0.00 |
| nm21 | 0.00 | 0.00 | -3.90 | 5.09 |
| nm12 | 0.00 | -3.00 | -2.49 | 0.00 |
| nm03 | 0.00 | 0.00 | 1.92 | -6.07 |
| nh1 | 28.42 | 3.21 | 2.63 | 4.88 |
| nh2 | 17.41 | 4.11 | 3.62 | -13.64 |
| nh3 | -12.88 | -2.78 | -3.23 | 7.64 |
| nh4 | -9.23 | 7.75 | 7.08 | 5.15 |
| nh5 | 0.00 | 7.26 | 5.94 | 4.90 |
| nh6 | 0.00 | 9.19 | 7.64 | -1.41 |
| nh7 | 0.00 | 0.00 | 2.08 | 0.00 |
| theta | 0.00 | 0.00 | -0.65 | 0.00 |
| cos | 10.00 | 10.00 | 9.79 | 10.00 |
| sin | 0.00 | 0.00 | -2.03 | 0.00 |
| aspect | 12.94 | 3.83 | 3.42 | -7.92 |
| fixed aspect | 12.94 | 4.10 | 3.62 | 0.04 |
| spread | 29.05 | 0.81 | 0.01 | 3.07 |

Only the usual location and size normalization is done. Rotation and aspect normalization, while definitely worthwhile in practice, are avoided here so that the reader can see the effect of rotation on moment variables. The *transformed* raw moments appear at the top of the list. The basic width and height measures, nm20 and nm02, are computed with Equation (8-18). The remaining five, all involving odd powers, are computed with Equation (8-19). The seven Hu variables are based on Equations (8-31) through (8-37). *Theta* is 6.3662 times the angle of the major principal axis, as discussed in the section *Measuring Orientation* starting on page 325. The associated variables *sin* and *cos* are ten times the respective trigonometric functions of twice the angle, as also discussed in that section. *Aspect* is the rotation-invariant aspect ratio based on the eigenvalues of the moment matrix, as shown in Equation (8-21). *Fixed aspect* is based on the raw width-to-height ratio and is defined in Equation (8-22). Finally, *spread* measures how much the shape is spread out. It is defined in Equation (8-23).

The transformations are designed so that two properties are obtained. A perfect circle forms the reference point. For raw variables that can be positive or negative, they take the value 0 for a circle. This property is preserved in the transformation, as is the sign of the raw variable. The transformation only compresses the range of the raw variable, allowing the neural network to focus on the most common values. For raw variables that are bounded below, the transformations are designed so that the transformed value is a moderate negative number at that minimum, typically –10 to –20. This minimum invariably is obtained for a perfect circle. The transformed value is 0 at the approximate median of the raw variable, and increasingly positive as the raw variable grows. With this general rule in mind, let us briefly examine these variables for some common shapes.

Start with the most basic shape, a circle. The variables based on raw moments are all 0, the center of their range. The first four transformed Hu variables are at their negative limit, as they can only increase as circularity is lost. The other three are 0, their central value. The major and minor principal axes are undefined, so *theta, sin,* and *cos* cannot be computed. The rotation-invariant aspect is at its minimum, while the fixed aspect is at its central value. The *spread* is also at its minimum for a circle. The *aspect* and *spread* are deliberately quite large negative numbers. They head toward 0 rapidly as circularity is lost, so they are effectively sensitive in that common "almost circular" situation.

Moving on to the first ellipse, moderately flattened and horizontal, we see a few changes. The basic extent variables, *nm20* and *nm02*, reflect the wider width and lower height. Several of the Hu variables also change. *Theta* is 0, indicating that the major axis is horizontal. *Cos* does the same, achieving its maximum value in this position. *Sin* is 0 because the figure is not inclined in either direction. The two aspect variables reflect the inequality of the principal axes. *Spread* is also moderately positive because the shape is considerably spread out.

Many variables for the other two ellipses are nearly equal to those of the first ellipse. All seven Hu variables, which are rotation invariant, are the same. (The small observed differences are due to slight differences in the discrete digitization process.) The *aspect* behaves similarly. The differences between these three identical-but-rotated ellipses are reflected strongly in the first three moments. The roles of *nm20* and *nm02* are reversed for the horizontal and vertical ellipses, and intermediate for the middle ellipse, which is inclined thirty degrees. The covariance moment, *nm11*, picks up strongly on the inclination of the middle ellipse. *Theta*, *sin*, and *cos* also indicate the orientation of the longest axis. *Fixed aspect* also responds to the changes in width versus height. Finally, *spread* should theoretically be the same for all three ellipses. However, it has responded with considerable sensitivity to differences in how the inclined ellipse is digitized relative to the other two.

The pencil figure is an extreme shape, unlikely to ever be encountered in real applications. It is a two-pixel-thick horizontal line. Thus, we should not be surprised that it produces extreme values for many of the shape variables. It generally behaves similarly to the first ellipse, but with everything pushed to the limit.

Controlled skewness is introduced with the egg shape. It is nearly identical to the first ellipse, except that some of its weight is shoved to the left, causing it to skew to the right. This is most reflected in the horizontal skewness variable, *nm30*. The Hu variables also change. Most of the other variables remain about the same, except for *spread* which shrinks in response to the egg being less spread out than the ellipse.

The last two shapes are reminiscent of a truck and an airplane. Note how the squat truck affects *nm20*, *nm02*, *aspect*, and *fixed aspect*. The engine produces some skewness to the right, as shown by *nm30*. It is only moderately spread out, so *spread* is near its central value of 0. The truck is the only object here that has no axis of symmetry, so it is the only one that produces a nonzero value for *nh7*. The airplane has downward skewness, which is reflected in *nm03*.

There is one special point of interest about the airplane. The width and height variables, *nm20* and *nm02*, are nearly equal, with the covariance, *nm11*, equal to 0. Looking back at Equation (8-9), we see that the eigenvalues will be nearly equal. In other words, the major principal axis is not well defined. It is defined well enough that *theta*, *sin*, and *cos* can be found correctly. But the reader should be a little frightened. If this airplane could appear in any orientation, and rotation normalization will be performed with Equation (8-13), the angle computed with Equation (8-12) will be subject to wide variation. As a result, many of the normalized moments will vary at the whim of noise and other random errors. This will drive a neural network crazy. If the anticipated shapes include such indeterminate cases, it would probably be better to use the fundamentally invariant measures discussed in the section *Moment Invariants and Their Relatives*, starting on page 327.

# Tone and Range Moments

Up until now, we have defined shapes using a binary variable. Each pixel is either a member of the figure or is not a member. We now generalize this idea to *weighted moments*, in which the membership function can take on more than just two values. That generalization should not be unfamiliar to statisticians. At the very beginning of this chapter, we made an analogy between shape moments and the statistical moments of a probability distribution. It was originally a uniform density defined over an irregular domain. We now allow the density function to depart from uniformity. Certain areas of the domain may be more densely populated than other areas. Or it may be that certain areas are brighter than other areas. Or some areas may be nearer than other areas. Regardless of the reason for the nonuniformity, we now allow the membership function to take on values other than 0 and 1.

One of the most common situations in which a nonbinary weight function is used is for handling range information. For each pixel in the object, we know how much distance separates that pixel from the observer. This is wonderful supplementary information, and we will make good use of it. There are some complications, though. In many applications, especially radar and sonar, all that we see is the front surface of the object. We can't see any part of the object behind what we do see. We may be looking at the end of a log, or we may be looking at a pancake, and not know which it is. We don't even know anything about the distribution of the mass of the object behind what we see. We obviously must make some assumptions. There is an infinite number of ways that this problem could be approached. It is in our best interest to choose assumptions in such a way that our task will be as easy as possible. Keep them simple. Probably the simplest assumption is that the object extends straight away from the sensor, then stops suddenly at some distance. In other words, its back is flat and lies on a plane that is parallel to the $x$-$y$ plane. The cross-section of the back of the object is identical to the cross-section of the part that we can see. That assumption is obviously not realistic in a physical sense. Few real-life objects look like that. But it is very nice from a mathematical point of view. And as long as we apply that same assumption to all figures, we should be safe. When we do this, it is traditional to let the weight function be the distance separating each pixel from the flat back of the object. Scaling that distance and positioning the hypothetical back of the object will be discussed in detail later. For now, understand that whenever range applications are discussed throughout the remainder of this chapter, the assumptions just mentioned will be assumed.

For the reader's convenience, let us repeat the defining formula for discrete central moments. We will also take the liberty of using $M$ to refer to these central moments, as that is the usual practice in the literature. When this formula was developed early in this chapter, $U$ was used to help the reader distinguish between raw moments and central moments. Now that we know that *only* central moments are ever used, we can stick with $M$.

$$M(p, q) = \sum_x \sum_y (x - \mu_x)^p (y - \mu_y)^q f(x, y) \qquad (8\text{-}38)$$

There is an annoying number of decisions that must be made in regard to scaling of both the defining function, $f(x, y)$, and the computed moments. In many cases, these decisions are disturbingly arbitrary. In case of doubt, it pays to try several alternatives and to choose the method that gives the best performance. We will now discuss several of the alternatives that are possible. Applications of these techniques are presented in [Reeves *et al.*, 1988], [Kuhl *et al.*, 1986], [Taylor *et al.*, 1992], and [Wang *et al.*, 1992].

## What Does the Weight Function Mean?

The meaning of the weight function is clear when it is binary. It simply tells us whether or not a pixel is a member of the figure being studied. When the weight function can take on more than two values, its meaning may not always be so clear. And the meaning can have a profound effect on the applicability of weighted moments. Also, the meaning of the function can impact some of the decisions that will need to be made concerning scaling and origin of the function. It is worth putting some thought into the function and perhaps generating a revised function that is more suitable to the task at hand.

Things are easiest if the function has a quantitative value with an absolute physical meaning that is relevant to the shape. Perhaps a shotgun blast has perforated a target. We may divide the target into a grid and count the number of pellet holes in each grid cell. The overall outline of the damage pattern, a binary function defined by the outermost boundary of the collection of holes, is of some interest. But that same shape, weighted by the number of holes in each grid cell, is surely more useful than just a simple function that tells us nothing about the distribution of damage. The hole count in each pixel is a quantitative variable that has an absolute meaning. Weighted moments will work well.

Another example of a quantitative variable that is well suited for shape analysis by weighted moments is the density of an object. Perhaps we take an X-ray of a moderately opaque object like a tumor. The image could be normalized so that areas surrounding the tumor have a reference value of 0, while the tumor takes on positive values. Thicker or more dense areas of the tumor would have larger values than thinner or less dense areas. Such a weighting function is ideal for analysis with moments.

The situation becomes a bit more tricky when the weight function is range. Perhaps the image is derived from a stereo camera pair or maybe from a laser radar targeting system. A target object has been segmented, and we have a set of measurements reflecting the distance separating the sensor from each pixel on the target. We can process this sort of function with weighted moments, but we must be careful. The scale in the $z$ (distance) dimension must be scaled in conjunction with the $x$ and $y$ distances in some way that is consistent for all objects. The measurements need not be exactly to scale, but they must be done the same way at all times. An even worse problem stems from the fact that we generally cannot see hidden parts of the object. We do not have a complete 3-D representation of the object. All we have is what is up front. We cannot locate the centroid in the $z$ dimension, so we must use some arbitrary technique to set the origin in that dimension. And last but not least, we have no idea what the shape of the object is behind the occluding part. We will return to these problems later.

The worst situation is when the data has no absolute quantitative meaning at all. For example, suppose that we photograph a human face, then separate the face from the background. We want to identify faces based on their shape, using the tone values in the photograph as the weight function. That is dangerously close to nonsense. Does bright equate to large weight? Or is it dark areas that should be weighted most heavily? Or perhaps we should apply an edge detection filter and assign the highest weight to pixels that lie on an edge. The problem is evident. We are trying to count apples using a yardstick. A way might be found to make it work, but it would be klutzy at best. If your application resembles this scene, it is time to do some rethinking. If your weight function does not have a direct quantitative relationship to the shape, beware.

## Binary or Weighted Normalization?

One important decision concerns whether the normalizations are based on the binary profile moments or on the weighted moments. This issue cannot be sidestepped by avoiding normalization. At the very minimum, immunity against change in location must be obtained through the use of central moments. This entails computation of the centroid. Also, size normalization is nearly always required, which requires computation of the total area or mass. Rotation, reflection and aspect normalization are often optional, and often not. So there's no getting around it. We need to compute some parameters to use in the normalization formulae. But how? At first glance, it would seem obvious that we simply compute the moments based on the defining function, as shown in Equation (8-38), having gotten the centroid from Equation (8-3); then compute the normalization parameters as discussed earlier in this chapter. In fact, that is often the best method. But extensive experience indicates that a simple alternative often provides superior results. Treat the figure as a binary shape and compute the normalization parameters in the traditional way. Then apply those parameters to the weighted moments in order to normalize them. Let us briefly examine the difference between these two methods using just one parameter, the centroid.

What is the centroid of a two-dimensional shape? Probably the most intuitive definition is that the centroid is the figure's balancing point. If the figure were cut out of a sheet of cardboard and laid out horizontally, sitting atop the point of a pin, the object would lie balanced if and only if the pin were located at the figure's centroid. If the cardboard were of uniform thickness, the centroid would be at its visual center. If, however, the cardboard's thickness varied across its extent, the centroid might not be at the visual center. If the cardboard were heavier toward its right side, the centroid would have to lie to the right of the visual center in order for it to balance. When we compute the central moments, we need the centroid as computed by Equation (8-3). But which centroid do we use? If the raw moments in that equation are based on the weighted membership function, the physical centroid will result. If they are based on a binary membership function, the visual centroid will result. Which is better? There is no universal answer. The plain truth is that some applications favor one and some favor the other. Guidance can be had by trying to assess the stability of each centroid estimator. If the profile of the figure is clean and well defined, but the weight function within that profile is noisy, then the binary centroid will probably be better. On the other hand, the boundary of the object may not be clear everywhere, while heavily weighted interior areas may be stable. In that case, the weighted centroid will probably be the better choice. If neither of those situations is dominant, then the best bet is to try both.

Similar considerations apply to the other normalization parameters. We may need to estimate the orientation of the major principal axis using Equation (8-12) in order to do rotation normalization. That angle may be computed from either the binary moments or the weighted moments. Since the angle will be used to normalize the weighted moments using Equation (8-13), it may seem appropriate to use weighted moments to estimate the angle. But the fact of the matter is that very often the binary moments provide a more stable estimate of the angle. The same goes for aspect and reflection normalization. The only fairly hard-and-fast rule is that either one or the other is used to estimate the parameters for *all* normalizations. One does not use binary moments for some and weighted moments for others. Then again...

## Scaling the Weight Function

If we choose the option of determining the normalization constants using the binary moments, *and* if the weight function has an absolute 0 origin, we do not need to worry about scaling the weight function. Multiplicative scaling will be ignored when computing binary moments. Further, look back at the moment definition, Equation (18-7). It should be obvious that if $f(x, y)$ is multiplied by a constant, that constant will factor out in such a way that all moments will be multiplied by it. We, in general, will no longer have the weighted area, $M_w(0, 0)$, equal to 1, because the normalization was done to set the binary moment equal to 1. But that's no big deal. All of the weighted moments will be multiplied by the same amount, whatever it is. When the moments are transformed and scaled for presentation to the neural network, that constant will be summarily jettisoned. The only important consideration is that the scaling be identical for all cases during both training and ultimate use. But that goes without saying, right?

We do have something to keep in mind if we normalize with binary moments, but we do not have an absolute 0 origin for the weight function. In that case, the task of finding a suitable origin will interact with scaling. That subject will be discussed in depth in the next section.

Finally, if we choose to normalize based on parameters computed from the weighted moments, there is one small point to which we must pay attention. It involves the concept of size. For binary moments, size was straightforward. It was just the pixel count, that being the number of square pixels encompassed by the figure. But what is size when $M_w(0, 0)$ is computed with a function that can take on values other than 0 and 1? It resembles something more like a volume in many applications. Therein lies the problem.

To review how we normalize against change in size, look back on page 310. We compute a scale factor as the reciprocal of the square root of the area, then multiply the moments by powers of that factor. Second-order moments are multiplied by the fourth power of that factor, and third-order moments are multiplied by its fifth power. Things can obviously get out of hand if the factor is very small or very large. Suppose that the weighting function is of an order of magnitude of 100. The third-order moments will by divided by a factor on the order of 100,000. That is above and beyond the division by a factor to normalize against the spread of the object. That factor of 100,000 is due to the scaling of the weight function alone. The magnitude of the moments will be disparate. There are two ways to compensate for this effect. The author prefers to be ready to use both. The first step is to scale the weight function so that it tends to center around 1.0. (This can interact with the choice of origin, so see the next section also.) That will often be sufficient to keep the magnitudes of the moments in a range that is suitable for the standard transformations described earlier in this chapter. However, that is no absolute guarantee of uniformity. The second step is to actually examine the moments for a representative collection of cases. If the variables that will be sent to the neural network are not up to spec, it will be necessary to do custom modifications of the transformation functions. If this step is neglected, performance may suffer, and training time certainly will suffer.

Regardless of what method is used for normalization, and whether or not the origin is an absolute 0, there is one vital point regarding scaling. *Whatever it is, it must be fixed in advance and used consistently for all cases.* The scaling must not be done on an individual basis. One must not, for example, look at the minimum and maximum weight function for each case and rescale it to a fixed range based on the original range. That would prevent us from discriminating between cases based on the absolute variation in weight function across the figure. There are a few rare applications when that is precisely what we want to do. It may be that the degree of variation is a random variable, unrelated to the important information. If that is so, then scale each case individually. But those situations are rare. In the vast majority of applications, we must determine a scale factor in advance, then stick with it for every case.

Sometimes there is a known correspondence between the weight function and the *x-y* coordinates of each pixel. We may have a range sensor mounted at a fixed distance from the objects being analyzed. Thus, we may have in our possession a scale factor that allows us to express the sensed range to each point on the target in the same units as those that separate the pixels in the image plane. When we have this information at our disposal, it seems a shame to ignore it. There are at least two ways to preserve true scaling. If we are normalizing with weighted moments, we could treat the object as having volume rather than area as was the case for Equations (8-5) through (8-7). The rescaling would now be done with Equation (8-39). $M(0, 0)$ in that equation is the sum of the heights of all pixels, the volume of the object. (Remember that the height is the distance from the back of the object toward the observer and is inversely related to the range from the sensor to the pixel.)

$$V(p, q) = \frac{U(p, q)}{\sqrt[3]{M(0, 0)}^{(p+q+3)}} \qquad (8\text{-}39)$$

If we are normalizing with binary moments, $M(0, 0)$ is the number of pixels in the figure, the area of its base. Thus, the scale factor is its square root rather than its cube root. However, we still have the 3 in the normalizing equation, as that is the effect on the moments of scaling all three coordinate axes simultaneously. The normalization is shown in Equation (8-40).

$$V(p, q) = \frac{U(p, q)}{\sqrt{M(0, 0)}^{(p+q+3)}} \qquad (8\text{-}40)$$

## Setting the Z-Axis Origin

Let us start by saying that this section is irrelevant if the weight function intrinsically possesses an origin, typically 0. The shotgun example presented earlier has an absolute origin of 0 pellet holes in a grid cell. The tumor density function has an absolute origin of 0 density relative to normal tissue. Those sorts of weight functions give us no problem in choosing an origin, for they already have one that is perfectly excellent. In this section we will focus on the fairly common application of using range-to-target to define the weight function, and normalizing based on binary moments. Naturally, the methods discussed here are also applicable to weight functions other than range.

We need to decide where to place the z-axis origin. We want to place it in such a way that some agreed-upon normalization criterion is met. Thus, there are two tasks that face us. One is to choose a normalization criterion. The other is to determine the effect that the position of the origin has on that criterion. Once we have solved those two problems, we can move the z-axis origin until the criterion is satisfied. The assumptions noted on page 336 make our tasks a little easier. However, the method for setting the origin depends on whether we are normalizing the object based on binary moments or on weighted moments. We will discuss the latter option first.

We normally do not want to complicate our lives by trying to employ anything but the simplest techniques for setting the origin if we are normalizing based on weighted moments. Why? If we are using weighted moments, our choices for origin are limited by the profound effect that the location of the origin has on the normalization. Recall that size normalization depends on $M(0, 0)$, the area (or mass, here) of the figure. It is the sum of the weight function over all pixels. The position of the origin dramatically affects that sum. In fact, if the origin happens to be set at the mean range, the mass will be 0! Of course, we would never do that. It distresses the intuition to have negative values for the weight function. By tradition, that is almost never done, although the author knows of no sound mathematical reason for strictly forbidding it. But the point is, when weighted moments are used for normalization, the z-axis origin interacts with size normalization, and hence, with all moments, in large and complicated ways. It is best by far to fix a point in some simple, stable way. The most common method is to place the origin at the furthest point in the object, and let the function increase for closer points. The principal argument against this choice is that the sensed distance to that point may be subject to significant random variability. That error will be magnified to serious proportions through its effect on size normalization, not to mention through direct effects. If such error is expected, some other method for fixing the origin should be adopted. Perhaps a high percentile of the ranges, plus a small constant, could be used. Or we may be lucky and have a sound physical means for fixing a maximum range that will always touch the front or rear of the target, so we can use the distance from that maximum as the weight function. If we absolutely cannot come up with a stable origin, we may want to rethink our decision to base normalization on weighted moments. Basing normalization on binary moments is a totally respectable option, and one for which the task of setting the origin is much easier.

Suppose that we do base our normalization on binary moments. Since the $z$ origin now does not affect normalization, it is easy to see the effect that the origin has on the weighted moments. Let $f(x, y)$ be the raw measured weight function, and let $M(p, q)$ represent its family of weighted moments, as defined in Equation (8-38). In order to avoid assuming that the weight function is 0 outside the figure, let us rather assume that the summation in that equation and in the equations to follow is taken over the figure only. Let $M_{BIN}(p, q)$ be the binary moments of the figure. Offset the weight function by adding $a$ to it everywhere. We can calculate the moments of the offset figure, $M_a(p, q)$, by Equation (8-41). For simplicity, that equation does not include subtraction of the centroid for location normalization. Assume that it has been done to the $x$ and $y$ coordinates already.

$$
\begin{aligned}
M_a(p, q) &= \sum_x \sum_y x^p y^q \, (f(x, y) + a) \\
&= \sum_x \sum_y x^p y^q f(x, y) + \sum_x \sum_y x^p y^q a \qquad (8\text{-}41) \\
&= M(p, q) + a \, M_{BIN}
\end{aligned}
$$

That equation does not in itself provide a magical solution to the problem of fixing the origin. But it is an extremely powerful tool. We can now come up with some definition of what "normalized" means and use that equation to choose a value of $a$ that achieves it.

We already saw that when weighted moments were used for size normalization, a good way to normalize was to set the volume, $M(0, 0)$, equal to 1. That is just as great an idea when we normalize with binary moments. That size normalization has already set the surface area, $M_{BIN}(0, 0)$, equal to 1. The volume is, of course, no longer equal to 1.0 except by unlikely coincidence. But we can easily adjust the volume to insure that $M(0, 0) = 1$. Just slide the origin (the back of the object) until that happens. The amount to slide it, $a$, is found by setting the left-hand side of Equation (18-10) equal to 1, then trivially solving for $a$. That same equation is then used to revise all of the other moments to account for the origin shift.

Let's think about that operation a little and prepare a defense against an unlikely but ugly possibility. Suppose that the weight function has been scaled so that it never exceeds 1. Also suppose that there are $n$ pixels in the object. The raw value of $M_{BIN}(0, 0)$ is $n$. Then it is divided by $n$ for binary size normalization, giving it a value of 1. The raw value of $M(0, 0)$ before size normalization is the sum of those weights. If every value of the weight function is less than or equal to 1, then that sum will be less than or equal to $n$. Thus, $M(0, 0)$ after size normalization is less than or equal to 1. In other words, the volume of the size-normalized object is less than or equal to 1. Our normal course of action will be to push the back of the object away from us, adding the volume due to the cross-sectional area to $M(0, 0)$ until it is brought up to 1. That makes simple intuitive sense.

Unfortunately, we may not always have it so easy. In practice, we may not be scaling the weight function from 0 to 1. We may be preserving known scale by means of Equation (8-40). If the object's maximum height does not exceed its breadth, everything is still fine. But if the object is much taller than it is wide, its volume can exceed 1. In this case we will find ourselves in the uncomfortable position of moving the hypothetical back of the object up into the object itself. In extreme cases, the implied negative volume can even cause pure even moments to become negative! The transformation functions for preparing the moments for neural networks will not be pleased. If we are adamant about sticking with true scaling, there is only one way out of this dilemma. There is nothing magical about a value of 1 for the normalized volume. Increase that constant. Establish the convention of setting $M(0, 0)$ equal to 2, or 3, or whatever is needed to avoid $a$ being negative. Just try to keep it as small as possible so that the height information in the moments does not get washed out by the binary moments that are brought in with the addition of the cross-section volume at the back of the figure.

## Summary of Scale, Offset, and Normalization

By now, the reader is probably totally confused over the number of issues raised regarding scaling the weight function, choosing an origin, scaling the moments for unit size, and so on. The harsh reality of the matter is that it truly *is* a confusing state of affairs. There are few rigid rules to follow, and a lot of fairly arbitrary choices need to be made. The one point that should be remembered is this: *If in doubt, try several methods and choose the best*. We will close with a few guidelines that will hopefully be of some assistance.

- If the weight function has a physically meaningful origin of absolute 0, use it unless there is a compelling reason to do otherwise. One such reason is if the bulk of the data is far from that origin. In that case we would want to move the origin a fixed distance closer to the minimum possible data value.

- If the perimeter of the figure is more stable than the distribution of the weights within the figure, normalize based on binary moments. If the perimeter is relatively vulnerable to noise, normalize based on weighted moments.

- If the weight function is the height of an object measured from its hypothetical back, scale it so that its height is in the same units as its base, if that information is available. Otherwise, scale it so that its height hovers around 1 or less.

- If the weight function is the height of an object measured from its hypothetical back, and the origin is not absolute, try to normalize using binary moments so that the origin can be set by fixing the volume at a constant such as 1. If the data is very clean, so that the distance to the furthest point is reliable, that point can be used as the origin, in which case we may normalize with weighted moments if desired. The data must be *very* clean for this to be a good choice.

• Never normalize with weighted moments if the origin is subject to noise.

• If normalization is done with weighted moments, either the weight function must be scaled so that its average value is not extremely different from 1, or individual moment scaling must be done to compensate for size disparity.

# 9

Tone/Texture Descriptors

In the previous chapter, we showed how moment-based variables could be used to describe image features. However, the methods of that chapter demand that the image be segmented. Those feature variables can only be computed for objects in the image that are clearly defined in terms of the pixels that comprise the object. This chapter takes a looser approach. No segmentation is needed. We will be examining circular or rectangular areas of the image and characterizing those areas in terms of the distribution of tones of the pixels in that area. There are many uses for the techniques presented in this chapter. Some common examples are as follows:

- We may work with satellite imagery. A trained photo-interpreter moves a rectangle across a computer display of an image, identifying areas as to vegetation type, terrain type, and so forth. That creates a training set. The neural network is then able to classify new images.

- We may work with medical imagery. A technician locates normal and abnormal areas so that a neural network can be trained to identify these areas on similar images.

- We may work with X-ray diffraction patterns of a manufactured product. Members of a collection of samples are known to be good or defective. A neural network is trained to differentiate between the two.

The variables discussed in this chapter are somewhat related to the two-dimensional FFT, Gabor, and wavelet variables already discussed. Both families of variables respond to relatively small areas of an image, where those areas have a predefined shape. No segmentation is necessary for computing the variables. Their values depend on the distribution of tones throughout the area of interest. However, there is an important difference that must be kept in mind when selecting variables for a particular application. The variables discussed earlier all respond to *frequency* information in the area. The variables of this chapter operate more in the spatial domain. If the application involves information that is periodic, the former set of variables is probably

345

better. If no periodic waves are expected, this family is suggested. But always remember, if in doubt, use both! Let the neural network pick and choose what it finds best. That is their strong point. Take advantage of it.

# To Window, or Not to Window

The frequency-domain variables discussed earlier in this book all used a window that had its peak at the center of the area of interest and tapered down to nearly zero sensitivity some distance away. That windowing was intrinsically necessary to prevent harmful artifacts, like leakage from nonintegral frequency bands, from contaminating the computed variables. Windowing is not so strictly required for the variables of this chapter. However, it can still be useful. The author uses it frequently.

In order to decide whether or not windowing is appropriate, one must consider the relative importance of the pixels near the border of the area of interest. A data window reduces the impact of these distant pixels, while highlighting the pixels near the center. Suppose that we are examining enlarged photographs of the surface of sandpaper in order to verify proper manufacturing. It is likely that all areas of the image are equally important. In this case we would not use a window. Now suppose that we are dealing with infrared images, searching for vehicles. We have located a suspicious spot, and our area of interest is centered on that spot. This area is presumably somewhat larger than the maximum expected size of the vehicle, so that we are sure of encompassing the vehicle entirely. Therefore, the furthest reaches of the area of interest are probably background and should not weigh heavily in the decision-making process. A window is definitely called for.

There is an infinite number of choices for the shape of the data window. The author's favorite is a simple Gaussian, similar to what is used in the Gabor transform. No mathematical optimality of any sort is claimed for this shape when used in conjunction with the tone/texture variables of this chapter. It is simply convenient. All of the routines in this chapter use the Gaussian window when a window is requested.

It would not do to force the exact same window shape for all applications. Sometimes we want the importance of the distant perimeter to be virtually zero, while other times we only want to moderately reduce its importance. An easy way to parameterize the width of the Gaussian window is to specify its height at pixels that are at the maximum distance from the center in one dimension and centered in the other dimension. This is what is done here. For example, we may have an area of interest that extends six pixels to the left and right of the center pixel, and five pixels above and below. The total rectangle size is 11 rows by 13 columns. We may perhaps specify that the height of the data window is 0.1 (with a maximum of 1.0 at the center) at the four extremes: center row at plus and minus six columns, and center column at plus and minus five rows. At the four corners of the rectangle, the height of the window will drop even lower than 0.1.

# Foundation Code

The variables presented in this chapter are all computed via member functions of a general tone/texture class. This allows us to compute and save the expensive window function and a few important constants. A member function called getpix is used to copy the pixels from a rectangular area of interest into a local work area. All variable computation will be based on the data in that local area. This avoids the time-consuming array processing that would be necessary if the image itself were used every time.

There is one inefficiency inherent in this code. Most variables require looping through the data. Since each variable is computed in a separate member function, the loop overhead is repeated. This gives the user full versatility in selecting which variables are desired and which are not. But it significantly slows operation. Readers whose applications must run as fast as possible would do well to combine computation as much as possible.

In this section we present the code for the constructor, destructor, and pixel retriever. The constructor allocates memory for saving the data window if it is needed. We know that no window is to be used if the tapered value at the extreme positions is essentially equal to one. It also allocates memory to save the contents of the rectangle, both in its original order and sorted order. It computes and saves the mean (across all pixels in the rectangle area) of the distance from the center pixel. One variable, rel_tone, needs this. Finally, the data window is computed. The destructor simply frees these arrays. The getpix function copies the data from the area of interest in the image to the local area. It also sets flags to indicate that some basic parameters of the area, needed by many functions, have not yet been computed. The code, along with the class declaration, now follows.

```
class ToneText {

public:
   ToneText ( int rows , int cols , double taper , int *ok ) ;
   ~ToneText () ;
   void getpix ( int nrows , int ncols , unsigned char *img ,
            int rcent , int ccent ) ;
   double mean () ;
   double mean_dev () ;
   double std_dev () ;
   double skewness () ;
   double ncon1 () ;
   double ncon2 () ;
   double rel_tone () ;
   double hgrad () ;
   double vgrad () ;
   double median () ;
   double range () ;
```

```
    double iq_range () ;
    double skew10 () ;

private:
    int nr ;                                // Row radius of area
    int nc ;                                // Column radius of area
    int ntot ;                              // Number of pixels in area
    float *window ;                         // Weight window stored here
    double wt_tot ;                         // Total of all weights
    double mean_rad ;                       // Mean radius of area
    int *unsorted ;                         // Raw pixels from image put here
    int *sorted ;                           // Sorted here if needed
    int sort_done ;                         // Have they been sorted?
    double p_mean ;                         // Mean tone in array, -1.0 if not done
    double p_stddev ;                       // Std dev tone in array, -1.0 if not done
    void sort () ;                          // Local routine to sort data
} ;


ToneText::ToneText (
    int rows ,                              // Row radius (above and below center)
    int cols ,                              // Column radius (left and right of center)
    double taper ,                          // Fraction of max at (row, 0), (0, col) etc.
    int *ok                                 // Memory allocation go ok?
    )
{
    int r, c ;
    float *wptr ;
    double dr, dc, winfac, wt ;

    window = NULL ;
    unsorted = NULL ;
    sorted = NULL ;
    *ok = 0 ;

    nr = rows ;
    nc = cols ;

    ntot = (2 * nr + 1) * (2 * nc + 1) ;  // Pixels in rectangle

/*
    Allocate memory
*/
```

```
   if (taper <= 0.99999999) { // No window if flat
     if ((window = (float *) malloc ( ntot * sizeof(float) )) == NULL)
       return ;
     }

   unsorted = (int *) malloc ( ntot * sizeof(int) ) ;
   if (unsorted != NULL)
     sorted = (int *) malloc ( ntot * sizeof(int) ) ;

   if (sorted == NULL) {
     if (window != NULL) {
       free ( window ) ;
       window = NULL ;  // So destructor doesn't try to free it
       }
     if (unsorted != NULL) {
       free ( unsorted ) ;
       unsorted = NULL ;
       }
     return ;
     }

   *ok = 1 ;  // Signal caller that memory allocs ok

/*
   Compute the mean radius of the area.  This is needed by rel_tone.
*/

   mean_rad = 0.0 ;
   for (r=-nr ; r<=nr ; r++) {
     for (c=-nc ; c<=nc ; c++)
       mean_rad += sqrt ( r * r + c * c ) ;
     }
   mean_rad /= ntot ;

/*
   Compute window if needed.  No window if flat taper.
*/

   if (window == NULL)
     return ;

   if (taper <= 0.0)    // Protection from silly user
     taper = 1.e-30 ;
```

```
        winfac = log ( taper ) ;
        wptr = window ;
        wt_tot = 0.0 ;
        for (r=-nr ; r<=nr ; r++) {
          dr = (double) r / (double) nr ;
          dr = dr * dr ;
          for (c=-nc ; c<=nc ; c++) {
            dc = (double) c / (double) nc ;
            dc = dc * dc ;
            wt = exp ( winfac * (dr + dc) ) ;   // Weight for this position
            *wptr++ = wt ;                        // Save it
            wt_tot += wt ;                        // And cumulate sum
            }
          }
}


ToneText::~ToneText ()
{
  if (window != NULL)
    free ( window ) ;
  if (unsorted != NULL)
    free ( unsorted ) ;
  if (sorted != NULL)
    free ( sorted ) ;
}


void ToneText::getpix (
    int nrows ,                      // Number of rows in image
    int ncols ,                      // And columns
    unsigned char *img ,             // Image
    int rcent ,                      // Row (origin 0) where area centered
    int ccent                        // And column
    )
{
  int r, c, *uptr ;
  unsigned char *iptr ;

/*
  Make sure we do not overrun the image.
*/
```

```
       if ((rcent < nr)  ||  (rcent >= nrows-nr))
          return ;
       if ((ccent < nc)  ||  (ccent >= ncols-nc))
          return ;

       uptr = unsorted ;
       for (r=rcent-nr ; r<=rcent+nr ; r++) {
          iptr = img + r * ncols + ccent - nc ;
          c = 2 * nc + 1 ;
          while (c--)
             *uptr++ = *iptr++ ;
          }
/*
   Flag some local variables as not being found yet
*/
       sort_done = 0 ;
       p_mean = p_stddev = -1.0 ;
    }
```

## Basic Tone and Texture

We will start out with the most basic tone and texture variables. These are all based on the raw, unsorted pixel values in the rectangle. The most basic of them all is the mean tone. Since that value is needed by many other functions, it is preserved in a private area. This avoids having to recompute it. The function getpix initialized p_mean to a negative number as a flag that the mean has not yet been computed. The computation of the mean tone is done in one of two ways. If no data window is used, it is just the sum of the individual tones, divided by the number of pixels. If a window is used, the tones are weighted and the sum is divided by the sum of the weights. Code for computing the mean is shown here.

```
double ToneText::mean ()
{
   int i, *uptr ;
   float *wptr ;

   if (p_mean >= 0.0)                    // Already computed?
      return p_mean ;

   p_mean = 0.0 ;                        // Will cumulate and save mean here
   i = ntot ;                            // This many pixels
```

```
    uptr = unsorted ;                   // They are here

    if (window == NULL) {               // If no window just sum
      while (i--)
        p_mean += *uptr++ ;
      p_mean /= (double) ntot ;
      }

    else {
      wptr = window ;                   // Window weights
      while (i--)
        p_mean += *uptr++  *  *wptr++ ;
      p_mean /= wt_tot ;
      }

    return p_mean ;
}
```

The most basic texture measure is the mean deviation. This is the average absolute difference between each pixel and the mean of the area. Code for computing the mean deviation is now given. Note that the first line of code calls mean to verify that the mean has been computed. That value is assigned to mdev. That assignment is not needed. If it is omitted, many compilers complain that the returned function value has been ignored. This assignment avoids those otherwise useful warnings.

```
double ToneText::mean_dev ()
{
  int i, *uptr ;
  float *wptr ;
  double mdev ;

  mdev = mean () ;                    // Need mean, mdev just to silence warnings

  mdev = 0.0 ;                        // Will cumulate and save mean_dev here
  i = ntot ;                          // This many pixels
  uptr = unsorted ;                   // They are here

  if (window == NULL) {               // If no window just sum
    while (i--)
      mdev += fabs ( (double) *uptr++  -  p_mean ) ;
    mdev /= (double) ntot ;
    }
```

```
   else {
     wptr = window ;                     // Window weights
     while (i--)
       mdev += fabs ( (double) *uptr++  -  p_mean )  *  *wptr++ ;
     mdev /= wt_tot ;
     }

   return mdev ;
}
```

A closely related texture variable is the standard deviation.  This squares each tone difference, rather than taking its absolute value.  To return to linear scaling, the square root of the mean squared deviation is then taken.  The code is largely redundant but is listed here for completeness.

```
double ToneText::std_dev ()
{
   int i, *uptr ;
   float *wptr ;
   double diff ;

   if (p_stddev >= 0.0)                  // Already computed?
     return p_stddev ;

   diff = mean () ;                      // Need mean, diff just to silence warnings

   p_stddev = 0.0 ;                      // Will cumulate and save std_dev here
   i = ntot ;                            // This many pixels
   uptr = unsorted ;                     // They are here

   if (window == NULL) {                 // If no window just sum
     while (i--) {
       diff = *uptr++  -  p_mean ;
       p_stddev += diff * diff ;
       }
     p_stddev = sqrt ( p_stddev / (double) ntot ) ;
     }

   else {
     wptr = window ;                     // Window weights
     while (i--) {
       diff = *uptr++  -  p_mean ;
       p_stddev += diff * diff * *wptr++ ;
       }
```

```
    p_stddev = sqrt ( p_stddev / wt_tot ) ;
    }

  return p_stddev ;
}
```

An often useful piece of information about the distribution of a set of tones is its skewness. Do the outliers tend to be above or below the mean, or are they symmetrically distributed? If most of the unusual pixel tones are above the mean, the distribution is said to be positively skewed. If the unusual tones are primarily darker than the mean, it is negatively skewed. If they are equally distributed on both sides of the mean, the skewness is zero. Code for computing the skewness is shown below. It is just the mean of the cubed deviation of each pixel from the mean tone, divided by the cube of the standard deviation. Note that the kurtosis of the tones can be computed in a similar fashion. Instead of using the cube of the individual deviations and standard deviation, use the fourth power. This measures the heaviness of the tails in a symmetric manner. Many people subtract three from the quotient, as that is the value obtained for a normal distribution. The author has not found kurtosis to be particularly valuable, so it is not listed here. The reader can program it easily by following the patterns for skewness.

```
double ToneText::skewness ()
{
  int i, *uptr ;
  float *wptr ;
  double diff, std, skew ;

  std = std_dev () ;                   // Need standard deviation

  skew = 0.0 ;                         // Will cumulate and save skewness here
  i = ntot ;                           // This many pixels
  uptr = unsorted ;                    // They are here

  if (window == NULL) {                // If no window just sum
    while (i--) {
      diff = *uptr++  -  p_mean ;
      skew += diff * diff * diff ;
      }
    skew /= (double) ntot ;
    }

  else {
    wptr = window ;                    // Window weights
```

```
      while (i--) {
        diff = *uptr++  -  p_mean ;
        skew += diff * diff * diff * *wptr++ ;
        }
      skew /= wt_tot ;
      }

    return skew / (std * std * std) ;
    }
```

## Neighbor Contrasts

The variables presented so far are all based on the individual pixel tones, regardless of their positions in the rectangle of interest. We now advance in complexity by defining variables that depend on spatial position.

The simplest spatial variable is *first neighbor contrast*. It is the mean difference in the tone of adjacent pixels. In its most general form, given here, it examines both vertical and horizontal neighbors. The reader can very easily modify this routine to consider only one direction if that is more appropriate for an application. A closely related variable is *second neighbor contrast*. That variable looks at pixels that are separated by an intervening pixel. It is more sensitive to coarse textures. Code for these two variables is now given.

```
double ToneText::ncon1 ()
{
  int r, c, curr, prev, *uptr, nrb, ncb ;
  float *wptr ;
  double con, wt, wtsum, wprev, wcurr ;

  con = 0.0 ;                       // Will cumulate and save contrast here
  nrb = 2 * nr + 1 ;                // Number of rows in box
  ncb = 2 * nc + 1 ;                // And columns

  if (window == NULL) {             // If no window just sum
    for (r=0 ; r<nrb ; r++) {       // First, do left nbrs for all rows
      uptr = unsorted + r * ncb ;   // First pix in this row
      curr = *uptr++ ;              // This is its tone
      c = ncb ;                     // Number of columns in box
      while (--c) {                 // Do one less because pairs
        prev = curr ;               // Pixel to left of current
        curr = *uptr++ ;            // New current pixel
```

```
          con += abs ( curr - prev ) ;      // Cumulate contrast
          }
        }
      for (c=0 ; c<ncb ; c++) {              // Now, do top nbrs for all cols
        uptr = unsorted + c ;                // First pix in this column
        curr = *uptr ;                       // This is its tone
        r = nrb ;                            // Number of rows in box
        while (--r) {                        // Do one less because pairs
          prev = curr ;                      // Pixel above current
          uptr += ncb ;                      // Advance down one row
          curr = *uptr ;                     // New current pixel
          con += abs ( curr - prev ) ;       // Cumulate contrast
          }
        }
      con /= nrb * (ncb-1)  +  ncb * (nrb-1) ; // Number of pairs used
      }

   else {
      wtsum = 0.0 ;                          // Will sum derived weights here
      for (r=0 ; r<nrb ; r++) {              // First, do left nbrs for all rows
        uptr = unsorted + r * ncb ;          // First pix in this row
        wptr = window + r * ncb ;            // Ditto window weights
        curr = *uptr++ ;                     // This is its tone
        wcurr = *wptr++ ;                    // And weight
        c = ncb ;                            // Number of columns in box
        while (--c) {                        // Do one less because pairs
          prev = curr ;                      // Pixel to left of current
          wprev = wcurr ;                    // Ditto for weights
          curr = *uptr++ ;                   // New current pixel
          wcurr = *wptr++ ;                  // And weight
          wt = wcurr + wprev ;               // Contrast weight includes both
          con += abs ( curr - prev ) * wt ;  // Cumulate contrast
          wtsum += wt ;                      // And weights
          }
        }
      for (c=0 ; c<ncb ; c++) {              // Now, do top nbrs for all cols
        uptr = unsorted + c ;                // First pix in this column
        wptr = window + c ;                  // Ditto window weights
        curr = *uptr ;                       // This is its tone
        wcurr = *wptr ;                      // And weight
        r = nrb ;                            // Number of rows in box
        while (--r) {                        // Do one less because pairs
          prev = curr ;                      // Pixel above current
          wprev = wcurr ;                    // Ditto for weights
```

```
         uptr += ncb ;                    // Advance down one row
         wptr += ncb ;                    // Ditto weights
         curr = *uptr ;                   // New current pixel
         wcurr = *wptr ;                  // And weight
         wt = wcurr + wprev ;             // Contrast weight includes both
         con += abs ( curr - prev ) * wt ; // Cumulate contrast
         wtsum += wt ;                    // And weights
         }
      }
   con /= wtsum ;
   }

 return con ;
}


double ToneText::ncon2 ()
{
   int r, c, curr, prev, *uptr, nrb, ncb ;
   float *wptr ;
   double con, wt, wtsum, wprev, wcurr ;

   con = 0.0 ;                            // Will cumulate and save contrast here
   nrb = 2 * nr + 1 ;                     // Number of rows in box
   ncb = 2 * nc + 1 ;                     // And columns

   if (window == NULL) {                  // If no window just sum
     for (r=0 ; r<nrb ; r++) {            // First, do left nbrs for all rows
       uptr = unsorted + r * ncb ;        // First pix in this row
       c = ncb - 2 ;                      // Number of pairs in box
       while (c--) {                      // Do all pairs
         curr = *(uptr+2) ;               // One pixel
         prev = *uptr++ ;                 // Pixel 2 to left of current
         con += abs ( curr - prev ) ;     // Cumulate contrast
         }
       }
     for (c=0 ; c<ncb ; c++) {            // Now, do top nbrs for all cols
       uptr = unsorted + c ;              // First pix in this column
       r = nrb - 2 ;                      // Number of pairs in box
       while (r--) {                      // Do one less because pairs
         prev = *uptr ;                   // One pixel
         uptr += ncb ;                    // Advance down one row
         curr = *(uptr + ncb) ;           // Pixel 2 rows beyond other
```

```
                con += abs ( curr - prev ) ;       // Cumulate contrast
              }
            }
          con /= nrb * (ncb-2)  +  ncb * (nrb-2) ; // Number of pairs used
          }

      else {
        wtsum = 0.0 ;                          // Will sum derived weights here
        for (r=0 ; r<nrb ; r++) {              // First, do left nbrs for all rows
          uptr = unsorted + r * ncb ;          // First pix in this row
          wptr = window + r * ncb ;            // Ditto window weights
          c = ncb - 2 ;                        // Number of pairs in box
          while (c--) {                        // Do one less because pairs
            curr = *(uptr+2) ;                 // One pixel
            prev = *uptr++ ;                   // Pixel 2 to left of current
            wcurr = *(wptr+2) ;                // One weight
            wprev = *wptr++ ;                  // Weight 2 to left of current
            wt = wcurr + wprev ;               // Contrast weight includes both
            con += abs ( curr - prev ) * wt ;  // Cumulate contrast
            wtsum += wt ;                      // And weights
            }
          }
        for (c=0 ; c<ncb ; c++) {              // Now, do top nbrs for all cols
          uptr = unsorted + c ;                // First pix in this column
          wptr = window + c ;                  // Ditto window weights
          r = nrb - 2 ;                        // Number of pairs in box
          while (r--) {                        // Do one less because pairs
            prev = *uptr ;                     // One pixel
            uptr += ncb ;                      // Advance down one row
            curr = *(uptr + ncb) ;             // Pixel 2 rows beyond other
            wprev = *wptr ;                    // One pixel
            wptr += ncb ;                      // Advance down one row
            wcurr = *(wptr + ncb) ;            // Pixel 2 rows beyond other
            wt = wcurr + wprev ;               // Contrast weight includes both
            con += abs ( curr - prev ) * wt ;  // Cumulate contrast
            wtsum += wt ;                      // And weights
            }
          }
        con /= wtsum ;
        }

      return con ;
   }
```

We may also be interested in regular variation in tone throughout the rectangle. It may be that the tone is brightest at the center of the area of interest, becoming dimmer toward the edges. Or the opposite may be true. An extremely primitive way to detect this is to compute the mean tone of all pixels except the center, then subtract that mean from the center pixel. That method is too sensitive to noise to be of good general use. A more sophisticated method is to use linear regression. Compute the relative tone of pixels as a linear function of their distance from the center pixel. This is easy to do. For each pixel, compute its tone relative to the mean tone in the rectangle. Also compute its distance from the center relative to the mean distance. That latter quantity was computed by the constructor and saved as mean_rad. As can be found in any basic statistics text, the *beta* coefficient is the ratio of the cross-products of these two quantities divided by the sum of squares of the independent variable. The beta coefficient tells how much the tone changes per unit change in distance from the center. If the center is brighter than the surrounding area, this quantity as computed here will be positive. If the center is darker, it will be negative. It will be zero when the tone of a pixel does not depend on its distance from the center. Code for computing the beta coefficient is now given.

```
double ToneText::rel_tone ()
{
   int r, c, *uptr ;
   float *wptr ;
   double sxy, ssx, ymean, x ;

   ymean = mean () ;                        // Mean tone

   sxy = 0.0 ;                              // Will cumulate cross-product here
   ssx = 0.0 ;                              // Sum of squares of distance (x)
   uptr = unsorted ;                        // Pixel tones are here

   if (window == NULL) {                    // If no window just sum
      for (r=-nr ; r<=nr ; r++) {
         for (c=-nc ; c<=nc ; c++) {
            x = mean_rad - sqrt ( r * r + c * c ) ;
            sxy += x * ((double) *uptr++  -  ymean ) ;
            ssx += x * x ;
            }
         }
      }

   else {
      wptr = window ;                        // Window weights
      for (r=-nr ; r<=nr ; r++) {
         for (c=-nc ; c<=nc ; c++) {
            x = mean_rad - sqrt ( r * r + c * c ) ;
```

```
      sxy += x  *  ((double) *uptr++  -  ymean )  *  *wptr ;
      ssx += x * x * *wptr++ ;
      }
    }
  }

  return sxy / ssx ;
}
```

We can do a similar thing with horizontal and vertical directions. It may be that a characteristic of some objects is that they become progressively darker or lighter from left to right or top to bottom. The technique for measuring this is exactly the same as for rel_tone, except that we use horizontal or vertical position as the independent variable. Code for computing the horizontal gradient is shown below. The vertical gradient is computed in almost exactly the same way. The only difference is that the row variable, r, is used in the summing instead of the column variable, c.

```
double ToneText::hgrad ()
{
  int r, c, *uptr ;
  float *wptr ;
  double sxy, ssx, ymean ;

  ymean = mean () ;                 // Mean tone

  sxy = 0.0 ;                       // Will cumulate cross-product here
  ssx = 0.0 ;                       // Sum of squares of distance (x)
  uptr = unsorted ;                 // Pixel tones are here

  if (window == NULL) {             // If no window just sum
    for (r=-nr ; r<=nr ; r++) {
      for (c=-nc ; c<=nc ; c++) {
        sxy += c  *  ((double) *uptr++  -  ymean ) ;
        ssx += c * c ;
        }
      }
    }

  else {
    wptr = window ;                 // Window weights
    for (r=-nr ; r<=nr ; r++) {
      for (c=-nc ; c<=nc ; c++) {
        sxy += c  *  ((double) *uptr++  -  ymean )  *  *wptr ;
```

```
            ssx += c * c * *wptr++ ;
          }
        }
      }

    return sxy / ssx ;
}
```

# Variables Based on Sorted Tones

It is often the case that some tone values in the area of interest will be wildly unlike the majority of pixels. Measures of tone and texture that involve sums throughout the area will be unduly influenced by these outliers. We can avoid this problem by sorting the tones in the rectangle and avoiding the extremes. A *trimmed mean* is computed by finding the mean in all except a preordained extreme fraction of the data. For example, we may discard the minimum 10 percent and maximum 10 percent of the data and compute the mean of the remainder. The trimmed mean deviation, standard deviation, and skewness may also be computed this way. The author has not found these variables to be particularly useful, so functions for them are not included here. The interested reader can easily program them if desired. On the other hand, sorted data can be used in other ways. Some of the most common are now discussed.

The ToneText class contains several functions that need the data sorted. It would not do to sort it every time a different member function is called. So we keep a sorted copy that is computed the first time it is needed, and we also keep a flag that tells whether or not the sorting has been done. The private sorting function is listed now. A suitable quicksort was listed on page 79.

```
void ToneText::sort ()
{
  if (sort_done)
    return ;

  memcpy ( sorted , unsorted , ntot * sizeof(int) ) ;
  qsort ( 0 , ntot-1 , sorted ) ;
  sort_done = 1 ;
}
```

The most fundamental order statistic is the *median* of the sample. Half of the sample cases are greater than the median, and half are less. (There are standard rules for dealing with equality and breaking ties. These do not concern us here because ntot is always odd, but they can be found in any standard statistics text.) It should be obvious that the median is, in one sense, a more stable measure of the central tendency of the sample. If one or even many samples are extremely large or small, they will not significantly affect the median. They will, however, pull the mean strongly away from what it would otherwise be.

Here, we will compute the median by sorting the data and retrieving the middle element of the sorted array. In the event that the only order statistic used is the median, and speed is extremely critical, the reader should be aware that there is a faster way to compute the median than sorting the entire array. This algorithm can be found in most good computer texts, such as [Press *et al.*, 1992]. It will not be given now because we are more concerned with the general case.

```
double ToneText::median ()
{
  sort () ;
  return sorted[ntot/2] ;
}
```

There are two common measures of spread based on order statistics that can be used to gauge texture. One is the *range*. It is the difference between the maximum and minimum tones. This variable is rarely used because it is obviously sensitive to outliers. The other is the *interquartile range*. It is the tone that is three-quarters of the way up the sorted array minus the tone that is only one-quarter of the way up. In other words, the brightest 25 percent and the darkest 25 percent of the pixels are discarded. The range of the middle 50 percent is the interquartile range. This is a very good indicator of the spread of the tone values, yet it is not influenced by extremes. The trivial code for computing these two variables is now shown.

```
double ToneText::range ()
{
  sort () ;
  return sorted[ntot-1] - sorted[0] ;
}

double ToneText::iq_range ()
{
  sort () ;
  return sorted[3*ntot/4] - sorted[ntot/4] ;
}
```

We have already discussed the skewness variable. As previously defined, it is computed by summing across all pixels. Thus, it will be affected by outliers. In fact, since it involves cubing tone differences, it will be extremely affected by outliers. We can avoid this by discarding a few extreme values and basing the variable on order statistics. One common method is to compare the distance separating the median from a high percentile with the distance separating the median from a low percentile. If the distribution is symmetric, these distances will be the same. If it is positively skewed, the high percentile will be further from the median than the low percentile. The converse is also true. The exact percentile to use is fairly arbitrary. The further out we go, the more sensitive to both skewness and outliers we become. The author traditionally discards the upper and lower 10 percent. A log transformation is a good way to center the value at zero for symmetric distributions. Note that this variable is immune to change in both location and scale. Also note that we must make provision for degenerate samples.

```
double ToneText::skew10 ()
{
  int numer, denom, median ;

  sort () ;
  median = sorted[ntot/2] ;
  numer = sorted[9*ntot/10] - median ;
  denom = median - sorted[ntot/10] ;
  if (numer) {
    if (denom)
      return log ( (double) numer / (double) denom ) ;
    else
      return 10.0 ;
    }
  else {
    if (denom)
      return -10.0 ;
    else
      return 0.0 ;
    }
}
```

# Other Variables

There is a vast array of choices for variables that describe the spatial structure of the tone values in an area of interest. This chapter has focused on variables that have worked well for the author. It has been his experience that most of the other variables described in the literature are of more academic than practical interest. On the other hand, there are quite possibly some applications for which some of these alternatives may be of use. This section will provide a rough overview of the "standard" texture variables. Detailed information is left for other references. A classic paper on this subject is [Haralick, 1979]. A very rigorous summary is given in [Rosenfeld and Kak, 1982]. A somewhat less rigorous but extremely broad summary can be found in [Pratt, 1991]. Both of those latter two references contain excellent bibliographies for even more detailed information.

One large family of texture measures is based on second-order gray-level statistics. One computes matrices whose elements are counts of the number of times certain patterns of pixel pairs occur. For example, an element of a matrix may be the number of times a gray level of 135 occurs exactly two pixels above a gray level of 132. In practice, the gray levels are usually quantized into relatively few bins, rather than using all possible values. Then, one can generate a seemingly infinite number of statistics from those matrices. These statistics are most useful when the area of interest is large. In most neural network applications that the author has seen, the area of interest is too small to allow stable values for these variables.

Another approach is based on *autocorrelation*. One computes the correlation between the tones of pixels a certain displacement apart. For example, suppose that we have an image of a log raft, with the logs running from the top of the image to the bottom. The areas between the logs will probably be unusually dark. If the digitization rate were 10 pixels per log, we would expect to see a high correlation at a horizontal lag of 10. Certain textures, such as woven fabrics, exhibit distinct autocorrelation functions.

[Pratt, 1991] describes a complex family of texture measures based largely on Markov processes. These are of great theoretical interest. However, relatively complicated normalizations are needed to standardize them. Readers who are interested in texture studies will be fascinated by this family of variables. That book also describes an interesting method based on whitening by decorrelation.

Finally, that text describes an unusual way of characterizing the structure of an image array by treating it as a matrix and computing its singular value decomposition. If the structure is highly random, the singular values will tend to be approximately equal to one another. If the structure is more organized, the energy will be concentrated in relatively few singular values. The author is not convinced that the enormous computation involved is worth the results obtained.

# A Final Note

In general, the variables presented in this chapter are *not* suitable for direct input to a neural network. No attempt was made to define widely applicable transformations, as was done for shape-based moments. The reason for avoiding such generality is that it is simply not possible. There is too wide a variety of tone patterns that can be encountered in practice. However, the standard guidelines still apply. They are repeated here.

1. **Transform.** If the variable is strictly positive, then consider a square root transform if the tail is light, or a log if it is heavy. If the variable can be positive or negative, try a cube root to reduce the extent of the tails.

2. **Center.** Compute the mean or median of a representative sample (*after* any transformations have been done), and subtract this from all cases.

3. **Scale.** Compute the interquartile range of a representative sample and use this to scale all cases so that all variables have about the same dispersion.

# 10

## Using the MLFN Program

MLFN is a general purpose program for training and testing multiple-layer feedforward neural networks. The input, hidden, and output layers may be independently defined as operating in the real or complex domain. A hybrid of simulated annealing and conjugate gradient optimization is used as the primary training algorithm. After training, validation sets may be processed for the purpose of evaluating the network's performance. The trained network may also be executed on unknown data, with the results saved to a disk file. Finally, the weights may be saved in a compact file format to facilitate reuse at a later date.

Interactive operation is possible. The user may enter commands that are executed immediately. However, the more usual mode of control is to prepare in advance an ASCII command file. Each line of this file is a single command, identical to those that might be entered interactively. There are at least two reasons why this method is preferable to interactive control. First, many program runs will require an extended period of time: overnight or even longer. The use of a self-contained command file allows the user to engage in other activities while the program is executing. The other reason is that a command file is self-documenting. The user will have a hard record of the parameters that went into producing the obtained results. The command to read an ASCII command file is the following:

### COMMAND FILE = filename

Usually this command is typed by the user to initiate processing of the command file. However, it may also appear in a command file. In other words, recursive processing of command files is allowed.

Every command has one or more components. The mandatory first component is the action to be taken. Many commands will also have auxiliary information. In this case, the action is followed by an equal sign (=), which in turn is followed by the auxiliary information. Blanks may optionally appear before or after the equal sign, and case (upper/lower) is ignored.

Commands may be commented by placing a semicolon (;) after the relevant parts, then following the semicolon with any text. Lines that are only comments are implemented by using a semicolon as the first character on the line.

In order to exit the program and return to the operating system, the command **BYE** must be used. Like all other rules, this may be typed interactively or it may appear in a command file. A reminder that this is the means of exiting MLFN is given to the user at each interactive command prompt.

# Output Mode

The neural network models implemented in the MLFN program are intrinsically function mappers. Given an input, they compute a scaler or vector numeric output. However, it is more convenient for the user if some common interpretations of the outputs are handled internally by the program. For example, MLFN provides a CLASSIFICATION mode in which training and testing is implemented by using predefined output values of 0.9 when a case belongs to the class associated with the output neuron, and –0.9 otherwise. Also, MLFN provides an AUTOASSOCIATION mode in which the inputs are automatically associated with the outputs, saving the user from explicitly specifying outputs. Finally, fully general function mapping is supported with the MAPPING output mode. This mode can be used to implement a wide variety of paradigms. However, the other modes can be very convenient. This section will examine these modes in some detail. The command to activate a given mode is shown below as a section head, and an explanation of the mode follows.

### MODE = MAPPING

This mode is the fully general $\Re^n$ to $\Re^m$ function mapping model. If $n$ is the number of inputs and $m$ is the number of outputs, each training and validation case must consist of $n+m$ numbers. If the output activation function is linear, the range of the $m$ outputs is unrestricted. However, the default hyperbolic tangent function has theoretical limits of –1 to 1, and it has effective limits in a somewhat narrower range. It is futile to attempt to train the network to achieve outputs outside this range when this activation function is employed.

### MODE = AUTOASSOCIATION

Autoassociative output mode is similar to the above mode in that general function mapping is implemented. The difference is that the user does not specify outputs. Rather, the inputs define the outputs. The number of inputs in the network model equals the number of outputs. All training and validation files contain only this one data vector for each case.

As in MAPPING mode, outputs have unrestricted range if a linear activation function is specified for the output neurons. However, the user must remember that since the inputs equal the outputs in this mode, the use of the default hyperbolic tangent function implies that the inputs are restricted in their range.

### MODE = CLASSIFICATION

This mode is different from the others in that the user does not directly specify the output values. Rather, the program itself determines them according to the class of each case. The user specifies the number of outputs as the number of classes that will be used. The class of each case in the training and validation sets is specified by means of the **CLASS** command which will be described later. For each case, the activation of the output neuron corresponding to the class of that case is defined as 0.9, and the activations of all other neurons are defined as −0.9.

# Network Structure

Several variations on the MLFN are available. These are distinguished by the number of layers, the number of neurons in each layer, the domains of the neurons' activation functions, and the nature of the activation functions of the output neurons. The domain of the neurons is specified by the following rule:

### DOMAIN = domain

The possible domains are now described.

**REAL** — All inputs, outputs and hidden neurons are in the real domain. This is the traditional MLFN. There may be zero, one, or two hidden layers with this model.

**COMPLEX-INPUT** — Inputs are complex numbers. The hidden layer, if any, has activation functions whose domain is complex and whose range is real. The output layer has real outputs. This model seems to have no advantages relative to a REAL model and is not generally recommended. Users are invited to experiment, though.

**COMPLEX-HIDDEN** — Inputs are complex. All hidden neurons map complex to complex. Output neurons map complex to real outputs. This model is valid only if a single hidden layer is used. For classification, this is often the best model when the input data is inherently complex.

**COMPLEX** — Inputs, outputs, and all neurons are fully complex. This model cannot be used if there are two hidden layers, nor can it be used in CLASSIFICATION output mode.

The network can contain zero, one, or two hidden layers. (The current version of MLFN allows two hidden layers only for REAL models.) It must have at least one input neuron and one output neuron. The network architecture is defined by specifying the domain and giving the number of input, hidden, and output neurons. Five commands are required. For example:

> **DOMAIN = COMPLEX**
> **INPUTS = 8**
> **FIRST HIDDEN = 4**
> **SECOND HIDDEN = 0**
> **OUTPUTS = 2**

The above commands define a three-layer network. It has 8 complex inputs (16 numbers altogether) and 2 complex outputs (4 numbers). All internal operation is entirely in the complex domain.

A model with no hidden layer is specified by setting FIRST HIDDEN to zero. In this case, SECOND HIDDEN must also be set to zero.

When the mode is AUTOASSOCIATION, the user must specify the same number of inputs as outputs. Also, the COMPLEX-INPUT and COMPLEX-HIDDEN domains cannot be used in autoassociative mode, since in those domain models the inputs are complex and the outputs are real.

The number of outputs in classification mode is normally set equal to the number of classes that will be used. The COMPLEX domain model cannot be used to classify, as real outputs are required in this mode.

## Output Activation Function

All hidden neurons use the hyperbolic tangent (tanh) activation function. This cannot be changed by the user. By default, the output layer uses this same function. The identity function may be used for output neurons by giving the command

**OUTPUT ACTIVATION = LINEAR**

There are several advantages to doing this. Regression (as used in conjunction with simulated annealing) now provides exactly optimal output weights, rather than only approximations. Moreover, there are no restrictions on the range of outputs that can be learned. Compression of extremes, often observed in autoassociative filtering and time-series prediction as a consequence of the squashing effect of nonlinear activation functions, is less pronounced. Finally, learning may be slightly accelerated.

However, linear outputs cause some loss of immunity to noise. The squashing functions usually used for output neurons prevent generation of extreme values as a result of extreme inputs. Naturally, a considerable degree of limiting is still obtained from the nonlinear hidden layer. But a nonlinear output layer is often helpful in noisy situations.

The network may be returned to its default nonlinear (hyperbolic tangent) output activation functions with the command

**OUTPUT ACTIVATION = NONLINEAR**

## Building the Training Set

Before learning can commence, a training set must be in place. The training set is created by reading one or more disk files containing the training cases. Each time a file is read, any existing training set is preserved. The new data is appended to the old data. If the user wishes to start all over with a totally new training set, the following command may be used:

**ERASE TRAINING SET**

All MLFN data files share the same format. This includes the training set files currently under discussion, as well as validation and test files that will be described later. They are standard ASCII files that may be read and written by nearly any text editor. Each case occupies exactly one line in the file. That line contains as many numbers as are needed to define the case. In CLASSIFICATION and AUTOASSOCIATION modes, that will be the number of inputs. (Class membership, which is not in the data file, is discussed later in this section.) In MAPPING mode, the quantity of numbers will be the number of inputs plus the number of outputs, with the inputs appearing before the outputs on each line. Each number is separated from the other numbers by one or more spaces. A sign and a decimal point are optional. A training data file is read by means of the following command:

**CUMULATE TRAINING SET = filename**

Note that this command does not initiate any learning. It causes the named file to be read and appended to any existing training set. A path and extension may optionally appear as part of the filename.

For classification, the class membership information is not included in the data file. It is specified *before* the **CUMULATE TRAINING SET** command appears. This command is

**CLASS = integer**

The integer is the class number, from one through the number of classes. It may not exceed the number of outputs in the model. A reject category is implemented by specifying a class number of zero. When a case is in the reject category, the network will assume that all output neurons should be at their minimum activation. Once this command appears, cases in all subsequent **CUMULATE TRAINING SET** commands will be considered to be in the specified class. The **CLASS** command does not need to be repeated for each training file unless the class membership changes.

# Learning Methods

There are several training algorithms available. The one to use is specified with the following command:

**LEARNING ALGORITHM = Method**

Specifying the method with this rule does not cause learning to begin. That is effected with the COMPUTE WEIGHTS command discussed later. This rule only decrees the method that will be used at that time. Each of the possible methods will now be described.

## Simulated Annealing

This is specified with the following command:

**LEARNING ALGORITHM = SIMULATED ANNEALING**

The hidden weight vectors are computed by the author's version of simulated annealing. That algorithm is described in detail in [Masters, 1993]. Output weights are explicitly computed by regression. This will result in exactly optimal output weights (in the mean squared error sense) if the output activation function is LINEAR. Otherwise, they will be only approximately optimal, though virtually always far better than if they were found by simulated annealing alone.

This method of learning is generally inferior to other methods, and it is included for experimentation only. Total execution time is approximately proportional to the product of the number of temperatures times the number of iterations at each temperature. For a relatively reliable search for the global minimum, use few temperatures (3 or 4) and many iterations. For higher accuracy in finding a local minimum, use more temperatures (10 to 20 or more) and fewer iterations. A high value of the SETBACK parameter will slow execution but will improve accuracy.

The annealing parameters can be specified as follows:

**ANNEALING INITIALIZATION TEMPERATURES = integer** — This specifies the number of temperatures. It must be at least one, and it will rarely exceed 20 or so.

**ANNEALING INITIALIZATION ITERATIONS = integer** — The number of trials at each temperature is given with this command. It should really be at least several hundred for small problems, and thousands for large problems.

**ANNEALING INITIALIZATION SETBACK = integer** — This is the amount to set back the iteration counter each time improvement is had. A good value is about half of the number of iterations.

**ANNEALING INITIALIZATION START = number** — The starting temperature is the standard deviation of the weight perturbation when annealing starts. It should be about 3 to 10 if the data is mostly in single-digit numbers. When this parameter is used in conjunction with the ANNEALING_CJ method described later, this temperature should be much less than when it is used for pure simulated annealing.

**ANNEALING INITIALIZATION STOP = number** — The stopping temperature limits the accuracy of the final result. However, if it is much smaller than the starting temperature, the number of temperatures must be large to avoid getting stuck in a local minimum. A typical value might be 0.1

## Hybrid: Annealing Plus Conjugate Gradients

This is the standard workhorse method. It is specified with the following command:

**LEARNING ALGORITHM = ANNEALING_CJ**

It combines the global search strategy of simulated annealing with the powerful minimum-seeking conjugate gradient (CJ) algorithm.

Simulated annealing is used in two separate, independent ways in this method. First it is used in a high-temperature initialization mode, centered at weights of zero, to find a good starting point for CJ minimization. When the CJ algorithm subsequently converges to a local minimum, simulated annealing is called into play again. This time its goal is to escape from what may be a local minimum, so its temperature is usually lower. Also, annealing is centered about the best weights as found by the CJ algorithm. A complete flowchart for this operation is shown in Figure 10-1.

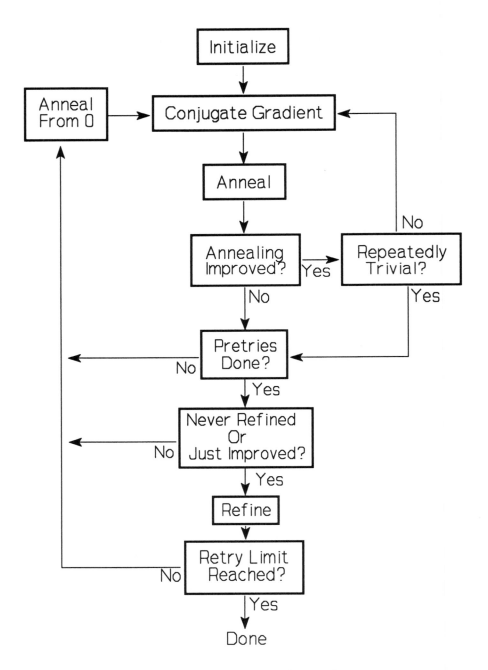

Fig. 10-1: MLFN network learning.

The parameters for the initial annealing around zero are set with the same commands as shown in the previous *Simulated Annealing* section. Parameters for escape from local minima are set with those same commands, except that **INITIALIZATION** is replaced with **ESCAPE**. For example, the following command would specify that three temperatures are to be used for escape from local minima:

**ANNEALING ESCAPE TEMPERATURES = 3**

There are several parameters that can be used to control some fine details of the learning process. The conjugate gradient step shown near the top of Figure 10-1 uses a relatively easily satisfied stopping criterion. Learning continues until the error cannot be reduced at CJ ACCURACY significant digits. The main loop, which is simulated annealing around zero for starting weights, followed by conjugate gradient optimization, followed by simulated annealing for escape from local minima, is repeated CJ PRETRIES times. Then, much slower refinement with the conjugate gradient algorithm is performed on the best of those tries, attempting to get an additional CJ REFINE significant digits improved. From that point on, this refinement is done whenever learning at the CJ ACCURACY level improves upon the previous record at that level. This algorithm saves the expensive refinement operation for only the most deserving candidates. In summary, the relevant parameters may be specified with the following commands:

**CJ ACCURACY = integer** gives the approximate number of error significant digits computed during the main retry loop. Smaller values will speed operation but make it less likely that a good set of weights will be found. The default may be found in the listing of MLFN.CPP. Reasonable values range from 5 to 10.

**CJ REFINE = integer** gives the additional approximate number of error significant digits computed after **CJ PRETRIES** tries are done and after any subsequent improvement. Smaller values will speed operation but make it less likely that a good set of weights will be found. The default may be found in the listing of MLFN.CPP. Reasonable values range from 1 to 4.

**CJ PRETRIES = integer** gives the number of retries done before refinement is performed. It is best to use as large a value as possible so that expensive refinement is postponed until a good collection of candidates is available. But remember that if operation is interrupted before refinement is done, the results may be less than excellent. The default may be found in the listing of MLFN.CPP. Reasonable values range from 1 (if you are in an unavoidable hurry) to 10 or more (if you can wait and are willing to save time in the long run).

Since learning with large networks or training sets can be excruciatingly slow, many users like to be kept informed of progress. This clutters the screen, but it is often a balm for sore nerves. Two commands control this supplementary output.

**CJ NO PROGRESS** — tells the conjugate gradient subroutine to remain silent until it has converged to a local minimum. This is the default, as it avoids confusing screen clutter.

**CJ PROGRESS** — tells the conjugate gradient algorithm to print (in parentheses) the current error after every 1000 internal iterations. This option helps the user keep track of progress for very large problems in which total silence for perhaps days can be daunting.

Finally, we should discuss the output that is written to the screen and methods by which the user can interrupt operations. Figure 10-1 should be studied during this discussion.

Every time outer-loop annealing is done, either at initialization or when retrying by annealing from zero, a new line of output will be written. This will show the temperature and will also show any improvement at that temperature. When annealing is complete, the best error so far will be printed at the start of a new line. During this annealing, ESCape may be pressed. Doing so will cause annealing to terminate and learning to proceed to the conjugate gradient step. Type *A* personalities will be severely tempted to press ESCape at this point, to the detriment of learning.

During conjugate gradient learning, no output will be produced unless the CJ PROGRESS command was given. In that case, regular reports on the current error will be written to the screen. When learning is complete to the CJ ACCURACY level, the attained error will be written to the screen on the same line. During this learning step, ESCape may be pressed. If so, learning will completely terminate and the program will enter interactive mode or proceed to the next command in the command file.

After the conjugate gradient step is complete, internal annealing will be performed. The purpose of this annealing is to escape from a local minimum if that is where the conjugate gradient step landed. The final result will be reported on the same line. Pressing ESCape during this step will terminate internal annealing and start a new retry.

If internal annealing improved the error, conjugate gradient learning will be done again. Its results will be reported on the same line.

When the time comes (often immediately) that internal annealing fails to improve the error, one of two things will happen. A new retry may be attempted by annealing from zero. In this case, reporting will begin on a new line as already stated. Or refinement may be performed, in which case its results are reported on the same line. The CJ PRETRIES parameter determines how many times the first action above, annealing from zero, is done. The second action, refinement, is done only after CJ PRETRIES complete loops have been done and after global improvement thereafter. If ESCape is pressed during refinement, it will be terminated and a new outer loop retry will commence.

The action of ESCape can now be summarized. Pressing ESCape will totally terminate learning only if done during normal conjugate gradient learning. If pressed during outer loop annealing from zero, annealing will terminate and conjugate gradient learning will commence. If pressed during inner loop annealing for escape from a local minimum, annealing will terminate and a new retry will commence. If pressed during refinement, a new retry will commence.

Results having to do with a particular retry, which includes its outer loop annealing, conjugate gradient learning, inner annealing, and any refinement, will be reported on the same line. A new retry will always start on a new line, with the temperature being written first. The only exception is that after outer loop annealing is complete, a new line will be started for the conjugate gradient learning. The first number written on this line is the try number, and it is immediately followed by the best error attained so far. Labeling included on the lines will make this confusing state of affairs far more clear.

### Regression-Initialized Conjugate Gradients

This excellent learning method is valid only if there are no hidden neurons. The input-to-output weights are computed using ordinary linear regression. If the output activation is linear (page 368) and mean square error is used for the error function, then nothing more is needed. The regression coefficients are already optimal. Otherwise, the conjugate gradient algorithm is used to trim up the weights. Simulated annealing is used to attempt to break out of a local minimum. No retries are ever done, as regression always gives the same initial weights. This learning method is requested with the following command:

**LEARNING ALGORITHM = REGRESS_CJ**

# Starting and Stopping Learning

Once the method and necessary parameters have been set, learning can commence. This is done with the following command:

**COMPUTE WEIGHTS**

There are several parameters related to the learning process which apply to many or to all of the learning methods. Once set, they will remain in effect for all COMPUTE WEIGHTS commands until explicitly reset. These commands, which relate to termination of learning, are now described.

**ALLOWABLE ERROR = number** can be used to specify a stopping criterion. If the learning progresses to the point that the error becomes this small, it is halted. Usually, we would set this parameter equal to zero, then press ESCape to halt training when patience expires. But setting it to larger, realistic values can be useful when running multiple experiments in a command file. This quantity corresponds to the error that is reported by the program, which is a constant multiple of that defined by equations in the text. See the next section for a detailed definition of error.

**MAXIMUM RESTARTS = integer** specifies the number of times that a recalcitrant learning session is retried by annealing around zero. At most this many retries (plus the first try) will be done if the error never drops to the quitting error level. This provides an alternative method of escape when running in batch mode. The ALLOWABLE ERROR could be set equal to zero, and the MAXIMUM RESTARTS set to a reasonably small number.

# Error Functions

By default, the traditional mean squared error is optimized during learning. There are several alternatives available. Many of them are worthless in practice and are included for education only. However, a few of them can be quite useful. The error type is specified with the following command:

**ERROR TYPE = Type**

These errors are defined starting on page 38. However, the errors that are reported by the MLFN program and those referred to by ALLOWABLE ERROR are actually constant multiples of the strict definitions. The reason is that some comparability between the different errors is thereby obtained. In each case, the constant is chosen so that (if possible) the maximum possible reported error using the hyperbolic tangent output activation function is 100. The constant used is specified in the definitions below.

**MEAN SQUARE** — The traditional mean squared error is used. This is the default. Reported is 25 times actual.

**ABSOLUTE** — The mean absolute error is used. All errors, regardless of their magnitude, are treated equally during learning. The default mean squared error favors reduction of large errors over small. Some applications, most notably some negative-feedback control situations, benefit from equal emphasis on all errors. Reported is 50 times actual.

**KK** — Kalman-Kwasny error is used. This measure is described in [Kalman and Kwasny, 1992]. It can be useful for classification models that have more than a few output neurons. It is not recommended for other applications. Reported is 25 times actual.

**CROSS ENTROPY** — Cross entropy is used as the error measure. It is valid only for CLASSIFICATION output models. This is a highly respected optimization criterion. Reported is 37.7 times actual.

**MAX** — The error is equal to the squared error of the output neuron having maximum error. This measure typically reduces the worst errors rapidly, then collapses due to numerical difficulties as ties for worst appear. Not recommended. Reported is 25 times actual.

**SIXTEENTH** — The mean across outputs of each error raised to the 16th power is optimized. Like MAX, this rapidly reduces the worst errors. However, even moderate errors are all but ignored. This method is not recommended. Reported is 1000 times actual.

**MEAN LOG** — This measure has essentially the opposite effect of MAX and SIXTEENTH. It reduces the mean across all outputs of the log of the error. The result is that for each case, some single output neuron already having small error is optimized even more, while other outputs are largely ignored. Never useful. Reported is a complicated function of actual. See GRADIENT.CPP for details.

**LOG MEAN** — This optimizes the log of the mean error across all outputs. This reversed tactic avoids the MEAN LOG problem of focusing on one output. However, it still gives excessive emphasis to some training cases at the expense of others. Not recommended. Reported is a complicated function of actual. See GRADIENT.CPP for details.

In summary, only a few of the above error measures are regularly useful. The default MEAN SQUARE is nearly always the best. A close second, and occasional winner, is CROSS ENTROPY. In some situations, ABSOLUTE may be best. The KK error tends to avoid some local minima when classifying a large number of classes. All other error measures are included for education only and most likely will fail miserably in practice.

# Confusion Matrices

One standard method for testing a classifier, neural or otherwise, is with the aid of a confusion matrix. This matrix portrays the patterns of misclassification that are obtained from a validation set. There is one row for each class, and as many columns as there are classes, plus one additional column to accommodate the reject category. The element in row $a$ and column $b$ is the number of cases that are truly members of class $a$ but that have been classified into class $b$. The element in the last column of row $a$ contains the number of cases in class $a$ whose activation did not attain the minimum threshold and so were tossed into the reject category. Ideally, we would want the last column to be entirely zero, and the rest of the matrix should be strictly diagonal. Off-diagonal quantities represent error.

MLFN evaluates and displays the confusion matrix one row at a time. It is the user's responsibility to keep track of the true class memberships and to assemble the rows into a matrix. Before computing a row of the confusion matrix, the user should issue the following command to zero the counters in all columns of the row that will be computed:

**ZERO CONFUSION ROW**

If desired, the user may also set the classification threshold. The command to do this is

**ACTIVATION THRESHOLD FOR CONFUSION = number**

The number is a percent (0–100) of full activation. In choosing a threshold, remember that the network was trained to achieve somewhat less than full activation for the true class. When the threshold is set, it will remain at that value unless it is set with this command again. The default may be found in MLFN.CPP, but it is highly recommended that the user explicitly set it. This threshold controls the ease with which cases are banished to the reject category. Larger values will force more marginal cases to land in the reject heap. Using a threshold of zero will let all cases be classified, so that the reject column will be entirely zero.

A single row of the confusion matrix is computed by means of one or more of the following commands:

**CLASSIFY WITH INPUT = filename**

The named data file has the same format as training files. In fact, the user may even want to use the training file to compute a confusion error measure to complement the reported error. Each case in the file is given to the network. The activation level of the maximally activated output neuron is compared to the confusion threshold. If the activation equals or exceeds the threshold, the corresponding element of the confusion row will be incremented. If it is less than the threshold, the reject counter which appears in the last column will be incremented.

The most recently computed row of the confusion matrix can be given to the user in one or both of two ways. The row can be printed on the screen, or it may be appended to the end of an ASCII text file. The named file will be created if it does not exist at the time this command is issued. These two commands are the following:

**LIST CONFUSION ROW**

**WRITE CONFUSION ROW TO FILE = filename**

## Testing in AUTOASSOCIATION and MAPPING Modes

The previous section discussed a straightforward method for testing when in CLASSIFI-CATION mode. This section covers the other two modes. The following command reads a data file and computes the mean squared error and RMS error for those cases.

**TEST NETWORK WITH INPUT = filename**

The format of the named data file is identical to that of training files. It is an ASCII text file having one line per case. If in AUTOASSOCIATION mode, each line contains as many numbers, with optional sign and decimal point, as there are network inputs. (If the input is complex, there will be two numbers for each input.) If in MAPPING mode, the lines also contain the desired outputs. The outputs follow the inputs, just as in a training file. Each case will be presented to the network, and the resulting outputs will be computed. Those outputs are compared to the target outputs (which are the inputs in AUTO mode), and the mean of the squared error is computed. This is the sum of squared errors divided by both the number of cases and the number of output neurons. The square root of this figure is the RMS error. Both errors are written to the display terminal.

There are two common uses for the TEST NETWORK WITH INPUT function. It allows us to test a validation set that is independent of the training set. In this way we can evaluate the performance of the network in an unbiased manner. The other use for this command is when we have trained the network using some error measure other than MEAN SQUARE, but we are interested in the mean squared error of the training set. We can use this command to reread the training set and compute its error.

## Saving Weights and Execution Results

We often want to preserve the weights of a trained network, particularly if the training required a long period of time. We may want to perform additional training starting with a weight set that is already good. Or we may simply want to use the network to perform some useful task. The weights are saved to and restored from a disk file with the following commands:

**SAVE NETWORK = filename**

**RESTORE NETWORK = filename**

Unlike data files, the weight file is not in a simple ASCII format. It is in a much more compact internal representation. This format can be found in WT_SAVE.CPP and in the wt_save and wt_restore functions in LAYERNET.CPP.

An ASCII text file containing the weights can be written with the following command:

**PRINT WEIGHTS = filename**

This file is for convenient visual inspection or for possible transfer to another program. It cannot be read with the RESTORE NETWORK command. This file is self-documenting, so its format is not described here. The code that writes this file is in wt_print in LAYERNET.CPP.

The MLFN program also provides a simple facility for actually using the trained network to perform tasks. It can read a file containing test cases, compute the activations attained in response to those cases, and write the results to a file. Two steps are required. The first step is to tell MLFN the name of the file to which outputs will be written. Outputs will be appended to that file if it exists. If it does not exist, it will be created when the first outputs are about to be written. The file is named with the following command:

**RUN OUTPUT = filename**

Then, the user issues one or more commands to read input data. The command to do this is

**RUN NETWORK WITH INPUT = filename**

In response to that command, all cases in the named file will be processed. This input file contains as many numbers as there are network inputs, one line per case. The generated output file will also contain one line per case, and it will have as many numbers as there are network outputs. If the default hyperbolic tangent output activation function is in use, the range of the output numbers will be –1 to 1. Otherwise, they may take on any values.

# Alphabetical Glossary of Commands

**ACTIVATION THRESHOLD FOR CONFUSION = number** — This is used in conjunction with computing confusion in CLASSIFICATION mode. It is the minimum percent activation (0–100) that the maximally activated output must attain to avoid that case being tossed into the reject category.

**ALLOWABLE ERROR = number** — When the error (see page 376) falls this low during learning, stop trying to improve it.

**ANNEALING ESCAPE ITERATIONS = integer** — This is the number of trials that will be performed at each temperature for simulated annealing local minimum escape.

**ANNEALING ESCAPE SETBACK = integer** — Each time improvement is had at a temperature, set back the iteration counter this amount to keep trying to do better. This is for simulated annealing local minimum escape.

**ANNEALING ESCAPE START = number** — The temperature is the standard deviation of the weight perturbation. This is the starting temperature for simulated annealing local minimum escape.

**ANNEALING ESCAPE STOP = number** — This specifies the final temperature for simulated annealing local minimum escape.

**ANNEALING ESCAPE TEMPERATURES = integer** — This is the number of temperatures for simulated annealing local minimum escape.

**ANNEALING INITIALIZATION ITERATIONS = integer** — This is the number of trials that will be performed at each temperature for simulated annealing weight initialization.

**ANNEALING INITIALIZATION SETBACK = integer** — Each time improvement is had at a temperature, set back the iteration counter this amount to keep trying to do better. This is for simulated annealing weight initialization.

**ANNEALING INITIALIZATION START = number** — The temperature is the standard deviation of the weight perturbation. This is the starting temperature for simulated annealing weight initialization.

**ANNEALING INITIALIZATION STOP = number** — This specifies the final temperature for simulated annealing weight initialization.

**ANNEALING INITIALIZATION TEMPERATURES = integer** — This is the number of temperatures for simulated annealing weight initialization.

**BYE** — Quit the MLFN program.

**CJ ACCURACY = integer** is the approximate number of error significant digits computed during the conjugate gradient part of the main retry loop.

**CJ NO PROGRESS** — tells the conjugate gradient subroutine to remain silent until it has converged to a local minimum.

**CJ PRETRIES = integer** is the number of retries done before refinement is performed.

**CJ PROGRESS** — tells the conjugate gradient algorithm to print (in parentheses) the current error after every 1000 internal iterations.

**CJ REFINE = integer** is the additional approximate number of error significant digits computed after **CJ PRETRIES** tries are done and after any subsequent improvement.

**CLASS = integer** — This sets the class number for subsequent CUMULATE TRAINING SET commands. Use CLASS = 0 for the reject category. This command is valid only in CLASSIFICATION mode.

**CLASSIFY WITH INPUT = filename** — After a network has been trained in CLASSIFICATION mode, this reads an ASCII text file of input data and cumulates a row of the confusion matrix.

**COMMAND FILE = filename** — This processes an ASCII text file of commands.

**COMPUTE WEIGHTS** — Train the network.

**CUMULATE TRAINING SET = filename** — This is the command used to build a training set. It reads an ASCII text file of training data. If a training set already exists, the new data will be appended.

**DOMAIN = COMPLEX** — Set the network to be entirely complex.

**DOMAIN = COMPLEX-HIDDEN** — Set the network to have complex inputs and hidden neurons, but real outputs. For classification tasks with complex data, this is almost always the preferred model.

**DOMAIN = COMPLEX-INPUT** — Set the network to have complex inputs, but real neurons for all subsequent layers. This model does not appear to have much to recommend it, but reader experimentation is encouraged.

**DOMAIN = REAL** — Set the network to be entirely real.

**ERASE TRAINING SET** — Erase the entire training set. This frees memory, so it should be done when the training set is no longer needed.

**ERASE NETWORK** — Erase learned weights.

**ERROR TYPE = ABSOLUTE** — The mean absolute error is minimized during learning. Occasionally useful.

**ERROR TYPE = CROSS ENTROPY** — The cross entropy is minimized during learning. Legal only for classification. Highly recommended alternative to mean squared error.

**ERROR TYPE = KK** — The Kalman-Kwasny error is minimized during learning. Appropriate only for classification, and apparently useful only when there are a large number of classes.

**ERROR TYPE = LOG MEAN** — The log of the mean squared error across outputs is minimized during learning. Not recommended.

**ERROR TYPE = MAX** — The maximum error across outputs is minimized during learning. Not recommended.

**ERROR TYPE = MEAN LOG** — The mean across outputs of each log error is minimized during learning. Not recommended.

**ERROR TYPE = MEAN SQUARE** — The traditional mean squared error is minimized during learning. This is the default and is a reliable all-purpose method.

**ERROR TYPE = SIXTEENTH** — The mean of the output errors raised to the 16th power is minimized during learning. Not recommended.

**FIRST HIDDEN = integer** — This defines the number neurons in the first hidden layer. If there is to be no hidden layer, set this to zero.

**INPUTS = integer** — This is the number of inputs. If the inputs are complex, there will actually be twice this many real numbers.

**LEARNING ALGORITHM = ANNEALING_CJ** — Use a hybrid of simulated annealing with conjugate gradients to train the network. This is the recommended method if there are any hidden layers.

**LEARNING ALGORITHM = REGRESS_CJ** — Use linear regression to initialize the weights, then continue with conjugate gradients to train the network. This can only be used if there is no hidden layer. In that case it is excellent.

**LEARNING ALGORITHM = SIMULATED ANNEALING** — Use pure simulated annealing to train the network. This is a slow technique that is included for educational purposes only.

**LIST CONFUSION ROW** — Write the most recently computed confusion row on the computer screen.

**MAXIMUM RESTARTS = integer** — This limits the number of times learning will restart by annealing around zero.

**MODE = AUTOASSOCIATION** — Set the mode to AUTOASSOCIATION.

**MODE = CLASSIFICATION** — Set the mode to CLASSIFICATION.

**MODE = MAPPING** — Set the mode to general function mapping.

**OUTPUT ACTIVATION = LINEAR** — The output neurons are to have an identity activation function.

**OUTPUT ACTIVATION = NONLINEAR** — The output neurons are to have the hyperbolic tangent activation function.

**OUTPUTS = integer** — This is the number of outputs. If the domain model is COMPLEX, there will actually be twice this many real numbers.

**PRINT WEIGHTS = filename** — This prints the weights to an ASCII file for convenient inspection. The file is self-documenting. This file cannot be read by the RESTORE NETWORK command.

**RESTORE NETWORK = filename** — This restores a trained network that was saved to a disk file. It sets the weights and fundamental structure items such as the size and domain of each layer and the output activation function. Learning parameters are not restored. (In fact, they weren't even saved.)

**RUN NETWORK WITH INPUT = filename** — This lets the user actually use the network to perform a task. It causes a file of input data to be read, and it writes the network's outputs to the current output file (as set by RUN OUTPUT).

**RUN OUTPUT = filename** — This names the ASCII text file to which subsequent RUN NETWORK WITH INPUT commands will write results.

**SAVE NETWORK = filename** — This saves a trained network to a disk file. It also saves model information, such as the size and domain of each layer and the output activation functions. The parameters that controlled the training of the network are not saved.

**SECOND HIDDEN = integer** — This specifies the number of neurons in the second hidden layer. If FIRST HIDDEN is zero, then this must also be zero.

**TEST NETWORK WITH INPUT = filename** — This reads a validation set (or the training set, if desired) and computes the mean squared error and RMS error. This command is valid only in AUTOASSOCIATION and MAPPING modes.

**WRITE CONFUSION ROW TO FILE = filename** — This appends to an ASCII disk file the most recently computed row of confusion.

**ZERO CONFUSION ROW** — This zeros all elements in the confusion row.

## Verification of Program Operation

This section lists a few test files for exercising MLFN. There are two purposes for this presentation. The first purpose is to aid the user in understanding the program by including some examples of realistic command files. A secondary purpose is to serve as a testing tool for readers who recompile the MLFN source code for their own system. Although these tests are not utterly exhaustive, they do provide a fairly rigorous test of most aspects of program operation. These files are also included on the program disk.

The command files will refer to two small data files. These files each contain eight cases, with each case consisting of six variables. The first four variables are, considering all 16 cases, the $2^4$ possible combinations of 0.7 and −0.7. Half of these 16 cases, those which correspond to even parity, are contained in one file. The other half are in the other file. The remaining two variables indicate the parity of the first four variables. In one file the indicators are exact negatives of each other. In the other file a bit of noise is introduced. This arrangement allows us to pursue a wide variety of difficult tests with this data. The two data files are now listed.

**ZERO4**

```
-.7  -.7  -.7  -.7      -.7   .72
-.7  -.7   .7   .7      -.71  .7
-.7   .7  -.7   .7      -.73  .7
-.7   .7   .7  -.7      -.7   .74
 .7  -.7  -.7   .7      -.7   .72
 .7  -.7   .7  -.7      -.71  .7
 .7   .7  -.7  -.7      -.75  .7
 .7   .7   .7   .7      -.7   .76
```

**ONE4**
```
-.7  -.7  -.7   .7        .7   -.7
-.7  -.7   .7  -.7        .7   -.7
-.7   .7  -.7  -.7        .7   -.7
-.7   .7   .7   .7        .7   -.7
 .7  -.7  -.7  -.7        .7   -.7
 .7  -.7   .7   .7        .7   -.7
 .7   .7  -.7   .7        .7   -.7
 .7   .7   .7  -.7        .7   -.7
```

Overall operation is supervised by one small command file that calls six other specialized command files. There is much unnecessary repetition within these files. For example, the training set is read, then erased, then reread many times. This is not always necessary. Also, the mode and domain are often set when they already have that value. This sort of repetition is done so that the reader can easily see what is happening. The supervising command file is listed below.

**VALID.CON**
```
ALLOWABLE ERROR = 0.0     ; Ask for the best
MAXIMUM RESTARTS = 1      ; But only give it two tries

COMMAND FILE = VALID1A.CON
COMMAND FILE = VALID2A.CON
COMMAND FILE = VALID3A.CON

COMMAND FILE = VALID1B.CON
COMMAND FILE = VALID2B.CON
COMMAND FILE = VALID3B.CON

BYE
```

The first command file uses an autoassociative model to train the network to reproduce its inputs. This command file takes two approaches. The first approach is trivial. It uses no hidden layer, so if the output activation function were linear, it could exactly learn these few unique cases. Hence, we make its life just a tiny bit more difficult by using nonlinear output activations. This way we can use regression to initialize the weights, followed by conjugate gradients to do the final tweaking. Observe just how little tweaking is needed.

The second approach in this file introduces a hidden layer that is too small to allow perfect autoassociation. Also, just for a little twist, we make the outputs linear. Learning is quite fast because the error surface is well behaved.

**VALID1A.CON**
; This is a trivial autoassociative task that can be easily solved.
; We use no hidden layer, so that regression initialization is available.
; This would give us a perfect fit if the output activation were linear.
; So we make it nonlinear and use conjugate gradients to trim things up.

DOMAIN = REAL
ERROR TYPE = MEAN SQUARE
MODE = AUTOASSOCIATION
OUTPUT ACTIVATION = NONLINEAR

INPUTS = 6
OUTPUTS = 6
FIRST HIDDEN = 0
SECOND HIDDEN = 0

CUMULATE TRAINING SET = zero4
CUMULATE TRAINING SET = one4

LEARNING ALGORITHM = REGRESS_CJ
COMPUTE WEIGHTS
ERASE TRAINING SET

PRINT WEIGHTS = 1a1.prn
SAVE NETWORK = 1a1.wts
RUN OUTPUT = 1a1.prn
RUN NETWORK WITH INPUT = zero4
RUN NETWORK WITH INPUT = one4
ERASE NETWORK

; We now limit the ability of the network to exactly duplicate the
; input data by imposing an insufficiently large hidden layer.
; We also make learning a little easier by linearizing the outputs.

DOMAIN = REAL
ERROR TYPE = MEAN SQUARE
MODE = AUTOASSOCIATION
OUTPUT ACTIVATION = LINEAR

INPUTS = 6
OUTPUTS = 6
FIRST HIDDEN = 4
SECOND HIDDEN = 0

```
CUMULATE TRAINING SET = zero4
CUMULATE TRAINING SET = one4

LEARNING ALGORITHM = ANNEALING_CJ
COMPUTE WEIGHTS
ERASE TRAINING SET

PRINT WEIGHTS = 1a2.prn
SAVE NETWORK = 1a2.wts
RUN OUTPUT = 1a2.prn
RUN NETWORK WITH INPUT = zero4
RUN NETWORK WITH INPUT = one4
ERASE NETWORK
```

The second command file that is called uses general function mapping to solve a parity problem. The first four variables of each case are the inputs. Each of the remaining two variables indicates the parity of the first four variables. This file starts out with an entirely real-domain network using the theoretical minimum number of hidden neurons. As usual when the architecture is so close to the theoretical edge, learning is slow. The error surface has many local minima, and the global minima (there are many) are at the bottom of narrow chasms.

This file then takes a second approach by switching the inputs and hidden neurons to the complex domain. Two hidden neurons are sufficient in this case, but finding those weights is nearly impossible. So we simplify things by using three hidden neurons. This network trains very rapidly, even though the pattern that must be learned is complicated.

**VALID2A.CON**
```
; Make the network learn a parity problem using  function mapping.
; Since we are using the minimum number of hidden neurons, this is a
; very difficult task to learn.  It may well fail in the tries allowed.

DOMAIN = REAL
ERROR TYPE = MEAN SQUARE
MODE = MAPPING
OUTPUT ACTIVATION = LINEAR

INPUTS = 4
OUTPUTS = 2
FIRST HIDDEN = 4
SECOND HIDDEN = 0

CUMULATE TRAINING SET = zero4
```

```
CUMULATE TRAINING SET = one4

LEARNING ALGORITHM = ANNEALING_CJ
COMPUTE WEIGHTS
ERASE TRAINING SET

PRINT WEIGHTS = 2a1.prn
SAVE NETWORK = 2a1.wts
RUN OUTPUT = 2a1.prn
RUN NETWORK WITH INPUT = zero4
RUN NETWORK WITH INPUT = one4
ERASE NETWORK

;  Now solve the same problem, but in the complex domain.
;  It can be solved with two complex hidden neurons, but the
;  solution is very difficult to find.  Using three makes it easy.

DOMAIN = COMPLEX
ERROR TYPE = MEAN SQUARE
MODE = MAPPING
OUTPUT ACTIVATION = LINEAR

INPUTS = 2
OUTPUTS = 1
FIRST HIDDEN = 3
SECOND HIDDEN = 0

CUMULATE TRAINING SET = zero4
CUMULATE TRAINING SET = one4

LEARNING ALGORITHM = ANNEALING_CJ
COMPUTE WEIGHTS
ERASE TRAINING SET

PRINT WEIGHTS = 2a2.prn
SAVE NETWORK = 2a2.wts
RUN OUTPUT = 2a2.prn
RUN NETWORK WITH INPUT = zero4
RUN NETWORK WITH INPUT = one4
ERASE NETWORK
```

The third command file also solves a parity problem.  This time it uses the built-in CLASSIFICATION mode to help the user.  Also, cross entropy error is minimized.  The most important aspect of this command file is that the learning scheme is very different from those of the previous files. Up until now we used the default, moderate annealing parameters and just one retry. That is fine in many problems. However, for problems in which the global minimum is difficult to find, an alternative usually works better. Initialize with just a small amount of annealing to save time. Then do conjugate gradient optimization with relatively few significant digits of error. This enables us to descend almost to the bottom of the nearest local minimum very quickly. Save the expensive optimization for deserving areas of the error surface. Combine this strategy with a large number of retries, and we have the makings of an excellent procedure.

**VALID3A.CON**
; Make the network learn a parity problem using classification mode.
; Since we are using the minimum number of hidden neurons in the
; first hidden layer, this is a very difficult task to learn.
; Thus, we use a different training scheme.
; Note that just to be cute, we stick in a second hidden layer.

DOMAIN = REAL
ERROR TYPE = CROSS ENTROPY
MODE = CLASSIFICATION
OUTPUT ACTIVATION = LINEAR

INPUTS = 4
OUTPUTS = 1
FIRST HIDDEN = 4
SECOND HIDDEN = 2

CLASS = 0
CUMULATE TRAINING SET = zero4

CLASS = 1
CUMULATE TRAINING SET = one4

LEARNING ALGORITHM = ANNEALING_CJ
ANNEALING INITIALIZATION TEMPERATURES = 2
ANNEALING INITIALIZATION ITERATIONS = 200
ANNEALING INITIALIZATION SETBACK = 50
ANNEALING INITIALIZATION START = 0.2
ANNEALING INITIALIZATION STOP = 0.1
ANNEALING ESCAPE TEMPERATURES = 2
ANNEALING ESCAPE ITERATIONS = 20
ANNEALING ESCAPE SETBACK = 10
ANNEALING ESCAPE START = 0.5
ANNEALING ESCAPE STOP = 0.05
ALLOWABLE ERROR = 0.05

```
MAXIMUM RESTARTS = 100
CJ ACCURACY = 5
CJ REFINE = 4
CJ PRETRIES = 6

COMPUTE WEIGHTS
ERASE TRAINING SET

PRINT WEIGHTS = class1.prn
SAVE NETWORK = class1.wts
RUN OUTPUT = class1.prn
RUN NETWORK WITH INPUT = zero4
RUN NETWORK WITH INPUT = one4
ACTIVATION THRESHOLD FOR CONFUSION = 50
ZERO CONFUSION ROW
CLASSIFY WITH INPUT = zero4
LIST CONFUSION ROW
ZERO CONFUSION ROW
CLASSIFY WITH INPUT = one4
LIST CONFUSION ROW
ERASE NETWORK

; Now we solve the parity problem in CLASSIFICATION mode using
; inputs and hidden neurons in the complex domain.
; The minimum number of complex hidden neurons is two, but that
; is an extremely difficult problem.  We use three to make it easier.

DOMAIN = COMPLEX-HIDDEN
ERROR TYPE = CROSS ENTROPY
MODE = CLASSIFICATION
OUTPUT ACTIVATION = LINEAR

INPUTS = 2
OUTPUTS = 1
FIRST HIDDEN = 3
SECOND HIDDEN = 0

CLASS = 0
CUMULATE TRAINING SET = zero4

CLASS = 1
CUMULATE TRAINING SET = one4

LEARNING ALGORITHM = ANNEALING_CJ
```

```
COMPUTE WEIGHTS
ERASE TRAINING SET

PRINT WEIGHTS = class2.prn
SAVE NETWORK = class2.wts
RUN OUTPUT = class2.prn
RUN NETWORK WITH INPUT = zero4
RUN NETWORK WITH INPUT = one4
ACTIVATION THRESHOLD FOR CONFUSION = 50
ZERO CONFUSION ROW
CLASSIFY WITH INPUT = zero4
LIST CONFUSION ROW
ZERO CONFUSION ROW
CLASSIFY WITH INPUT = one4
LIST CONFUSION ROW
ERASE NETWORK
```

The remaining three command files are simply for verification that MLFN is correctly saving and restoring networks. Examine the last few lines of each of the previous command files. The network is saved to a weight file for later restoration, and the weights are printed to an ASCII file. Then the outputs corresponding to the training cases are appended to that same ASCII file.

The following three command files restore the previously saved networks and pass the training data through them. The outputs are again appended to the very same files. The user can then examine the files with an ASCII text editor and verify that the output results are exactly the same. This is not a perfect test, but it should catch most errors.

**VALID1B.CON**
```
RESTORE NETWORK = 1a1.wts
RUN OUTPUT = 1a1.prn
RUN NETWORK WITH INPUT = zero4
RUN NETWORK WITH INPUT = one4
ERASE NETWORK

RESTORE NETWORK = 1a2.wts
RUN OUTPUT = 1a2.prn
RUN NETWORK WITH INPUT = zero4
RUN NETWORK WITH INPUT = one4
ERASE NETWORK
```

**VALID2B.CON**
```
RESTORE NETWORK = 2a1.wts
RUN OUTPUT = 2a1.prn
```

RUN NETWORK WITH INPUT = zero4
RUN NETWORK WITH INPUT = one4
ERASE NETWORK

RESTORE NETWORK = 2a2.wts
RUN OUTPUT = 2a2.prn
RUN NETWORK WITH INPUT = zero4
RUN NETWORK WITH INPUT = one4
ERASE NETWORK

**VALID3B.CON**
RESTORE NETWORK = class1.wts
RUN OUTPUT = class1.prn
RUN NETWORK WITH INPUT = zero4
RUN NETWORK WITH INPUT = one4
ERASE NETWORK

RESTORE NETWORK = class2.wts
RUN OUTPUT = class2.prn
RUN NETWORK WITH INPUT = zero4
RUN NETWORK WITH INPUT = one4
ERASE NETWORK

# Appendix

This appendix contains information concerning the files supplied on the accompanying disk. The most important code listed in the text is included on this disk so that the reader does not need to type it in manually. The disk also contains the complete source code and executable of a complex-domain neural network program.

## Disk Contents

The enclosed diskette contains an installation program that loads a fully functioning version of the MLFN neural network program along with all source files required to compile it. The C++ source files included in the MLFN directory should be able to be compiled using any ANSI C++ compiler. The author has tested them with Borland C++ 4.0 and Symantec C++ 6.1.

Included in a directory called VALIDATE are two data files and seven sample validation command files that can be used to test the MLFN program. Refer to Chapter 10 for information on how to construct a command file and run the validation suite.

The MISC directory contains a variety of programs that are listed in the text but that are not directly related to the MLFN program. These include subroutines for the Fourier, Gabor, and Morlet wavelet transforms. Moment and tone/texture variable subroutines, as well as some other miscellaneous tools, are also supplied in this directory.

The following is a listing of files included on the disk:

\SIGNAL
MLFN.EXE
READ.ME

\SIGNAL\MLFN
CLASSES.H
CONST.H
FUNCDEFS.H
ACTIVITY.CPP
ACT_FUNC.CPP
AN1.CPP
AN1_CJ.CPP
ANNEAL1.CPP
CONFUSE.CPP
CONJGRAD.CPP
CONTROL.CPP
DIRECMIN.CPP

DOTPROD.CPP
DOTPRODC.CPP
EXECUTE.CPP
FLRAND.CPP
GRADIENT.CPP
LAYERNET.CPP
MEM.CPP
MESSAGES.CPP
MLFN.CPP
PARSDUBL.CPP
REGRESS.CPP
REGRS_CJ.CPP
SHAKE.CPP
SVDCMP.CPP
TEST.CPP
TRAIN.CPP
WT_SAVE.CPP

**\SIGNAL\VALIDATE**
ONE4
ZERO4
VALID.CON
VALID1A.CON
VALID1B.CON
VALID2A.CON
VALID2B.CON
VALID3A.COIN
VALID3B.CON

**\SIGNAL\MISC**
DETREND.CPP
FASTFILT.CPP
FFT.CPP
GABOR.CPP
GABOR2D.CPP
MOM.CPP
MORLET.CPP
MORLET2D.CPP
QSORT.CPP
QSORT_I.CPP
SCALER.CPP
TT.CPP

# Hardware and Software Requirements

The MLFN program included on disk can be run on any standard IBM-compatible computer having an 80386 or higher processor. A math coprocessor is strongly recommended, but it is not mandatory. An ANSI C++ compiler is required if the user wishes to recompile the program.

# Making a Backup Copy

Before using the enclosed diskette, make a backup copy of the original. This backup is for personal use and will only be required in case of damage to the original. Any other use of the diskette violates copyright law. Assuming the floppy drive you will be using is drive A, please do the following:

1. Insert the original diskette included with the book into drive A.

2. At the A:> prompt, type DISKCOPY A: A: and press Return. You will be prompted to place the source diskette into drive A.

3. Press return and wait until you are prompted to place the target diskette in drive A.

4. Remove the original diskette and replace it with your blank backup diskette. Press return.

Continue to alternately insert the original (source) and backup (target) diskettes as prompted until the message COPY COMPLETE appears.

# Installing the Disk

The installation program included on the diskette contains 52 files in compressed format. The default installation settings will create a directory called SIGNAL and the three subdirectories MLFN, VALIDATE, and MISC. To install the files, please do the following:

1. Assuming you will be using drive A as the floppy drive for your diskette, at the A:> prompt type INSTALL.

2. Follow the instructions displayed by the installation program. At the end of the process you will be given the opportunity to review the README.TXT file for more information about the diskette.

# Bibliography

Aarts, E., and van Laarhoven, P. (1987). *Simulated Annealing: Theory and Practice.* John Wiley and Sons, New York.

Abe, S., Kayama, M., Takenaga, H., and Kitamura, T. (1992). "Neural Networks as a Tool to Generate Pattern Classification Algorithms." *International Joint Conference on Neural Networks*, Baltimore, MD.

Acton, Forman S. (1959). *Analysis of Straight-Line Data.* Dover Publications, New York.

Acton, Forman S. (1970). *Numerical Methods That Work.* Harper & Row, New York.

Anderson, James and Rosenfeld, Edward, eds. (1988). *Neurocomputing: Foundations of Research.* MIT Press, Cambridge, MA.

Austin, Scott (1990). "Genetic Solutions to XOR Problems." *AI Expert* (December), 52–57.

Avitzur, Ron (1992). "Your Own Handprinting Recognition Engine." *Dr. Dobb's Journal* (April), 32–37.

Azencott, R., ed. (1992). *Simulated Annealing: Parallelization Techniques.* John Wiley and Sons, New York.

Baba, Norio (1989). "A New Approach for Finding the Global Minimum of Error Function of Neural Networks." *Neural Networks*, **2**:5, 367–373.

Baba, N., and Kozaki, M. (1992). "An Intelligent Forecasting System of Stock Price Using Neural Networks." *International Joint Conference on Neural Networks*, Baltimore, MD.

Barmann, Frank and Biegler-Konig, Friedrich (1992). "On a Class of Efficient Learning Algorithms for Neural Networks." *Neural Networks*, **5**: 139–144.

Barnard, Etienne and Casasent, David (1990). "Shift Invariance and the Neocognitron." *Neural Networks*, **3**: 403–410.

Barr, Avron, Cohen, Paul R., and Feigenbaum, Edward A., eds. (vol. I, 1981; vol. II, 1982; vol. III, 1982; vol. IV, 1989). *The Handbook of Artificial Intelligence.* Addison-Wesley, Reading, MA.

Bartlett, E. B. (1991). "Chaotic Time-series Prediction Using Artificial Neural Networks." *Abstracts from 2nd Government Neural Network Applications Workshop* (September), Session III.

Birx, D., and Pipenberg, S. (1992). "Chaotic Oscillators and Complex-Mapping Feedforward Networks for Signal Detection In Noisy Environments." *International Joint Conference on Neural Networks*, Baltimore, MD.

Birx, D., and Pipenberg, S. (1993). "A Complex Mapping Network for Phase-Sensitive Classification." *IEEE Transactions on Neural Networks*, **4**:1, 127-135.

Blum, A. L., and Rivest, R. L. (1992). "Training a 3-Node Neural Network Is NP-Complete." *Neural Networks*, **5**:1, 117–127.

Blum, Edward, and Li, Leong (1991). "Approximation Theory and Feedforward Networks." *Neural Networks*, **4**: 511–515.

Booker, L. B., Goldberg, D. E., and Holland, J. H. (1989). "Classifier Systems and Genetic Algorithms." *Artificial Intelligence*, **40**: 235–282.

Box, George, and Jenkins, Gwilym (1976). *Time-series Analysis, Forecasting and Control.* Prentice Hall, Englewood Cliffs, NJ.

Bracewell, Ronald N. (1986). *The Fourier Transform and Its Applications.* McGraw-Hill, New York.

Brent, Richard (1973). *Algorithms for Minimization without Derivatives.* Prentice-Hall, Englewood Cliffs, NJ.

Brillinger, David R. (1975). *Time Series, Data Analysis and Theory.* Holt, Rinehart and Winston, New York.

Burgin, George (1992). "Using Cerebellar Arithmetic Computers." *AI Expert* (June), 32–41.

Cacoullos, T. (1966). "Estimation of a Multivariate Density." *Annals of the Institute of Statistical Mathematics* (Tokyo), **18**:2, 179–189.

Cardaliaguet, Pierre and Euvrard, Guillaume (1992). "Approximation of a Function and its Derivative with a Neural Network." *Neural Networks*, 5:2, 207–220.

Carpenter, Gail A., and Grossberg, Stephen (1987). "A Massively Parallel Architecture for a Self-Organizing Neural Pattern Recognition Machine." Academic Press *(Computer Vision, Graphics, and Image Processing)*, 37: 54–115.

Carpenter, Gail A., Grossberg, Stephen, and Reynolds, John H. (1991). "ARTMAP: Supervised Real-Time Learning and Classification of Nonstationary Data by a Self-Organizing Neural Network." *Neural Networks*, 4: 565–588.

Caruana, R. A., and Schaffer, J. D. (1988). "Representation and Hidden Bias: Gray vs. Binary Coding for Genetic Algorithms" in Laird, J. (ed.) *Proceedings of the Fifth International Congress on Machine Learning.* Morgan Kaufmann, San Mateo, CA.

Caudill, Maureen (1988). "Neural Networks Primer, Part IV—The Kohonen Model." *AI Expert* (August).

Caudill, Maureen (1990). "Using Neural Nets: Fuzzy Decisions." *AI Expert* (April), 59–64.

Chui, C. (1992). *An Introduction to Wavelets.* Academic Press, New York.

Cotter, Neil E., and Guillerm, Thierry J. (1992). "The CMAC and a Theorem of Kolmogorov." *Neural Networks*, 5: 221–228.

Cottrell, G., Munro, P., and Zipser, D. 1987. "Image Compression by Backpropagation: An Example of Extensional Programming." *ICS Report 8702*, University of California at San Diego.

Cox, Earl (1992). "Solving Problems with Fuzzy Logic." *AI Expert* (March), 28–37.

Cox, Earl (1992). "Integrating Fuzzy Logic into Neural Nets." *AI Expert* (June), 43–47.

Crooks, Ted (1992). "Care and Feeding of Neural Networks." *AI Expert* (July), 36–41.

Daubechies, Ingrid (1990). "The Wavelet Transform, Time-Frequency Localization, and Signal Analysis." *IEEE Transactions on Information Theory*, 36:5, 961–1005.

Davis, D. T., and Hwang, J. N. (1992). "Attentional Focus Training by Boundary Region Data Selection." *International Joint Conference on Neural Networks*, Baltimore, MD.

Davis, Lawrence (1991). *Handbook of Genetic Algorithms.* Van Nostrand Reinhold, New York.

Dracopoulos, D. and Jones, A. (1993). "Modeling Dynamic Systems." *World Congress on Neural Networks*, (Portland, OR).

Draper, N. R., and Smith, H. (1966). *Applied Regression Analysis.* John Wiley and Sons, New York.

Duffin, R. J., and Schaeffer, A. C. (1952). "A Class of Nonharmonic Fourier Series." *Transactions of the American Mathematical Society*, **72**: 341-366.

Eberhart, Russell C., and Dobbins, Roy W., eds. (1990). *Neural Network PC Tools, A Practical Guide.* Academic Press, San Diego, CA.

Fahlmann, Scott E. (1988). "An Empirical Study of Learning Speed in Backpropagation Networks." *CMU Technical Report CMU-CS–88–162* (June 1988).

Fakhr, W., Kamel, M., and Elmasry, M. I. (1992). "Probability of Error, Maximum Mutual Information, and Size Minimization of Neural Networks." *International Joint Conference on Neural Networks*, Baltimore, MD.

Finkbeiner, Daniel T., II (1972). *Elements of Linear Algebra.* W. H. Freeman, San Francisco, CA.

Foley, James D., van Dam, Andries, Feiner, Steven K., and Hughes, John F. (1990). *Computer Graphics: Principles and Practice* (Second Edition). Addison-Wesley, Reading, MA.

Forsythe, George E., Malcolm, Michael A., and Moler, Cleve B. (1977). *Computer Methods for Mathematical Computations.* Prentice-Hall, Englewood Cliffs, NJ.

Freeman, James A., and Skapura, David M. (1992). *Neural Networks: Algorithms, Applications, and Programming Techniques.* Addison-Wesley, Reading, MA.

Fu, K. S. , ed. (1971). *Pattern Recognition and Machine Learning.* Plenum Press, New York.

Fukunaga, Keinosuke (1972). *Introduction to Statistical Pattern Recognition.* Academic Press, Orlando, FL.

Fukushima, Kunihiko (1987). "Neural Network Model for Selective Attention in Visual Pattern Recognition and Associative Recall." *Applied Optics* (December), **26**:23.

Fukushima, Kunihiko (1989). "Analysis of the Process of Visual Pattern Recognition by the Neocognitron." *Neural Networks*, **2**: 413–420.

Gallant, Ronald, and White, Halbert (1992). "On Learning the Derivatives of an Unknown Mapping with Multilayer Feedforward Networks." *Neural Networks*, **2**: 129–138.

Gallinari, P., Thiria, S., Badran, F., and Fogelman-Soulie, F. (1991). "On the Relations between Discriminant Analysis and Multilayer Perceptrons." *Neural Networks*, **4**:3, 349–360.

Garson, David G. (1991). "Interpreting Neural-Network Connection Weights." *AI Expert* (April), 47–51.

Georgiou, G. (1993). "The Multivalued and Continuous Perceptrons." *World Congress on Neural Networks*, (Portland, OR).

Gill, Philip E., Murray, Walter, and Wright, Margaret H. (1981). *Practical Optimization*. Academic Press, San Diego, CA.

Glassner, Andrew S., ed. (1990). *Graphics Gems*. Academic Press, San Diego, CA.

Goldberg, David E. (1989). *Genetic Algorithms in Search, Optimization and Machine Learning*. Addison-Wesley, Reading, MA.

Gori, M., and Tesi, A. (1990). "Some Examples of Local Minima during Learning with Back-Propagation." *Third Italian Workshop on Parallel Architectures and Neural Networks (E. R. Caianiello, ed.)*. World Scientific Publishing Co.

Gorlen, Keith E., Orlow, Sanford M., and Plexico, Perry S. (1990). *Data Abstraction and Object-Oriented Programming in C++*. John Wiley & Sons, Chichester, England.

Grossberg, Stephen (1988). *Neural Networks and Natural Intelligence*. MIT Press, Cambridge, MA.

Guiver, John P., and Klimasauskas, Casimir, C. (1991). "Applying Neural Networks, Part IV: Improving Performance." *PC AI* (July/August).

Hald, A. (1952). *Statistical Theory with Engineering Applications*. John Wiley and Sons, New York.

Haralick, R. M. (1979). "Statistical and Structural Approaches to Texture." *Proceedings of the IEEE*, **67**: 786-804.

Harrington, Steven (1987). *Computer Graphics, A Programming Approach* (Second Edition). McGraw-Hill, New York.

Hashem, M. (1992). "Sensitivity Analysis for Feedforward Neural Networks with Differentiable Activation Functions." *International Joint Conference on Neural Networks*, Baltimore, MD.

Hastings, Cecil, Jr. (1955). *Approximations for Digital Computers*. Princeton University Press, Princeton, NJ.

Hecht-Nielsen, Robert (1987). "Nearest Matched Filter Classification of Spatiotemporal Patterns." *Applied Optics* (May 15), **26**:10.

Hecht-Nielsen, Robert (1991). *Neurocomputing*. Addison-Wesley, Reading, MA.

Hecht-Nielsen, Robert (1992). "Theory of the Backpropagation Network." *Neural Networks for Perception, vol. 2 (Harry Wechsler, ed.)* Academic Press, New York.

Hirose, A. (1992). "Proposal of Fully Complex-Valued Neural Networks." *International Joint Conference on Neural Networks*, (Baltimore, MD).

Hirose, A. (1993). "Simultaneous Learning of Multiple Oscillations of Recurrent Complex-Valued Neural Networks." *World Congress on Neural Networks*, (Portland, OR).

Hirose, Yoshio, Yamashita, Koichi, and Hijiya, Shimpei (1991). "Back-Propagation Algorithm Which Varies the Number of Hidden Units." *Neural Networks*, **4**:1, 61-66.

Hornik, Kurt, Stinchcombe, Maxwell, and White, Halbert (1989). "Multilayer Feedforward Networks are Universal Approximators." *Neural Networks*, **2**:5, 359–366.

Hornik, Kurt (1991). "Approximation Capabilities of Multilayer Feedforward Networks." *Neural Networks*, **4**:2, 251–257.

Howell, Jim (1990). "Inside a Neural Network." *AI Expert* (November), 29–33.

Hu, M. K. (1962). "Visual Pattern Recognition By Moment Invariants." *IRE Transactions on Information Theory*, **8**:2, 179–187.

IEEE Digital Signal Processing Committee, eds. (1979). *Programs for Digital Signal Processing*. IEEE Press, New York.

Ito, Y. (1991a). "Representation of Functions by Superpositions of a Step or Sigmoid Function and Their Applications to Neural Network Theory." *Neural Networks*, **4**:3, 385–394.

Ito, Y. (1991b). "Approximation of Functions on a Compact Set by Finite Sums of a Sigmoid Function without Scaling." *Neural Networks*, **4**:6, 817–826.

Ito, Y. (1992). "Approximation of Continuous Functions on $\mathbf{R}^d$ by Linear Combinations of Shifted Rotations of a Sigmoid Function with and without Scaling." *Neural Networks*, **5**:1, 105–115.

Kalman, B. L., and Kwasny, S. C. (1991). "A Superior Error Function For Training Neural Networks." *International Joint Conference on Neural Networks*, Seattle, WA.

Kalman, B. L., and Kwasny, S. C. (1992). "Why Tanh? Choosing a Sigmoidal Function." *International Joint Conference on Neural Networks*, Baltimore, MD.

Karr, Chuck (1991). "Genetic Algorithms for Fuzzy Controllers." *AI Expert* (February), 26–33.

Karr, Chuck (1991). "Applying Genetics to Fuzzy Logic." *AI Expert* (March), 39–43.

Kendall, M., and Stuart, A. (vol. I, 1969; vol. II, 1973; vol. III, 1976). *The Advanced Theory of Statistics*. Hafner, New York.

Kenue, S. K. (1991). "Efficient Activation Functions for the Back-Propagation Neural Network." *SPIE, Proceedings from Intelligent Robots and Computer Vision X: Neural, Biological, and 3-D Methods* (November).

Kim, M. W., and Arozullah, M. (1992). "Generalized Probabilistic Neural Network-Based Classifiers." *International Joint Conference on Neural Networks*, Baltimore, MD.

Klimasauskas, Casimir C. (1987). *The 1987 Annotated Neuro-Computing Bibliography*. NeuroConnection, Sewickley, PA.

Klimasauskas, Casimir C. (1992a). "Making Fuzzy Logic 'Clear'." *Advanced Technology for Developers*, 1 (May), 8–12.

Klimasauskas, Casimir C. (1992b). "Hybrid Technologies: more power for the future." *Advanced Technology for Developers*, 1 (August), 17–20.

Klir, George J., and Folger, Tina A. (1988). *Fuzzy Sets, Uncertainty, and Information*. Prentice Hall, Englewood Cliffs, NJ.

Knuth, Donald (1981). *Seminumerical Algorithms*. Addison-Wesley, Reading, MA.

Kohonen, Teuvo (1982). "Self-Organized Formation of Topologically Correct Feature Maps." *Biological Cybernetics*, **43**: 59–69.

Kohonen, Teuvo (1989). *Self-organization and Associative Memory*. Springer-Verlag, New York.

Kosko, Bart (1987). "Fuzziness vs. Probability." *Air Force Office of Scientific Research (AFOSR F49620-86-C-0070) and Advanced Research Projects Agency (ARPA Order No. 5794)*, (July).

Kosko, Bart (1988a). "Bidirectional Associative Memories." *IEEE Transactions on Systems, Man, and Cybernetics* (January/February), **18**:1.

Kosko, Bart (1988b). "Hidden Patterns in Combined and Adaptive Knowledge Networks." *International Journal of Approximate Reasoning*, vol. 1.

Kosko, Bart (1992). *Neural Networks and Fuzzy Systems*. Prentice Hall, Englewood Cliffs, NJ.

Kreinovich, Vladik Ya. (1991). "Arbitrary Nonlinearity Is Sufficient to Represent All Functions by Neural Networks: A Theorem." *Neural Networks*, **4**:3, 381–383.

Kuhl, Frank, Reeves, Anthony, and Taylor, Russell (1986). "Shape Identification With Moments and Fourier Descriptors." *Proceedings of the 1986 ACSM-ASPRS Convention* (March), 159–168.

Kurkova, Vera (1992). "Kolmogorov's Theorem and Multilayer Neural Networks." *Neural Networks*, **5**:3, 501–506.

Lawton, George (1992). "Genetic Algorithms for Schedule Optimization." *AI Expert* (May), 23–27.

Levin, A. (1993). "Predicting With Feedforward Networks." *World Congress on Neural Networks*, (Portland, OR).

Lim, Jae S. (1990). *Two-Dimensional Signal and Image Processing*. Prentice Hall, Englewood Cliffs, NJ.

Lo, Zhen-Ping, Yu, Yaoqi, and Bavarian, Behnam (1993). "Analysis of the Convergence Properties of Topology-Preserving Neural Networks." *IEEE Transactions on Neural Networks*, **4**:2, 207–220.

Lu, C. N., Wu, H. T., and Vemuri, S. (1992). "Neural Network Based Short-Term Load Forecasting." *IEEE/PES 1992 Winter Meeting, New York* (92 WM 125-5 PWRS).

Maren, Alianna, Harston, Craig, and Pap, Robert (1990). *Handbook of Neural Computing Applications*. Academic Press, New York.

Masters, Timothy (1993). *Practical Neural Network Recipes in C++*. Academic Press, New York.

Matsuba, I., Masui, H., and Hebishima, S. (1992). "Optimizing Multilayer Neural Networks Using Fractal Dimensions of Time-Series Data." *International Joint Conference on Neural Networks*, Baltimore, MD.

McClelland, James and Rumelhart, David (1988). *Explorations in Parallel Distributed Processing*. MIT Press, Cambridge, MA.

Meisel, W. (1972). *Computer-Oriented Approaches to Pattern Recognition*. Academic Press, New York.

Miller, J., Goodman, R., and Smyth, P. (1991). "Objective Functions for Probability Estimation." *International Joint Conference on Neural Networks*, (Seattle, WA).

Minsky, Marvin and Papert, Seymour (1969). *Perceptrons*. MIT Press, Cambridge, MA.

Mougeot, M., Azencott, R., and Angeniol, B. (1991). "Image Compression with Back Propagation: Improvement of the Visual Restoration Using Different Cost Functions." *Neural Networks*, **4**:4, 467–476.

Mucciardi, A., and Gose, E. (1970). "An Algorithm for Automatic Clustering in N-Dimensional Spaces Using Hyperellipsoidal Cells." *IEEE Sys. Sci. Cybernetics Conference*, (Pittsburgh, PA).

Musavi, M., Kalantri, K., and Ahmed, W. (1992). "Improving the Performance of Probabilistic Neural Networks." *International Joint Conference on Neural Networks*, Baltimore, MD.

Negoita, Constantin V., and Ralescu, Dan (1987). *Simulation, Knowledge-Based Computing, and Fuzzy Statistics*. Van Nostrand Reinhold, New York.

Nitta, T. (1993). "Three-Dimensional Backpropagation." *World Congress on Neural Networks*, (Portland, OR).

Nitta, T. (1993). "A Complex-Numbered Version of the Backpropagation Algorithm." *World Congress on Neural Networks*, (Portland, OR).

Pao, Yoh-Han (1989). *Adaptive Pattern Recognition and Neural Networks*. Addison-Wesley, Reading, MA.

Parzen, E. (1962). "On Estimation of a Probability Density Function and Mode." *Annals of Mathematical Statistics*, **33**: 1065–1076.

Pethel, S. D., Bowden, C. M., and Sung, C. C. (1991). "Applications of Neural Net Algorithms to Nonlinear Time Series." *Abstracts from 2nd Government Neural Network Applications Workshop* (September), Session III.

Polak, E. (1971). *Computational Methods in Optimization*. Academic Press, New York.

Polzleitner, Wolfgang and Wechsler, Harry (1990). "Selective and Focused Invariant Recognition Using Distributed Associative Memories (DAM)." *IEEE Transactions on Pattern Analysis and Machine Intelligence* (August), **12**:8.

Pratt, William K. (1991). *Digital Image Processing*. John Wiley and Sons, New York.

Press, William H., Flannery, B., Teukolsky, S., and Vetterling, W. (1992). *Numerical Recipes in C*. Cambridge University Press, New York.

Raudys, Sarunas J., and Jain, Anil K. (1991). "Small Sample Size Effects in Statistical Pattern Recognition: Recommendations for Practitioners." *IEEE Transactions on Pattern Analysis and Machine Intelligence* (March), **13**:3.

Reed, R., Oh, S., and Marks, R. J. (1992). "Regularization Using Jittered Training Data." *International Joint Conference on Neural Networks*, Baltimore, MD.

Reeves, A., Prokop, R., Andrews, S., and Kuhl, F. (1988). "Three-Dimensional Shape Analysis Using Moments and Fourier Descriptors." *IEEE Transactions on Pattern Analysis and Machine Intelligence* (November), **10**: 937–943.

Rich, Elaine (1983). *Artificial Intelligence*. McGraw-Hill, New York.

Rosenblatt, Frank (1958). "The Perceptron: A Probabilistic Model for Information Storage and Organization in the Brain." *Psychological Review*, **65**: 386–408.

Rosenfeld, A., and Kak, A. (1982). *Digital Picture Processing*. Academic Press, New York.

Rumelhart, David, McClelland, James and the PDP Research Group (1986). *Parallel Distributed Processing*. MIT Press, Cambridge, MA.

Sabourin, M., and Mitiche, A. (1992). "Optical Character Recognition by a Neural Network." *Neural Networks*, **5**: 843–852.

Samad, Tariq (1988). "Backpropagation Is Significantly Faster if the Expected Value of the Source Unit Is Used for Update." *1988 Conference of the International Neural Network Society*.

Samad, Tariq (1991). "Back Propagation with Expected Source Values." *Neural Networks*, **4**:5, 615–618.

Schioler, H., and Hartmann, U. (1992). "Mapping Neural Network Derived from the Parzen Window Estimator." *Neural Networks*, **5**:6, 903–909.

Schwartz, Tom J. (1991). "Fuzzy Tools for Expert Systems." *AI Expert* (February), 34–41.

Sedgewick, Robert (1988). *Algorithms*. Addison-Wesley, Reading, MA.

Shapiro, Stuart C., ed. (1990). *Encyclopedia of Artificial Intelligence*. John Wiley & Sons, New York.

Siegel, Sidney (1956). *Nonparametric Statistics for the Behavioral Sciences*. McGraw-Hill, New York.

Soulie, Francoise Fogelman, Robert, Yves, and Tchuente, Maurice, eds. (1987). *Automata Networks in Computer Science*. Princeton University Press, Princeton, NJ.

Specht, Donald (1990). "Probabilistic Neural Networks." *Neural Networks*, **3**: 109–118.

Specht, Donald (1992). "Enhancements to Probabilistic Neural Networks." *International Joint Conference on Neural Networks*, Baltimore, MD.

Specht, Donald F., and Shapiro, Philip D. (1991). "Generalization Accuracy of Probabilistic Neural Networks Compared with Back-Propagation Networks." *Lockheed Missiles & Space Co., Inc. Independent Research Project RDD 360*, I-887-I-892.

Spillman, Richard (1990). "Managing Uncertainty with Belief Functions." *AI Expert* (May), 44–49.

Stork, David G. (1989). "Self-Organization, Pattern Recognition, and Adaptive Resonance Networks." *Journal of Neural Network Computing* (Summer).

Strand, E. M., and Jones, W. T. (1992). "An Adaptive Pattern Set Strategy for Enhancing Generalization While Improving Backpropagation Training Efficiency." *International Joint Conference on Neural Networks*, Baltimore, MD.

Styblinski, M. A., and Tang, T.-S. (1990). "Experiments in Nonconvex Optimization: Stochastic Approximation with Function Smoothing and Simulated Annealing." *Neural Networks*, **3**: 467–483.

Sudharsanan, Subramania I., and Sundareshan, Malur K. (1991). "Exponential Stability and a Systematic Synthesis of a Neural Network for Quadratic Minimization." *Neural Networks*, **4**: 599–613.

Sultan, A. F., Swift, G. W., and Fedirchuk, D. J. (1992). "Detection of High Impedance Arcing Faults Using a Multi-Layer Perceptron." *IEEE/PES 1992 Winter Meeting, New York* (92 WM 207-1 PWRD).

Sussmann, Hector J. (1992). "Uniqueness of the Weights for Minimal Feedforward Nets with a Given Input-Output Map." *Neural Networks*, **5**:4, 589–593.

Tanimoto, Steven L. (1987). *The Elements of Artificial Intelligence.* Computer Science Press, Rockville, MD.

Taylor, Russell, Reeves, Anthony, and Kuhl, Frank (1992). "Methods For Identifying Object Class, Type, and Orientation, in the Presence of Uncertainty." *Remote Sensing Reviews*, **6**:1, 183–206.

Ulmer, Richard, Jr., and Gorman, John (1989). "Partial Shape Recognition Using Simulated Annealing." *IEEE Proceedings, 1989 Southeastcon.*

Unnikrishnan, K. P., and Venugopal, K. P. (1992). "Learning in Connectionist Networks Using the Alopex Algorithm." *International Joint Conference on Neural Networks*, Baltimore, MD.

van Ooyen, A., and Nienhuis, B. (1992). "Improving the Convergence of the Back-Propagation Algorithm." *Neural Networks*, **5**:3, 465–471.

Wallace, Timothy P., and Wintz, Paul A. (1980). "An Efficient Three-Dimensional Aircraft Recognition Algorithm Using Normalized Fourier Descriptors." *Computer Graphics and Image Processing*, **13**, 99-126.

Wang, Kaitsong, Gorman, John, and Kuhl, Frank (1992). "Spherical Harmonics and Moments for Recognition of Three-Dimensional Objects." *Remote Sensing Reviews*, **6**:1, 229–250.

Wayner, Peter (1991). "Genetic Algorithms." *BYTE* (January), 361–368.

Webb, Andrew R., and Lowe, David (1990). "The Optimized Internal Representation of Multilayer Classifier Networks Performs Nonlinear Discriminant Analysis." *Neural Networks*, **3**:4, 367–375.

Wenskay, Donald (1990). "Intellectual Property Protection for Neural Networks." *Neural Networks*, **3**:2, 229–236.

Weymaere, Nico and Martens, Jean-Pierre (1991). "A Fast and Robust Learning Algorithm for Feedforward Neural Networks." *Neural Networks*, **4**:3, 361–369.

White, Halbert (1989). "Neural-Network Learning and Statistics." *AI Expert* (December), 48–52.

Wiggins, Ralphe (1992). "Docking a Truck: A Genetic Fuzzy Approach." *AI Expert* (May), 29–35.

Wirth, Niklaus (1976). *Algorithms + Data Structures = Programs*. Prentice-Hall, Englewood Cliffs, NJ.

Wolpert, David H. (1992). "Stacked Generalization." *Neural Networks*, **5**: 241–259.

Yau, Hung-Chun and Manry, Michael T. (1991). "Iterative Improvement of a Nearest Neighbor Classifier." *Neural Networks*, **4**: 517–524.

Zadeh, Lotfi A. (1992). "The Calculus of Fuzzy If/Then Rules." *AI Expert* (March), 23–27.

Zeidenberg, Matthew (1990). *Neural Network Models in Artificial Intelligence*. Ellis Horwood, New York.

Zhang, Y., Chen, G. P., Malik, O. P., and Hope, G. S. (1992). "An Artificial Neural Network-Based Adaptive Power System Stabilizer." *IEEE/PES 1992 Winter Meeting, New York* (92 WM 018-2 EC).

Zhou, Yi-Tong and Chellappa (1992). *Artificial Neural Networks for Computer Vision.* Springer-Verlag, New York.

Zornetzer, Steven, Davis, Joel, and Lau, Clifford, eds. (1990). *An Introduction to Neural and Electronic Networks.* Academic Press, New York.

# Index